INTRACELLULAR SIGNAL TRANSDUCTION

ADVANCES IN
PHARMACOLOGY

VOLUME 36

INTRACELLULAR SIGNAL TRANSDUCTION

Edited by

Hiroyoshi Hidaka

Department of Pharmacology
Nagoya University School of Medicine
Nagoya, Japan

Angus C. Nairn

Laboratory of Molecular and Cellular Neuroscience
The Rockefeller University
New York, New York

Associate Editors

Edwin G. Krebs

Howard Hughes Medical Research Labs
University of Washington School of Medicine
Seattle, Washington

Philip Cohen

Department of Biochemistry
Medical Science Institute
University of Dundee
Dundee, Scotland,
United Kingdom

Tamio Yamakawa

Tokyo College of Pharmacy
Tokyo, Japan

Fumimaro Takaku

International Medical Center of Japan
Tokyo, Japan

Michio Ui

The Tokyo Metropolitan Institute of Medical Science
Tokyo, Japan

ADVANCES IN
PHARMACOLOGY

VOLUME 36

ACADEMIC PRESS

San Diego   New York   Boston   London   Sydney   Tokyo   Toronto

Contents

Cyclic Nucleotide Phosphodiesterases: Gene Complexity, Regulation by Phosphorylation, and Physiological Implications

Fiona Burns, Allan Z. Zhao, and Joseph A. Beavo

Structural Analysis of the MAP Kinase ERK2 and Studies of MAP Kinase Regulatory Pathways

Melanie H. Cobb, Shuichan Xu, Mangeng Cheng, Doug Ebert, David Robbins, Elizabeth Goldsmith, and Megan Robinson

Novel Protein Phosphatases That May Participate in Cell Signaling

Patricia T. W. Cohen, Mao Xiang Chen, and Christopher G. Armstrong

Protein Tyrosine Phosphatases and the Control of Cellular Signaling Responses

N. K. Tonks

Roles of the MAP Kinase Cascade in Vertebrates

Tetsuo Moriguchi, Yukiko Gotoh, and Eisuke Nishida

Signal Transductions of SH2/SH3: Ash/Grb-2 Downstream Signaling

Tadaomi Takenawa, Kenji Miura, Hiroaki Miki, and Kazutada Watanabe

Sphingolipid-Dependent Protein Kinases

Sen-itiroh Hakomori

G Protein-Coupled Receptor Kinase: Phosphorylation of Muscarinic Receptors and Facilitation of Receptor Sequestration

Tatsuya Haga, Kazuko Haga, Kimihiko Kameyama, and Hirofumi Tsuga

Molecular and Cellular Pharmacology of a Calcium/Calmodulin-Dependent Protein Kinase II (CaM Kinase II) Inhibitor, KN-62, and Proposal of CaM Kinase Phosphorylation Cascades

Hiroyoshi Hidaka and Hisayuki Yokokura

Intrasteric Regulation of Calmodulin-Dependent Protein Kinases

B. E. Kemp, J. A. Barden, B. Kobe, C. House, and M. W. Parker

Structure, Regulation, and Function of Calcium/ Calmodulin-Dependent Protein Kinase I

Marina R. Picciotto, Kent L. Nastiuk, and Angus C. Nairn

Gene Expression and CREB Phosphorylation Induced by cAMP and Ca^{2+} in Neuronal Cells

Masatoshi Hagiwara, Atsushi Shimomura, Kazuhiko Yoshida, and Junko Imaki

Contributors

Numbers in parentheses indicate the pages on which the authors' contributions begin.

J. Arias (1) Scripps Research Institute, La Jolla, California 92037

Christopher G. Armstrong (67) Medical Research Council Protein Phosphorylation Unit, Department of Biochemistry, The University, Dundee DD1 4HN, Scotland, UK

R. Armstrong (1) IDUN Pharmaceuticals, San Diego, California 92121

J. A. Barden (221) Department of Anatomy and Histology, University of Sydney, Sydney 2006, New South Wales, Australia

Joseph A. Beavo (29) Department of Pharmacology, University of Washington, Seattle, Washington 98195

P. Brindle (1) St. Jude Children's Research Hospital, Memphis, Tennessee 38101

Fiona Burns (29) Department of Pharmacology, University of Washington, Seattle, Washington 98195

Mao Xiang Chen (67) Medical Research Council Protein Phosphorylation Unit, Department of Biochemistry, The University, Dundee DD1 4HN, Scotland, UK

Mangeng Cheng (49) Department of Pharmacology, The University of Texas Southwestern Medical Center, Dallas, Texas 75235

Melanie H. Cobb (49) Department of Pharmacology, The University of Texas Southwestern Medical Center, Dallas, Texas 75235

Patricia T. W. Cohen (67) Medical Research Council Protein Phosphorylation Unit, Department of Biochemistry, The University, Dundee DD1 4HN, Scotland, UK

Philip Cohen (15) Medical Research Council Protein Phosphorylation Unit, Department of Biochemistry, The University, Dundee DD1 4HN, Scotland, United Kingdom

Doug Ebert (49) Department of Pharmacology, The University of Texas Southwestern Medical Center, Dallas, Texas 75235

K. Ferreri (1) VivoRx, Inc., Santa Monica, California 90404

Elizabeth Goldsmith (49) Department of Biochemistry, The University of Texas Southwestern Medical Center, Dallas, Texas 75235

Yukiko Gotoh (121) Department of Genetics and Molecular Biology, Institute for Virus Research, Kyoto University, Kyoto 606-01, Japan

Kazuko Haga (173) Department of Biochemistry, Institute for Brain Research, Faculty of Medicine, University of Tokyo, Hongo, Tokyo 113, Japan

Tatsuya Haga (173) Department of Biochemistry, Institute for Brain Research, Faculty of Medicine, University of Tokyo, Hongo, Tokyo 113, Japan

Masatoshi Hagiwara (277) Department of Anatomy, Nagoya University School of Medicine, Nagoya 466, Japan

Sen-itiroh Hakomori (155) Division of Biomembrane Research, Pacific Northwest Research Foundation, and Departments of Pathobiology and Microbiology, University of Washington, Seattle, Washington 98122

Hiroyoshi Hidaka (193) Department of Pharmacology, Nagoya University School of Medicine, Nagoya 466, Japan

C. House (221) St. Vincent's Institute of Medical Research, Fitzroy 3065, Victoria, Australia

Junko Imaki (277) Department of Anatomy, Nippon Medical School, Tokyo 113, Japan

Kimihiko Kameyama (173) Department of Biochemistry, Institute for Brain Research, Faculty of Medicine, University of Tokyo, Hongo, Tokyo 113, Japan

B. E. Kemp (221) St. Vincent's Institute of Medical Research, Fitzroy 3065, Victoria, Australia

B. Kobe (221) St. Vincent's Institute of Medical Research, Fitzroy 3065, Victoria, Australia

Hiroaki Miki (139) Department of Biochemistry, Institute of Medical Science, University of Tokyo, Shirokanedai, Minato-ku, Tokyo 108, Japan

Kenji Miura (139) Department of Biochemistry, Defense Medical College, Namiki, Tokorozawa, Saitama 359, Japan

M. Montminy (1) The Salk Institute, The Clayton Foundation Laboratories for Peptide Biology, La Jolla, California 92037

Tetsuo Moriguchi (121) Department of Genetics and Molecular Biology, Institute for Virus Research, Kyoto University, Kyoto 606-01, Japan

Angus C. Nairn (251) Laboratory of Molecular and Cellular Neuroscience, The Rockefeller University, New York, New York 10021

Kent L. Nastiuk (251) Laboratory of Molecular and Cellular Neuroscience, The Rockefeller University, New York, New York 10021

Eisuke Nishida (121) Department of Genetics and Molecular Biology, Institute for Virus Research, Kyoto University, Kyoto 606-01, Japan

M. W. Parker (221) St. Vincent's Institute of Medical Research, Fitzroy 3065, Victoria, Australia

Marina R. Picciotto (251) Department of Psychiatry, Yale University School of Medicine, New Haven, Connecticut 06508

David Robbins (49) University of California, San Francisco, Hooper Foundation, San Francisco, CA 94143

Megan Robinson (49) Department of Pharmacology, The University of Texas Southwestern Medical Center, Dallas, Texas 75235

Atsushi Shimomura (277) Department of Anatomy, Nagoya University School of Medicine, Nagoya 466, Japan

Tadaomi Takenawa (139) Department of Biochemistry, Institute of Medical Science, University of Tokyo, Shirokanedai, Minato-ku, Tokyo 108, Japan

N. K. Tonks (91) Cold Spring Harbor Laboratory, Cold Spring Harbor, New York 11724

Hirofumi Tsuga (173) Department of Biochemistry, Institute for Brain Research, Faculty of Medicine, University of Tokyo, Hongo, Tokyo 113, Japan

Kazutada Watanabe (139) Department of Experimental Biology, Tokyo Metropolitan Institute of Gerontology, Sakae-cho, Itabashi-ku, Tokyo 173, Japan

Shuichan Xu (49) Department of Pharmacology, The University of Texas Southwestern Medical Center, Dallas, Texas 75235

Hisayuki Yokokura (193) Department of Pharmacology, Nagoya University School of Medicine, Nagoya 466, Japan

Kazuhiko Yoshida (277) Department of Ophthalmology, Hokkaido University School of Medicine, Sapporo 060, Japan

Allan Z. Zhao (29) Department of Pharmacology, University of Washington, Seattle, Washington 98195

M. Montminy*
P. Brindle†
J. Arias‡
K. Ferreri§
R. Armstrong‖

*The Salk Institute
The Clayton Foundation Laboratories for Peptide Biology
La Jolla, California 92037

† Department of Biochemistry
St. Jude Children's Research Hospital
Memphis, Tennessee 38101

‡Scripps Research Institute
La Jolla, California 92037

§ VivoRx, Inc.
Santa Monica, California 90404

‖ IDUN Pharmaceuticals
San Diego, California 92121

Regulation of Somatostatin Gene Transcription by cAMP

I. Introduction

A number of hormones and growth factors stimulate target cells through receptors that are coupled to second messenger pathways. The second messenger cAMP, for example, mediates a wide variety of cellular responses to hormonal signals, including changes in intermediary metabolism, cellular proliferation, and cellular motility. In mammalian cells, all these biological responses are triggered by the activation of the cAMP-dependent protein kinase (PKA), a heterotetramer consisting of paired catalytic (C) and regulatory (R) subunits. Upon hormonal stimulation, cAMP binds tightly to the R-subunits, thereby liberating C-subunits and promoting the phosphorylation of cellular substrates. In the liver, cAMP functions as a starvation-state signal, mediating hormonal cues from the pancreas and adrenal to stimulate

Advances in Pharmacology, Volume 36
Copyright © 1996 by Academic Press, Inc. All rights of reproduction in any form reserved.

glucose production. cAMP stimulates glucose production, in part, by regulating transcription of the gene for phosphoenol pyruvate carboxykinase (PEPCK), a rate-limiting enzyme in gluconeogenesis (Short *et al.*, 1986). Following hormonal stimulation, cAMP induces PEPCK gene expression 10-fold within 20–30 min, and this induction appears to be independent of new protein synthesis.

II. Characterization of the cAMP Response Unit

Previous studies in our (Montminy *et al.*, 1986) and other (Comb *et al.*, 1986; Short *et al.*, 1986) laboratories have identified a consensus cAMP response element (CRE), usually positioned within 100 nucleotides of the transcriptional initiation site, which mediates transcriptional induction of PEPCK and other cAMP-responsive genes. Characterized by a core motif, which often contains the palindromic sequence 5′-TGACGTCA-3′, the CRE can confer cAMP inducibility when placed upstream of a heterologous promoter and can function in both distance- and orientation-independent contexts. Remarkably, mutant cell lines deficient in PKA cannot support cAMP-responsive transcription from a CRE reporter gene (Montminy *et al.*, 1986), suggesting that this kinase might phosphorylate proteins that bind to the CRE and thereby activate transcription of cAMP-responsive genes. Toward this end, we purified a 43-kDa CRE-binding protein from PC12 and brain nuclear extracts using CRE–oligonucleotide affinity chromatography (Montminy and Bilezikjian, 1987; Yamamoto *et al.*, 1988). The purified CREB protein was highly phosphorylated by PKA *in vitro*, and primary sequencing of a CREB tryptic phosphopeptide revealed a consensus PKA site: (R)-R-P-S-Y-R. Using primary sequence information provided by other tryptic fragments of CREB, we obtained cDNAs encoding a 341-aa protein, which contained the PKA phosphoacceptor site at Ser133 (Gonzalez *et al.*, 1989). cDNAs for CREB were also characterized independently by Hoeffler *et al.* (1988).

Previous work showing that most, if not all, of the biological effects of cAMP are mediated by the cAMP-dependent PKA prompted us to test whether CREB activity in response to cAMP might be dependent on phosphorylation by PKA. We found that CREB is indeed phosphorylated by PKA at a single serine phosphoacceptor site Ser133 (Gonzalez and Montminy, 1989), and this phosphorylation is induced by hormonal stimuli that signal through the cAMP pathway. Remarkably, mutant cell lines that are deficient in PKA activity cannot stimulate CREB phosphorylation at Ser133, nor can they induce CRE-dependent transcription. To determine whether PKA phosphorylation of CREB at Ser133 was critical for cAMP-responsive transcription *in vivo*, we compared wild-type and mutant forms of CREB containing substitutions at the PKA phosphoacceptor site (Gonzalez and Montminy, 1989). When transfected into F9 teratocarcinoma cells, a wild-type CREB expression plasmid can stimulate CRE–CAT reporter activity greater than 200-fold in the presence of PKA. In contrast, a mutant CREB plasmid containing a Ser133 to Ala133 substitution was completely unable to respond to PKA induction. Acidic substitution mutants (Ser133 to Asp133 or Glu133) were also inactive, suggesting that the negative charge provided by phosphorylation is not sufficient to stimulate transcription.

The presence of phosphorylation sites for other kinases, most notably for casein kinase II, prompted us to test whether these might also regulate CREB activity (Gonzalez *et al.*, 1991; Hagiwara *et al.*, 1992). When treated with casein kinase II (CKII), recombinant CREB protein was phosphorylated at several Ser phosphoacceptor sites, including Ser111, -114, -117, and -156, as determined by primary sequencing of tryptic peptides (not shown). In contrast with the Ser133 phosphoacceptor site, however, mutagenesis of any or all of the CKII sites did not reduce but rather increased CREB activity somewhat in the presence of cAMP, suggesting that a single phosphorylation event may be both necessary and sufficient to induce transcription of cAMP-responsive genes. To test this hypothesis, we performed microinjection experiments with purified recombinant CREB protein (Alberts *et al.*, 1994). When coinjected with a CRE–lacZ reporter plasmid, unphosphorylated CREB had no effect on lacZ reporter activity in microinjected rat embryo fibroblasts. However, phospho-CREB markedly stimulated CRE-dependent transcription, suggesting that PKA-mediated phosphorylation at Ser133 might indeed be sufficient to induce expression of cAMP-regulated genes.

III. Kinetic Profile of the Transcriptional Response to cAMP

Like other signaling pathways, cAMP stimulates transcription of somatostatin and other target genes with burst-attenuation kinetics: transcription usually peaks within 20–30 min of induction and gradually declines over the next 4–8 hr (Fig. 1). The kinetics of PKA-dependent CREB phosphorylation closely parallel the changes in transcription of cAMP-responsive genes by run-on assay (Fig. 1). Nuclear translocation of PKA, visualized by microinjection of fluorescently labeled PKA holoenzyme, appears to represent the rate-limiting step in CREB phosphorylation and transcriptional activation (Hagiwara *et al.*, 1992, 1993).

To follow the stoichiometry of CREB phosphorylation in response to hormonal stimulation, we developed a phospho-Ser133-specific CREB antiserum (Hagiwara *et al.*, 1993). Maximal activation of the cAMP pathway with forskolin (10 μM) causes 40% of total CREB protein to be phosphorylated. By varying the strength of hormonal stimulus, we found that the degree of CREB phosphorylation and CRE-dependent transcription were directly proportional to the extent of PKA activation. Rather than simply

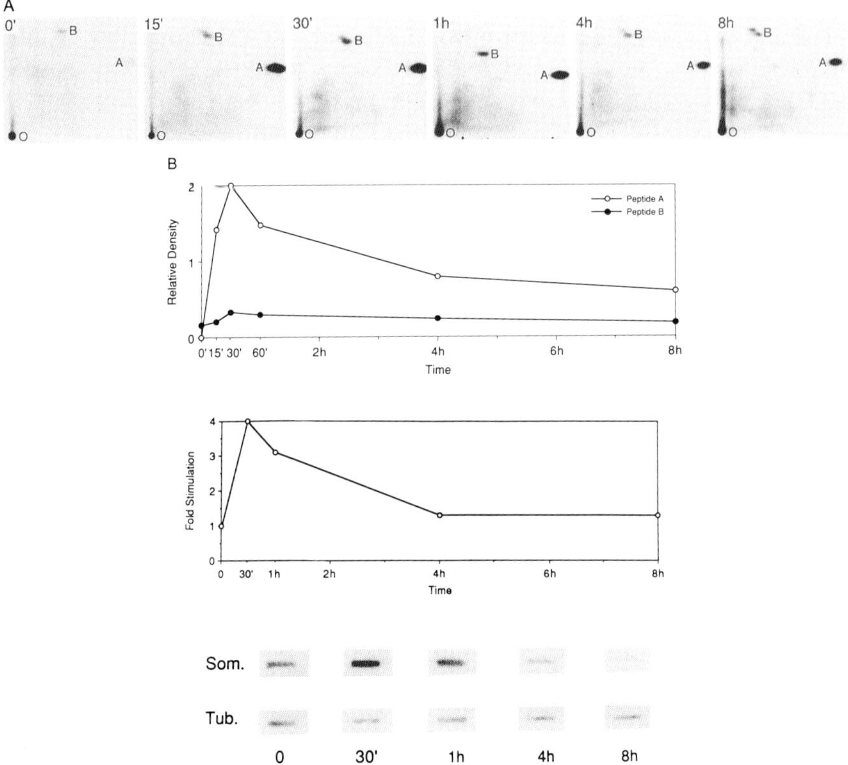

FIGURE I   (A) Two-dimensional phosphotryptic mapping of ^{32}P-labeled CREB in PC12 cells after treatment with forskolin. Spot A contains PKA phosphorylation site (Ser133). Spot B contains casein kinase II (Ser156) phosphorylation site. (B) Run-on transcription of the cAMP-responsive somatostatin gene in stable lines of NIH-3T3 cells transfected with the somatostatin gene. Tub., tubulin mRNA transcription.

being turned on or off, peak levels of cAMP-responsive transcription are proportional to signal strength.

The second phase in cAMP-responsive transcription is an attenuation phase during which CREB phosphorylation and CRE-dependent transcription decrease in parallel over a 4-hr period. During that time, levels of total CREB protein remain unchanged, suggesting that a CREB phosphatase may be critical in down-regulating cAMP-induced transcription (Hagiwara *et al.*, 1992). Indeed, CREB phosphorylation at Ser133 and CRE–CAT reporter activity in NIH-3T3 cells, at late times (4 hr) after cAMP induction, were markedly augmented by the phosphatase inhibitors okadaic acid (ID$_{50}$ = 20 nM) and the inhibitor-1 (I-1) protein. Of the four Ser/ Thr phosphatases characterized to date, only protein phosphatase-1 (PP-1) can be inhibited by both okadaic acid and I-1. When transfected into F9 teratocarcinoma cells, an expression vector for PP-1, but not the PP-2A catalytic subunit, could inhibit cAMP-dependent transcription from a CRE–CAT reporter plasmid. Furthermore, microinjection of purified PP-1 but not PP-2A protein severely diminished CRE–lacZ reporter expression in rat embryo fibroblasts, suggesting that cAMP-responsive transcription may not only depend on the level of PKA activation but also on cellular levels of PP-1. Indeed, microinjection of a constitutively active I-1 protein increases CRE-dependent transcription and stimulates CREB phosphorylation *in situ.* In contrast, the SV40 t antigen, which binds and specifically inhibits PP-2A, has no effect on either CRE-inducible transcription or on CREB phosphorylation.

IV. Mechanism of Phosphorylation-Dependent Induction

Phosphorylation may regulate nuclear factors at several levels, including DNA binding and nuclear targeting activities. To determine whether phosphorylation might induce CREB DNA binding activity, we performed gel mobility shift and footprinting assays using recombinant purified CREB protein. These assays, combined with filter-binding studies, revealed that CREB and PKA-phosphorylated CREB both bind the CRE with k_d of 10^{-9} M (Gonzalez and Montminy, 1989; Hagiwara *et al.,* 1993). If PKA phosphorylation does promote CREB DNA binding activity, we reasoned that CREB fusion proteins containing a heterologous binding domain (e.g., the yeast Gal4 DNA binding domain) may not respond to PKA induction. When transfected into F9 or PC12 cells, Gal4–CREB nevertheless induced a Gal4–CAT reporter 50- to 100-fold in response to PKA (Hagiwara *et al.,* 1992). Moreover, mutagenesis of the Ser133 phosphoacceptor site completely abrogated PKA-inducible activity, suggesting that the Gal4–CREB protein may function like the wild-type CREB protein on a heterologous DNA binding domain.

To determine whether phosphorylation may affect nuclear entry of CREB, we performed immunocytochemical studies with CREB-specific antisera. We found that endogenous CREB protein was nuclear in both unstimulated and forskolin-treated cells (Gonzalez and Montminy, 1989). Moreover, mutant forms of CREB lacking the Ser133 phosphoacceptor site were also targeted correctly to nuclei of transfected cells, suggesting that PKA-mediated phosphorylation may not induce nuclear transport of CREB.

Having eliminated CRE binding activity, dimerization, or nuclear translocation as plausible mechanisms by which PKA might stimulate CREB activity, we hypothesized that phosphorylation might directly regulate the transactivation potential of CREB. In this regard, we determined that a 70-aa kinase-inducible domain (KID) spanning aa 90–160 in CREB can modulate transcription of target genes in response to cAMP (Brindle *et al.,* 1993). The

KID region may not function alone, but must be paired with a constitutive activator to modulate transcription. In this regard, the CREB protein contains a glutamine-rich activator region termed Q2 with properties similar to those of glutamine-rich regions in the nuclear factor SP1. Because deletion of Q2 completely abrogates CREB activity in response to cAMP, we have speculated that KID must cooperate with Q2 to stimulate transcription in response to PKA phosphorylation.

To determine whether KID and Q2 could function as independent activators, we attached these regions to the Gal4 DNA binding domain and examined their activities on a 5× Gal4–CAT reporter plasmid following transfection into F9 teratocarcinoma cells (Brindle *et al.*, 1993) (Fig. 2A). The Q2–Gal4 fusion protein stimulated reporter activity 15-fold but did not respond to PKA induction, and a KID–Gal4 effector plasmid showed very low constitutive activity, which was stimulated 3-fold in response to PKA. When cotransfected together (Fig. 2B), Q2–Gal4 and KID–Gal4 effect-ors could reconstitute full PKA-inducible transcription, suggesting that

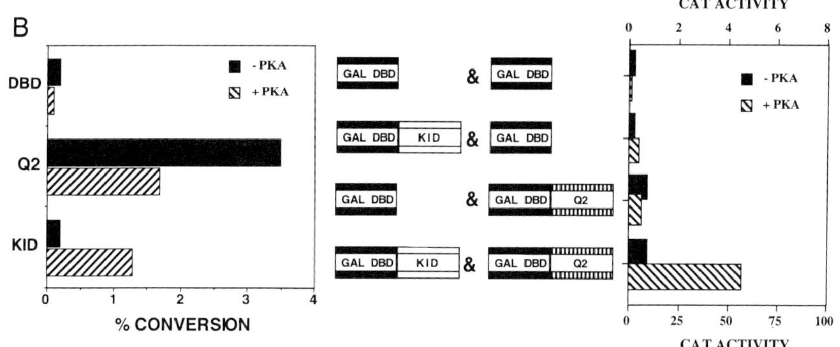

FIGURE 2   (A) The transcription factor CREB is organized in modular fashion. Q1 and Q2, hydrophobic, glutamine-rich domains; α, alternately splice 14-amino-acid sequence, which forms α-helical structure *in vitro*; KID, kinase-inducible domain containing PKA phosphoacceptor site at Ser133; DBD/LZ, basic DNA binding/leucine zipper dimerization domain. (B) The KID and Q2 domains in CREB function independently as modulatory and constitutive activation domains, respectively. Left, CAT activity derived from F9 cells following cotransfection with Gal4 DNA binding domain (Gal DBD), Gal4–Q2, or Gal4–KID fusion protein expression vectors as listed; PKA (− or + as indicated) plus Gal4–CAT reporter plasmid. Right, the KID and Q2 activators can synergize in *trans* to stimulate transcription of cAMP-responsive genes. F9 cells were cotransfected with expression vectors as indicated; PKA (− or + as indicated) plus Gal4–CAT reporter plasmid.

Q2 and KID are indeed independent activators that synergize to stimulate transcription in response to cAMP.

To characterize factors that functionally interact with CREB, we employed an *in vitro* transcription assay system using purified wild-type or mutant CREB polypeptides (Ferreri *et al.*, 1994). When added to crude Hela nuclear extracts, the unphosphorylated CREB protein stimulated transcription from the CRE-containing somatostatin promoter but not from a control α-globin template. In contrast, a mutant CREB polypeptide, lacking the glutamine-rich Q2 region, was transcriptionally inactive despite having wild-type DNA binding affinity *in vitro*. To test whether the Q2 domain was sufficient to stimulate transcription, we expressed and purified a fusion protein containing the Q2 region attached to the Gal4 DNA binding domain. When evaluated on a somatostatin promoter template containing two Gal4 recognition sites, the Gal4–Q2 protein stimulated somatostatin promoter activity in a dose-dependent manner, but the Gal4 DNA binding domain alone was minimally active, suggesting that the Q2 domain does indeed function as a constitutive activator both *in vitro* and *in vivo* (Brindle *et al.*, 1993).

V. Interactions between CREB and the Transcriptional Apparatus

On the basis of mutagenesis data presented previously, we hypothesized that CREB would form both constitutive and inducible interactions with proteins that are associated with the transcriptional apparatus. To characterize "constitutive" interactions between Q2 and the basal transcription complex, we prepared partially purified fractions of the basal transcription factors and monitored their ability to complement Q2 activity (Ferreri *et al.*, 1994). Transcription reactions reconstituted with all basal factors except the TFIID fraction showed minimal activity from either the somatostatin or control adenovirus-2 major late promoters. Addition of recombinant TBP restored basal but not Gal4–Q2 activator-dependent activity to both templates. Only the holo-TFIID fraction could fully reconstitute Gal4–Q2 transactivation, suggesting that some factor in this fraction, in addition to TBP, is required for Q2 function. Within the TFIID fraction, TBP is found associated with a number of factors (TAFs) that appear to have coactivator activity (Dynlacht *et al.*, 1991; Hoey *et al.*, 1993).

Inspection of sequences within Q2 revealed an intriguing resemblance to the hydrophobic and glutamine-rich Sp1 activation domain B (Gill *et al.*, 1994). Because mutational analysis indicated that this region was important for interaction with the *Drosophila* dTAF$_{II}$110 protein, we tested for similar interactions between Q2 and dTAF$_{II}$110. CREB stimulates transcription of target genes comparably in *Drosophila* and mammalian nuclear extracts

(M. Montminy, unpublished observations), thereby permitting us to evaluate interactions between Q2 and the recently cloned *Drosophila* TAFs. Toward this end, we prepared CRE oligonucleotide affinity resins bound by purified CREB or mutant ΔQ2 proteins. We observed that ^{35}S-labeled dTAF$_{II}$110 protein from reticulocyte lysates programmed with dTAF$_{II}$110 RNA could indeed bind to CRE resin containing wild-type CREB, albeit with low efficiency (10%). However, no such binding was observed between dTAF$_{II}$110 protein and ΔQ2 mutant CREB protein, suggesting that the interaction we detected was specific for the Q2 domain. Indeed, dTAF$_{II}$110 was also retained on resins containing Gal4–Q2 protein but not Gal4 DNA binding domain (DBD) alone.

To confirm these *in vitro* assays and begin to map the important sequences in Q2 that mediate interaction with dTAF$_{II}$110, we employed the yeast two-hybrid system (Chien *et al.*, 1991) (Fig. 3). A yeast expression

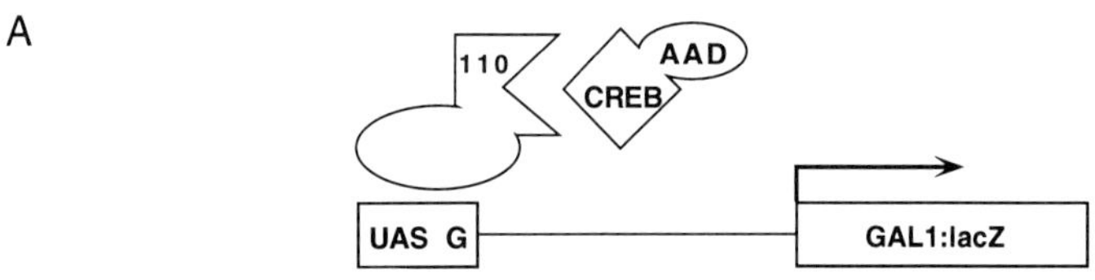

	Fusion to GAL4 AAD	Fusion to GAL4 DBD	β-galactosidase Activity
	NO	dTAF$_{11}$ 110	<1
Sp1B		dTAF$_{11}$ 110	51
Q2		dTAF$_{11}$ 110	19
Q2[Δ204-213]		dTAF$_{11}$ 110	4
Q2[Δ204-224]		dTAF$_{11}$ 110	<1
Q2[Δ204-256]		dTAF$_{11}$ 110	<1
Q2[Δ185-256]		dTAF$_{11}$ 110	<1
Q2		NO	<1

FIGURE 3 Interaction between dTAF110 and the CREB Q22 requires sequences in Q2 that are homologous to Sp1. (A) Diagram of yeast two-hybrid assay showing that recruitment of Gal4 acidic activator requires sequences in CREB that interact with Gal4–dTAF110 and promote transcription from a Gal4–lacZ reporter. (B) β-Galactosidase activity obtained from wild-type and mutant Q2 regions shown with Sp1 activity included for comparison.

plasmid encoding the Q2 domain fused to the acidic activation domain of Gal4 was cotransformed into a yeast strain with a second vector expressing a dTAF$_{II}$110–Gal4 DBD fusion protein. Interaction of CREB–Q2 with dTAF$_{II}$110, monitored by the recruitment of the acidic activation domain onto the promoter of a *Gal4–lacZ* reporter gene, was readily observed. In similar assays, CREB–Q2 did not appear to interact with other components of the TFIID complex such as TBP, TAF40, and TAF80.

Based on sequence similarity with Sp1 (aa 375–378) (Gill *et al.*, 1994), we postulated that the putative dTAF$_{II}$110 interaction motif in CREB is located at aa 204–208. This motif is absent from the CREM-α,-β, and -ε repressors, which lack sequences in Q2 extending from aa 183 to 244 of CREB. Similar deletion mutants of CREB (Δ204–214 and Δ204–224) exhibit far lower constitutive and PKA-dependent activity than the wild-type protein in cotransfection assays using a somatostatin–CAT reporter vector (Brindle *et al.*, 1993; Ferreri *et al.*, 1994). Consistently, dTAF$_{II}$110 association with these and other inactive CREB deletion mutants was markedly decreased compared to the wild-type Q2 domain (Fig. 3), demonstrating that Q2 activity was indeed correlated with its ability to bind dTAF$_{II}$110.

Dr. Goodman's group has characterized a CREB-binding protein (CBP) that specifically recognizes sequences within the Ser133 phosphorylated form of CREB (Chrivia *et al.*, 1993). CBP does not regulate the DNA binding, dimerization, or nuclear targeting properties of CREB. Rather, CBP binds selectively to the KID of CREB, a 60-amino-acid transactivation domain that is critical for PKA-inducible transcription. To characterize the functional properties of CBP, we developed an antiserum directed against aa 634–648 within the CREB binding domain of CBP (Arias *et al.*, 1994). Using this antiserum, we detected a 265-kDa polypeptide on Western blots (Fig. 4A, lane 1) as predicted from the cDNA (Chrivia *et al.*, 1993), which coincided with the predominant phospho-CREB binding activity in Hela nuclear extracts by "far-Western" blot assay (Fig. 4A, lane 2). An identical phospho-CREB binding activity was also found in NIH-3T3 cells (not shown). This phospho-CREB binding protein appeared to be specific for Ser133-phosphorylated CREB because no such band was detected with CREB labeled to the same specific activity at a nonregulatory phosphoacceptor site (Ser156) by CKII (Hagiwara *et al.*, 1992) (Fig. 4B). To further demonstrate that the major phospho-CREB binding protein in Hela and 3T3 cells is specifically bound by the anti-CBP antibody, we prepared immunoprecipitates from crude nuclear extracts using the CBP antiserum. Far-Western analysis of these immunoprecipitates (Fig. 4C) revealed a 265-kDa band in samples incubated with CBP antiserum but not with control IgG. To examine whether the phosphorylation-dependent interaction between CREB and CBP was critical for cAMP-responsive transcription, we employed a microinjection assay using CBP antiserum that would be predicted to impair formation of a CREB–CBP complex (Fig. 5). Following microinjection into nuclei of

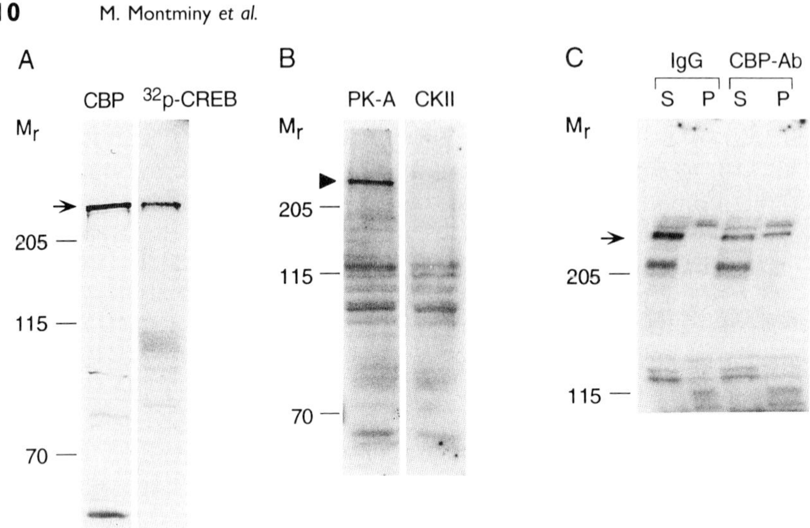

FIGURE 4 CBP is the predominant phospho-Ser133 CREB binding activity in nuclear extracts. (A) Western (CBP) and far-Western ([^{32}P]CREB) blot analyses of Hela nuclear extract following SDS–PAGE and transfer to nitrocellulose. (B) Far-Western blot of crude Hela nuclear extract using [^{32}P]CREB phosphorylated with PKA or casein kinase II (CKII). In A–C, arrow points to major 265-kDa band. M$_r$, relative mass in kDa. (C) Far-Western blot analysis of immunoprecipitates prepared from Hela nuclear extracts with control IgG (IgG) or affinity-purified CBP antiserum. CREB binding activity was detected with [^{32}P]CREB phosphorylated with PKA. S and P, supernant and pellet fractions, respectively.

NIH-3T3 cells, a CRE–lacZ reporter was markedly induced by treatment with 8-Br–cAMP plus IBMX. Coinjection of CBP antiserum with the CRE–lacZ plasmid inhibited cAMP-dependent activity in a dosage-dependent manner (Fig. 5A), but control IgG had no effect on this response (Fig. 5B).

To determine whether CBP antiserum inhibited cAMP-responsive transcription by binding specifically to CBP, we performed peptide-blocking experiments (Arias *et al.*, 1994) (Fig. 5B). CBP antiserum, preincubated with synthetic CBP peptide, was unable to recognize the 265-kDa CBP product on a Western blot (not shown) and could not inhibit CRE–lacZ reporter activity on microinjection into NIH-3T3 cells (Fig. 5B, top). However, antiserum treated with an unrelated synthetic peptide (ILS) retained full activity in both Western and microinjection assays, suggesting that the ability of the antiserum to bind CBP was critical for its inhibitory effect on cAMP-dependent transcription.

To determine whether CBP activity may be restricted to a subset of promoters, we tested several constitutively active reporter constructs (Arias *et al.*, 1994): cytomegalovirus, Rous sarcoma virus, and SV40. When examined in NIH-3T3 cells by transient transfection assay, each of these constructs had comparable basal activity relative to the cAMP-stimulated CRE reporter plasmid (not shown), thereby permitting us to compare the effects of CBP

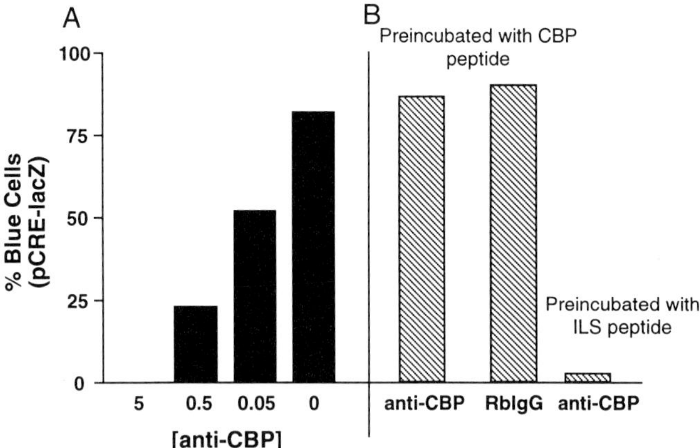

FIGURE 5   CBP is required from cAMP-inducible transcription. (A) Representative field of NIH-3T3 cells injected with CRE–lacZ reporter plasmid plus 5, 0.5, and 0.05 mg/ml of affinity-purified CBP antiserum. Total antibody concentration in microinjected cells was maintained at 5 mg/ml by adjusting with control rabbit IgG. Injected cells were stimulated with 0.5 mM 8Br–cAMP plus IBMX for 4 hr then fixed and assayed for lacZ activity (β-Gal) as well as antibody content (Texas Red-anti-Rb). (B) Effect of CBP antiserum on CRE–lacZ reporter activity following pretreatment of CBP antiserum with synthetic CBP peptide (anti-CBP + CBP) or unrelated peptide (anti-CBP + ILS). Rb IgG + CBP, control rabbit IgG pretreated with CBP peptide. NIH-3T3 cells were injected with CRE–lacZ reporter plus various CBP antisera, stimulated with 0.5 mM 8Br–cAMP plus IBMX for 4 hr, and assayed for lacZ activity. Cells expressing the lacZ gene product form a blue precipitate that quenches immunofluorescent detection of the injected antibody.

antiserum on these reporters directly. Although coinjected CBP antiserum could block cAMP-stimulated activity from a CRE–lacZ reporter in contemporaneous assays, no inhibition was observed on basal expression from any of the constitutive promoter constructs we tested, even when 10-fold lower amounts of reporter plasmid were employed. These results suggest that CBP can indeed discriminate between basal and signal-dependent activities *in vivo*.

With these factors in hand, we can now begin the process of reconstituting PKA-dependent transcription *in vitro* using well-characterized proteins such as CREB, TAF 110, and CBP. The assembly of such factors on cAMP-regulated promoters like somatostatin may thereby permit responsiveness to a variety of hormonal stimuli that employ cAMP as their second messenger.

Acknowledgments

This work was supported by an NIH Grant #GM37828 and the Foundation for Medical Research, Inc. M. M. is a Foundation for Medical Research investigator. This article was previously published for The CIBA Foundation Symposium volume 190, *Somatostatin and its receptors* (Derek J. Chadwick and Gail Cardew, eds.), John Wiley & Sons Ltd., Chichester, U.K.

References

Alberts, A., Arias, J., Hagiwara, M., Montminy, M., and Feramisco, J. (1994). Recombinant cyclic AMP response element binding protein (CREB) phosphorylated on Ser-133 is transcriptionally active upon its introduction into fibroblast nuclei. *J. Biol. Chem.* **269**, 7623–7630.

Arias, J., Alberts, A., Hua, X., Smeal, T., Karin, M., Feramisco, J., and Montminy, M. (1994). Activation of cAMP and mitogen responsive genes relies on a common nuclear factor. *Nature* **370**, 226–228.

Brindle, P., Linke, S., and Montminy, M. (1993). Analysis of a PK-A dependent activator in CREB reveals a new role for the CREM family of repressors. *Nature* **364**, 821–824.

Chien, C., Bartel, P., Sternglanz, R., and Fields, S. (1991). The two hybrid system: A method to identify and clone genes for proteins that interact with a protein of interest. *Proc. Natl. Acad. Sci. U.S.A.* **88**, 9578–9582.

Chrivia, J. C., Kwok, R. P., Lamb, N., Hagiwara, M., Montminy, M. R., and Goodman, R. H. (1993). Phosphorylated CREB binds specifically to the nuclear protein CBP. *Nature* **365**, 855–859.

Comb, M., Burnberg, N. C., Seascholtz, A., Herbert, E., and Goodman, H. M. (1986). A cyclic-AMP- and phorbol ester-inducible DNA element. *Nature* **323**, 353–356.

Dynlacht, B., Hoey, T., and Tjian, R. (1991). Isolation of co-activators associated with the TATA-binding protein that mediate transcriptional activation. *Cell* **66**, 563–576.

Ferreri, K., Gill, G., and Montminy, M. (1994). The cAMP regulated transcription factor CREB interacts with a component of the TFIID complex. *Proc. Natl. Acad. Sci. U.S.A.* **91**, 1210–1213.

Gill, G., Pascal, E., Tseng, Z., and Tjian, R. (1994). A glutamine hydrophobic patch in transcription factor Sp1 contacts the dTAF$_{II}$110 component of the Drosophila TFIID complex and mediates transcriptional activation. *Proc. Natl. Acad. Sci. U.S.A.* **91**, 192–196.

Gonzalez, G. A., Menzel, P., Leonard, J., Fischer, W. H., and Montminy, M. R. (1991). Characterization of motifs which are critical for activity of the cyclic AMP-responsive transcription factor CREB. *Mol. Cell. Biol.* **11**(3), 1306–1312.

Gonzalez, G. A., and Montminy, M. R. (1989). Cyclic AMP stimulates somatostatin gene transcription by phosphorylation of CREB at Serine 133. *Cell* **59**, 675–680.

Gonzalez, G. A., Yamamoto, K. K., Fischer, W. H., Karr, D., Menzel, P., Biggs, W., III, Vale, W. W., and Montminy, M. R. (1989). A cluster of phosphorylation sites on the cyclic AMP-regulated nuclear factor CREB predicted by its sequence. *Nature* **337**, 749–752.

Hagiwara, M., Alberts, A., Brindle, P., Meinkoth, J., Feramisco, J., Deng, T., Karin, M., Shenolikar, S., and Montminy, M. (1992). Transcriptional attenuation following cAMP induction requires PP-1-mediated dephosphorylation of CREB. *Cell* **70**, 105–113.

Hagiwara, M., Brindle, P., Harootunian, A., Armstrong, R., Rivier, J., Vale, W., Tsien, R., and Montminy, M. R. (1993). Coupling of hormonal stimulation and transcription via cyclic AMP-responsive factor CREB is rate limited by nuclear entry of protein kinase A. *Mol. Cell. Biol.* **13**, 4852–4859.

Hoeffler, J. P., Meyer, T. E., Yun, Y., Jameson, J. L., and Habener, J. F. (1988). Cyclic-AMP-responsive DNA-binding protein: Structure based on a cloned placental cDNA. *Science* **242**, 1430–1432.

Hoey, T., Weinzieri, R., Gill, G., Chen, J., Dynlacht, B., and Tjian, R. (1993). Molecular cloning and functional analysis of Drosophila TAF110 reveal properties expected of coactivators. *Cell* **72**, 247–260.

Montminy, M. R., and Bilezikjian, L. M. (1987). Binding of a nuclear protein to the cyclic-AMP response element of the somatostatin gene. *Nature* **328**, 175–178.

Montminy, M. R., Sevarino, K. A., Wagner, J. A., Mandel, G., and Goodman, R. H. (1986). Identification of a cyclic-AMP responsive element within the rat somatostatin gene. *Proc. Natl. Acad. Sci. U.S.A.* **83,** 6682–6686.

Short, J. M., Wynshaw-Boris, A., Short, H. P., and Hanson, R. W. (1986). Characterization of the phosphoenolpyruvate carboxykinase (GTP) promoter-regulatory region. II. Identification of cAMP and glucocorticoid regulatory domains. *J. Biol. Chem.* **261,** 9721–9726.

Yamamoto, K. K., Gonzalez, G. A., Biggs, W. H., III, and Montminy, M. R. (1988). Phosphorylation-induced binding and transcriptional efficacy of nuclear factor CREB. *Nature* **334,** 494–498.

Philip Cohen

MRC Protein Phosphorylation Unit
Department of Biochemistry
The University
Dundee, Scotland, United Kingdom

Dissection of Protein Kinase Cascades That Mediate Cellular Response to Cytokines and Cellular Stress

I. The First Two Protein Kinase Cascades

Almost every aspect of cell life is controlled by the reversible phosphorylation of proteins. About one in every three proteins in mammalian cells is now thought to contain at least some covalently bound phosphate, and the genes encoding protein kinases, protein phosphatases, and their regulatory subunits may account for perhaps 5% of all human genes. To control the biological functions of proteins, it is therefore critical to be able to regulate the activities of protein kinases. Some protein kinases are controlled by allosteric effectors, such as the second messengers, cyclic AMP, calcium ions, and diacylglycerol, but others are themselves regulated by phosphorylation catalyzed by "kinase kinases" that operate in protein kinase "cascades."

The first protein kinase cascade to be identified was that involved in the adrenergic regulation of glycogenolysis in skeletal muscle (Fig. 1). In this pathway, discovered by Edmond Fischer and Edwin Krebs over the

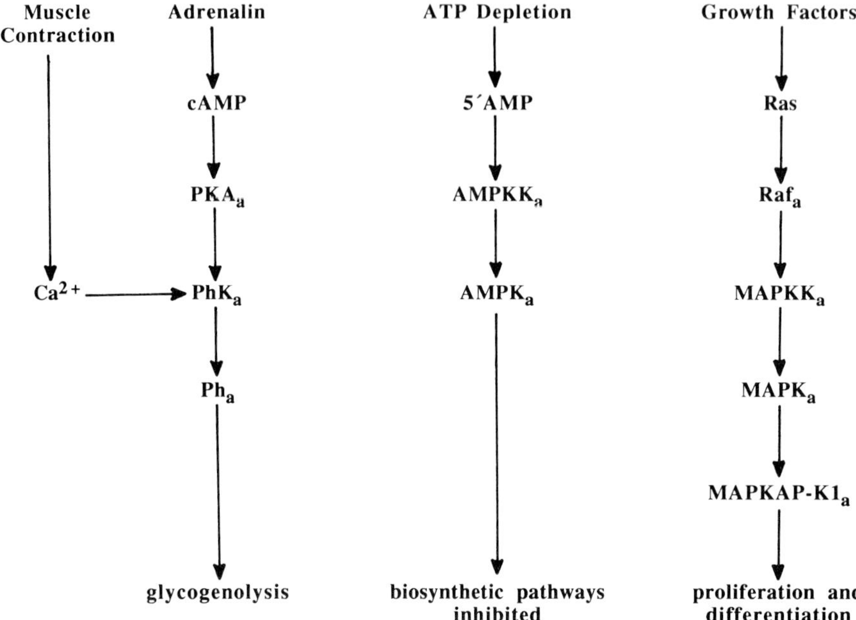

FIGURE 1   The first three protein kinase cascades. Abbreviations: a, activated; cAMP, cyclic AMP; PKA, cyclic AMP-dependent protein kinase; PhK, phosphorylase kinase; Ph, glycogen phosphorylase; AMPK, 5′AMP-activated protein kinase; AMPKK, AMPK kinase; MAPK, mitogen-activated protein kinase; MAPKK, MAPK kinase; MAPKAP-K1, MAPK-activated protein kinase-1.

period 1955–1968, cyclic AMP-dependent protein kinase (PKA) activates the calcium/calmodulin-dependent protein kinase, termed phosphorylase kinase (PhK), which then activates glycogen phosphorylase (Ph). This kinase cascade plays several roles. First, the PKA-catalyzed phosphorylation of PhK not only increases its catalytic activity, but also its sensitivity to Ca^{2+}, thereby integrating the effects of cyclic AMP and Ca^{2+}. Second, the cascade serves to amplify the effect of adrenalin because the molar ratios PKA:PhK:Ph are 1:10:240 in skeletal muscle. Third, the different molar ratios of each enzyme in the cascade allow the system to respond to a number of allosteric effectors whose intracellular concentrations also vary widely. Thus, the concentration of Ph in skeletal muscle (0.1 mM) is appropriate for activation by 5′AMP and inhibition by glucose-6-phosphate, but not for activation by cyclic AMP whose intracellular concentration rarely exceeds 1 μM. Similarly, the intracellular concentrations of PKA (0.2 μM) and PhK (2.5 μM) are appropriate for regulation by cyclic AMP and calcium ions, respectively.

The second kinase cascade, the AMP-activated protein kinase cascade (Fig. 1), was elucidated in the late 1970s and the 1980s (reviewed by Corton *et al.,* 1995). In this pathway, the AMP-activated protein kinase (AMPK) is activated about 5-fold by 5′AMP and a further 10- to 20-fold by phosphorylation catalyzed by AMP-activated kinase kinase (AMPKK), which is itself activated by 5′AMP. As a result, the AMPK is activated 100-fold or more under conditions in which 5′AMP is elevated, i.e., when the intracellular concentration of ATP decreases. The AMPK then conserves cellular ATP levels by inactivating major ATP-utilizing systems. In particular, it inhibits the rate-limiting enzymes in biosynthetic processes such as fatty acid, cholesterol, glycogen, and protein synthesis.

II. The Classical MAP Kinase Pathway

Over the past 5 years, work in a number of laboratories has largely elucidated the MAP kinase (MAPK) cascade (Fig. 1) that plays a critical role in mediating several of the intracellular actions of growth factors. In this pathway, the interaction of growth factors with their receptors triggers a series of protein–protein interactions at the plasma membrane that culminate in activation of the GTP-binding protein Ras. GTP–Ras then interacts with the protein kinase Raf, recruiting it to the plasma membrane where it is activated by an as yet unidentified mechanism. The activation of Raf is followed by the sequential activation of three additional kinases: MAP kinase kinase (MAPKK), MAPK, and MAPK-activated protein (MAPKAP) kinase-1 (also known as p90^{rsk}). The prolonged (30–60 min) activation of MAPK and MAPKAP kinase-1 leads to their translocation from the cytosol to the nucleus where they then regulate the activities of transcription factors, and it is now clear that sustained activation of the MAPK cascade is not only required, but it is sufficient to trigger the proliferation of some cells and the differentiation of others (reviewed by Marshall, 1995).

The activation of MAPKK, MAPK, and MAPKAP kinase-1 results from the phosphorylation of serine/threonine residues located in the same region of the catalytic domain just before the conserved "APE" motif between subdomains VII and VIII (Fig. 2). The activation of MAPKK by Raf involves the phosphorylation of Ser217 and Ser221 (Alessi *et al.,* 1994). Interestingly, only the diphosphorylated species phosphorylated at both Ser217 and Ser221 is observed *in vitro* or *in vivo,* even when the degree of activation of MAPKK is low, indicating that phosphorylation of the first residue triggers an extremely rapid phosphorylation of the second. Furthermore, phosphorylation of either Ser217 or Ser221 is sufficient for maximal activation, and dephosphorylation of both residues is required to inactivate MAPKK. These features are probably important for enabling significant activation of MAPKK to occur, even when the degree of activation of Raf is low. Moreover, only 10–20% activation of MAPKK is sufficient to activate MAPK maximally (Saito *et al.,* 1994).

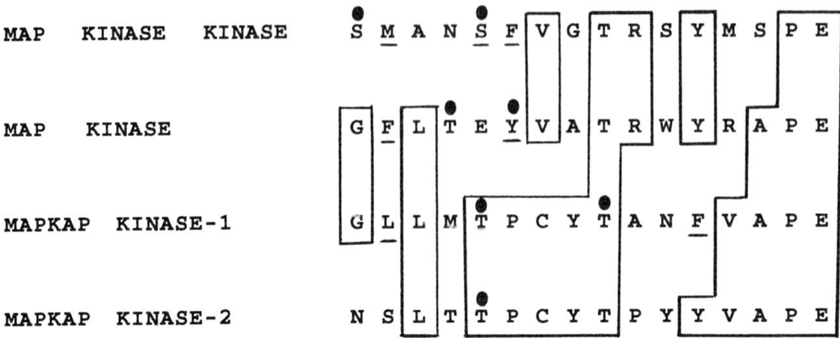

FIGURE 2   The activating phosphorylation sites in MAP kinase kinase, MAP kinase, and MAPKAP kinases are all located just N terminal to the "APE motif" between subdomains VII and VIII. Identities are boxed and conservative substitutions underlined. Phosphorylation sites are denoted by filled circles. The threonine residue in MAPKAP kinase-1 located five residues N terminal to the APE motif is thought to be introduced by autophosphorylation, whereas that located nine residues N terminal to the APE motif is phosphorylated by MAP kinase (Sutherland *et al.*, 1993).

Thus, the MAPK cascade plays a critical role in amplifying the initial stimulus in a situation in which classical second messengers like cyclic AMP and calcium ions may not exist.

The activation of MAPK by MAPKK, like the activation of MAPKK by c-Raf, involves the phosphorylation of two residues, a threonine and a tyrosine (Payne *et al.*, 1991). In contrast to the activation of MAPKK, however, phosphorylation of both residues is required to activate MAPK and inactivation can be achieved by the dephosphorylation of either residue (Anderson *et al.*, 1990). The reason why MAPK is activated by such a dual threonine/tyrosine phosphorylation mechanism is still a matter of debate, although one attractive possibility is to permit dephosphorylation by several types of protein phosphatase, thereby providing sophisticated control over the inactivation process. Recent work suggests that MAPK may be inactivated in the cytosol by protein phosphatase 2A (PP2A), which dephosphorylates the threonine residue specifically and by an as yet unidentified protein tyrosine phosphatase (PTPase). Nuclear MAPK is probably inactivated by a class of "dual specificity" phosphatases (termed CL100), which are structurally related to protein tyrosine phosphatases and dephosphorylate the threonine and tyrosine residues at similar rates (Alessi *et al.*, 1995a). These enzymes are immediate early genes and induced within an hour when cells are stimulated with growth factors or subjected to cellular stresses. However, the induction of these dual specificity phosphatases is not always accompanied by the inactivation of MAPK, suggesting that additional tiers of regulation may operate in many cells.

III. The Identification of MAPKAP Kinase-2

Several years ago, we identified a novel protein kinase in skeletal muscle that was inactivated by incubation with PP2A and reactivated by p42 MAP kinase. This enzyme could be distinguished from MAPKAP kinase-1 in several ways, in particular by its chromatographic properties, substrate specificity, and resistance to the protein kinase inhibitor H7; we therefore termed it MAPKAP kinase-2 (Stokoe *et al.*, 1992a). MAPKAP kinase-2 was purified from rabbit skeletal muscle and the sequences of several tryptic peptides were used to clone its cDNA (Stokoe *et al.*, 1993). The amino acid sequence of MAPKAP kinase-2 revealed a proline-rich N-terminal domain, a striking bipartite nuclear localization sequence at the C terminus, and a catalytic domain most similar to that of the family of calcium/calmodulin-dependent protein kinases (although calcium/calmodulin does not affect the activity of the dephosphorylated or phosphorylated forms of MAPKAP kinase-2). Northern blotting experiments revealed similar levels of the MAPKAP kinase-2 transcript in every tissue examined. Early work identified Thr334 as one phosphorylation site, although recent studies have revealed that Thr222 and Ser272 are also phosphorylated by p42 MAP kinase (Benlevy *et al.*, 1995). Like all serine and threonine residues phosphorylated by MAPK, Thr222, Ser272, and Thr334 are each followed by a proline residue. Thr222 lies in the same region of the catalytic domain as the activating phosphorylation sites in MAPKK, MAPK, and MAPKAP kinase-1, and the amino acid sequence surrounding Thr222 is similar to that surrounding the threonine residue in MAPKAP kinase-1 phosphorylated by p42 MAPK (Fig. 2).

Studies with synthetic peptide substrates have revealed that MAPKAP kinase-2 phosphorylates serine residues in the sequence H-x-R-x-x-S-x-x, where H is a bulky hydrophobic residue (Stokoe *et al.*, 1993). The serine cannot be replaced by threonine and the arginine at position n−3 cannot be substituted by lysine. Like MAPKAP kinase-1, MAPKAP kinase-2 cannot tolerate proline at position n+1, which may ensure that the specificities of these enzymes never overlap with MAPK.

The small heat shock protein (HSP27) is phosphorylated at serine residues that lie in H-x-R-x-x-S-x-x sequences in response to heat shock and other extracellular stimuli reported to activate MAPK in some cells. These observations led us to study the phosphorylation of HSP27 by MAPKAP kinase-2 *in vitro*. We found that both sites on HSP27 phosphorylated *in vivo* are phosphorylated by MAPKAP kinase-2 *in vitro*. Moreover, HSP27 kinase activity copurified with MAPKAP kinase-2 activity throughout the purification of the latter from skeletal muscle, suggesting that HSP27 might be a physiological substrate of MAPKAP kinase-2 (Stokoe *et al.*, 1992b).

IV. MAPKAP Kinase-2 Is Activated *in Vivo* by Reactivating Kinase (RK), a Novel MAPK Homolog Activated by Cellular Stress

It came as a great surprise in late 1993, when we found that nerve growth factor (NGF) and epidermal growth factor (EGF), which potently activate p42 and p44 MAPKs in PC12 cells and A431 cells, respectively, failed to activate MAPKAP kinase-2. However, incubation of PC12 or A431 cells with arsenite (an agonist that mimics many of the effects of heat shock) or subjecting HeLa cells to heat shock potently activated MAPKAP kinase-2 (Rouse *et al.*, 1994). Because arsenite did not activate p42 MAPK or p44 MAPK in PC12 cells, this implied that the activation of MAPKAP kinase-2 by arsenite must be mediated by a distinct protein kinase. This led us to discover a novel protein kinase that reactivated MAPKAP kinase-2 and that is activated when PC12 cells are stimulated with arsenite or when HeLa cells are subjected to heat shock. This enzyme, RK, is eluted from Mono Q at a higher concentration of NaCl than p42 or p44 MAPK (Rouse *et al.*, 1994). Thus, although MAPKAP kinase-2 was originally identified as an enzyme that is activated by p42 or p44 MAPK, the effects of growth factors, arsenite, and heat shock indicate that the RK is the enzyme that activates MAPKAP kinase-2 *in vivo*. This conclusion is further supported by the effects of a specific inhibitor of the RK described under Section VI.

The RK could be inactivated by PP2A, a protein serine/threonine phosphatase, by protein tyrosine phosphatase (PTP)-1B, or by the dual specificity phosphatase CL100 indicating that, like p42 and p44 MAPKs, its activation requires the "dual phosphorylation" of serine/threonine and tyrosine residues (Rouse *et al.*, 1994). We also identified an activator of the RK termed the RK kinase (RKK) in arsenite-stimulated PC12 cells. Like MAPKK, the RKK was inactivated by PP2A but not by PTP-1B. However, unlike MAPKK, it could not be activated by Raf. The RKK could not activate p42 or p44 MAPK, while MAPKK1 could not activate the RK. Moreover, the RK could not activate MAPKAP kinase-1 (Rouse *et al.*, 1994) and MAPKAP kinase-1 could not phosphorylate HSP27 (Stokoe *et al.*, 1992b). Thus, the RK cascade (Fig. 3) appears to be quite separate from the classical MAPK cascade (Fig. 1).

The pharmaceutical company SmithKline Beecham (SB) has developed a novel class of pyridinyl imidazoles called CSAIDS that inhibit the production of the cytokines interleukin-1 (IL-1) and tumor necrosis factor (TNF-α) in monocytes by bacterial lipopolysaccharide (LPS). They found that this drug binds specifically to one protein in monocytes, termed CSBP (CSAID-binding protein), which they isolated, cloned, and sequenced and showed was a novel MAPK homolog (Lee *et al.*, 1994a). I learned about this work in February 1994, when I joined an Advisory Board of SB, and I realized that the stimulation of monocytes by LPS (i.e., bacterial infection) could be

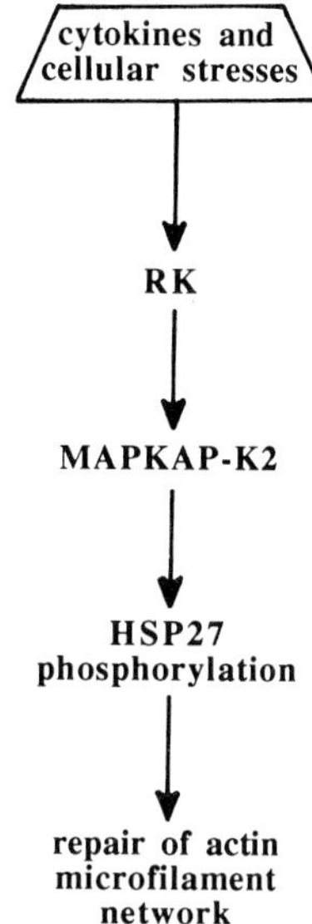

FIGURE 3 The protein kinase cascade that leads to the phosphorylation of heat shock protein 27 in response to cytokines and cellular stresses. The cytokines interleukin-1 and TNF-α, and cellular stresses such as heat shock, osmotic shock, bacterial infection or ultraviolet radiation, trigger the activation of reactivating kinase (RK), which then activates MAPKAP kinase-2 (MAPKAP-K2), the enzyme responsible for phosphorylating HSP27. The phosphorylation of HSP27 is thought to be involved in the repair of the actin microfilament network that is disrupted during cellular stress, thereby aiding cell survival.

regarded as just another cellular stress like heat shock. John Lee and Peter Young at SB generously supplied us with one of the most potent CSAIDs (SB 203580) and we found that it inhibited the RK with an IC_{50} value of 0.6 μM, with essentially complete inhibition occurring at 10 μM (Cuenda *et al.*, 1995). Moreover, after incubation with MgATP and RKK, CSBP (or its *Xenopus* homolog) could now activate MAPKAP kinase-2 (Rouse *et al.*, 1994) and it was inhibited by the same concentrations of SB 203580 that inactivated the RK from arsenite-stimulated PC12 cells or heat-shocked HeLa Cells (Cuenda *et al.*, 1995). In addition, antibodies raised against the *Xenopus* homolog of CSBP recognized the RK from arsenite-stimulated PC12 cells in both immunoblotting and immunoprecipitation experiments (Rouse *et al.*, 1994), establishing that the RK was either identical or very closely related to CSBP.

The amino acid sequence of the RK is about 50% identical to p42 or p44 MAPK (Rouse *et al.*, 1994; Lee *et al.*, 1994a), but an important distinguishing feature is the TGY sequence present in the activation domain instead of the TEY sequence found in p42 and p44 MAPKs (Fig. 2). The RK is most similar to a MAPK homolog in *S. cerevisiae* termed HOG1 (for high osmolarity glycerol), which also contains the TGY sequence and the RK and HOG1 are unique among MAPK homologs in that kinase subdomain VII is separated from the activation domain sequence in subdomain VIII by only 6 residues, whereas this gap is 8 residues in JNK and >12 in any other MAPK homolog (Rouse *et al.,* 1994).

Genetic evidence has revealed that HOG1 lies in a signaling pathway that restores the osmotic gradient across the yeast cell membrane in response to high external osmolarity (i.e., to osmotic stress) by stimulating the synthesis of enzymes that produces glycerol. We found that exposure of mammalian cells to 0.5 *M* sorbitol or 0.3 *M* NaCl for a few minutes activates the RK and MAPKAP kinase-2 to levels as high, or higher, than those observed after heat shock or stimulation with arsenite (Rouse *et al.,* 1994; Cuenda *et al.,* 1995).

V. What Lies Upstream of the RK?

The RK is activated in response to arsenite, heat shock, or osmotic shock, but also when cells are exposed to ultraviolet radiation or stimulated with LPS (Han *et al.,* 1994) or interleukin-1 (Freshney *et al.,* 1994). These findings raise the question of how cells sense these diverse stimuli and trigger activation of the RKK. The protein kinase lying upstream of HOG1 in the yeast osmoregulation pathway is a MAPKK homolog termed PBS2. The enzyme that activates PBS2 is unknown, but genetic experiments have revealed that two further genes, termed SLN1 and SSK1, are required even higher up the pathway (Maeda *et al.,* 1994). Intriguingly, SLN1 and SSK1 are eukaryotic homologs of bacterial "two-component" systems that sense changes in the external environment. By analogy with bacterial two-component systems, it is likely that SLN1 phosphorylates itself at a histidine residue and then transfers the phosphate to particular aspartyl residues present in both SLN1 and SSK1. It has been speculated that the phosphorylation of SSK1 prevents it from inhibiting the kinase that lies upstream of PBS2, thereby turning on the pathway. It would be of great interest to know whether ultraviolet radiation, heat, and osmotic stress activate mammalian homologs of bacterial two-component systems. However, LPS is known to trigger activation of this pathway by binding to the CD14 receptor (Han *et al.,* 1994), whereas signaling by interleukin-1 is also triggered by binding to a specific receptor. It is therefore most unlikely that activation of the RK pathway by LPS and interleukin-1 involves a two-component system. Probably a component lying upstream of the RKK can be activated by several distinct mechanisms, as occurs in the classical MAPK pathway.

Two mammalian MAPKK homologs have been identified, termed MKK3 and MKK4, which resemble yeast PBS2 more closely than they resemble MAPKK1 of the classical MAPK pathway (Sanchez *et al.,* 1994; Derijard *et al.,* 1995). These enzymes have been shown to activate the RK *in vitro* or when overexpressed in mammalian cells. MKK3 when overexpressed in cells was shown to be activated by osmotic stress, ultraviolet radiation, interleukin-1, and tumor necrosis factor-α (Derijard *et al.*, 1995).

Interestingly, MKK4 (also known as SEK1) is also capable of activating a further MAPK homolog, termed JNK or stress-activated protein kinase (SAPK) (Sanchez *et al.*, 1994; Derijard *et al.*, 1995), which is implicated in activation of the transcription factors c-*jun* and *ATF2* (Gupta *et al.*, 1994). JNK also phosphorylates the transcription factor *p53 in vitro* and *in vivo* (Milne *et al.*, 1995), although the biological consequence of the latter event is unknown. JNK is activated *in vivo* by stimuli similar to those that activate the RK, consistent with the notion that both enzymes may be activated by a common "upstream" kinase (such as MKK4). However, MKK3 is apparently unable to activate JNK (Derijard *et al.*, 1995). Moreover, MKK4, but not MKK3, is activated by MEK kinase *in vitro* or when coexpressed in cells with MEK kinase (Yan *et al.*, 1994). Further work is clearly needed to establish whether MKK3, MKK4, and MEK kinase are really the enzymes that regulate the RK and JNK *in vivo*, to identify the enzyme that activates MKK3, and to discover the mechanisms by which MEK kinase and the MKK3 activator are regulated (Fig. 4).

VI. What Lies Downstream of the RK?

SB 203580 appears to be a remarkably specific inhibitor of the RK *in vitro* and *in vivo*. It does not inhibit 12 other protein kinases, including p42 MAPK and JNK, even at 200-fold higher concentrations than those that inhibit the RK by 50% (Cuenda *et al.*, 1995). SB 203580 is a small cell-permeant molecule that does not inhibit activation of the RKK or JNK by arsenite (PC12 cells) or interleukin-1 (KB cells), nor does it inhibit the activation of p42 MAPK by NGF in PC12 cells or by EGF in KB cells. However, we found that SB 203580 prevents the activation of MAPKAP kinase-2 by arsenite in PC12 cells, by LPS in THP1 monocytes, and by arsenite, osmotic stress, or interleukin-1 in KB cells (Cuenda *et al.*, 1995). These observations establish that MAPKAP kinase-2 is a physiological substrate of the RK.

Stimulation of KB cells with arsenite leads to phosphorylation of the threonine and tyrosine residues in the TGY sequence of RK (Doza *et al.*, 1995) and to the phosphorylation of MAPKAP kinase-2 at Thr222, Ser272, and Thr334 within minutes. The RK, like p42 MAPK, phosphorylates Thr222, Ser272, and Thr334 *in vitro*, and SB 203580 prevents the phosphorylation of these three sites *in vivo* (Benlevy *et al.*, 1995). Ser272 is resistant to dephosphorylation by PP2A *in vitro*, unlike Thr222 and Thr334. Because phosphorylation of Ser272 is not sufficient to activate MAPKAP kinase-2 but essential for activation in conjunction with the phosphorylation of either

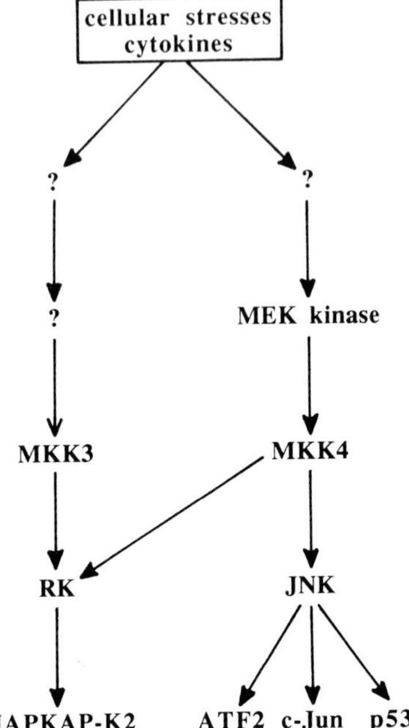

FIGURE 4   Protein kinases that may mediate activation of the MAP kinase homologs RK and JNK in response to cytokines and cellular stress. MKK3 and MKK4 are MAP kinase kinase homologs that are more closely related to PBS2 (the enzyme located "upstream" of HOG1 in the yeast osmoregulation pathway) than MAP kinase kinase-1 (MKK1) of the classical MAP kinase pathway (Fig. 1). MKK4 is activated *in vitro* by MEK kinase and phosphorylates RK and c-Jun kinase (JNK) *in vitro*. MKK3 is not activated by MEK kinase and reported to activate RK, but not JNK. Three physiological substrates for JNK are the transcription factors c-*jun*, *ATF2*, and *p53*.

Thr222 or Thr334, the presence of phosphate in Ser272 could represent a "memory" of a previous stress and perhaps prime the system to respond more rapidly should cells be subjected to a second stress soon after the first.

SB 203580 also prevents the phosphorylation of HSP27 induced by arsenite in PC12 cells and by arsenite, osmotic stress, or interleukin-1 in KB cells (Cuenda *et al.*, 1995). The concentration of SB 203580 required to prevent phosphorylation of HSP27 is similar to that which inhibits the RK *in vitro*. These results establish that HSP27 is a physiological substrate of MAPKAP kinase-2. HSP27 is an actin-binding protein and evidence has been presented that HSP27 phosphorylation promotes the polymerization of actin (Lavoie *et al.*, 1995). This may help to repair the actin microfilament network that is disrupted by cellular stress, thereby aiding cell survival (Fig. 3).

Recent studies employing SB 203580 have helped to identify additional physiological roles of the RK pathway (Fig. 5). SB 203580 and its analogs

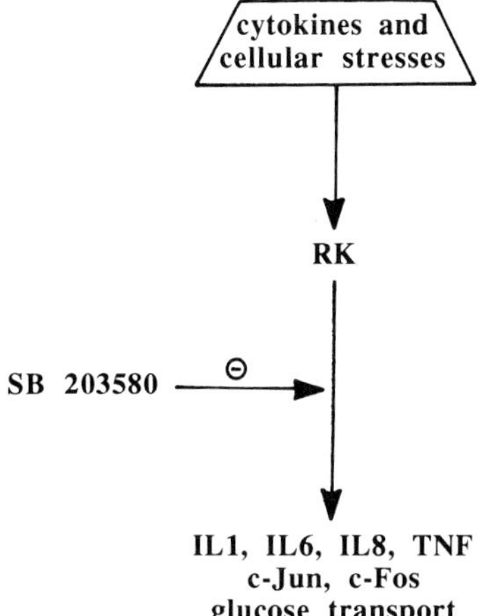

FIGURE 5 New functions of the RK cascade revealed by the use of the inhibitor SB 203580. Abbreviation: IL, interleukin.

were originally identified by their ability to prevent the synthesis of interleukins-1, -6, and -8 and TNF-α in monocytes (Lee *et al.*, 1994a,b), and it has been shown that SB 203580 also inhibits the production of IL-6 and GM–CSF by TNF-α in L929 cells (R. Beyaert and W. Fiers, personal communication). Thus, the RK pathway is not only involved in mediating the synthesis, but also some of the actions of interleukin-1 and TNF-α, and it is clearly involved in the production of many cytokines. However, SB 203580 does not inhibit the activation of NF-κB by TNF-α in L929 cells, illustrating the specificity of SB 203580. These observations suggest that NF-κB is activated by a distinct signaling pathway or by a component of the same pathway lying upstream of RKK.

SB 203580 also inhibits the stimulation of glucose transport in KB cells induced by interleukin-1, but not the stimulation of glucose transport induced by EGF or phorbol esters (G. Gould, personal communication). Similarly, the synthesis of the transcription factor c-*fos* induced by incubating fibroblasts with the protein synthesis inhibitor anisomycin (a potent activator of the RK pathway) is blocked by SB 203580 (L. Mahadevan, personal communication). However, SB 203580 does not prevent the induction of c-*fos* by EGF or phorbol esters (which is mediated by the classical MAPK pathway), again emphasizing the specificity of SB 203580.

The vast array of protein kinases in mammalian cells that are all members of the same superfamily had led many to doubt whether compounds that are really specific inhibitors of a single protein kinase can be synthesized. However, the identification of rapamycin as a specific inhibitor of the activation of p70 S6 kinase (Chung *et al.*, 1992), the identification of compounds that inhibit the EGF receptor protein kinase very potently (Fry *et al.*, 1994) and that block the activation of MAP kinase kinase (Alessi *et al.*, 1995b), together with the identification of SB 203580 have demonstrated that specific inhibitors of protein kinases can indeed be developed. It is likely that the further exploitation of SB 203580 will uncover many further roles of the RK/MAPKAP kinase-2 signaling pathway.

Acknowledgments

I thank the many members of the Medical Research Council (MRC) Protein Phosphorylation Unit who made the discoveries reviewed in this chapter and the MRC, Royal Society, and British Diabetic Association for financial support.

References

Alessi, D., Saito, Y., Campbell, D. G., Cohen, P., Sithanadam, G., Marshall, C. J., and Cowley, S. (1994). Identification of the sites in rabbit MAP kinase kinase-1 phosphorylated by p74raf-1. *EMBO J.* **13**, 1610–1619.

Alessi, D., Gomez, N., Moorhead, G., Lewis, T., Keyse, S. M., and Cohen, P. (1995a). Inactivation of p42 MAP kinase by protein phosphatase 2A and a protein tyrosine phosphatase. *Curr. Biol.* **5**, 283–295.

Alessi, D., Cuenda, A., Cohen, P., Dudley, D., and Saltiel, A. (1995b). PD 098059 is a specific inhibitor of the activation of MAP kinase kinase-1 in vitro and in vivo. *J. Biol. Chem.* **270**, 27489–27494.

Anderson, N., Maller, J. L., Tonks, N. K., and Sturgill, T. W. (1990). Requirement for integration of signals from two distinct phosphorylation pathways for activation of MAP kinase. *Nature* **343**, 651–653.

Benlevy, R., Leighton, I. A., Doza, Y., Attwood, P., Marshall, C. J., and Cohen, P. (1995). Identification of novel phosphorylation sites required for activation of MAPKAP kinase-2. *EMBO J.* **14**, 101–111.

Chung, J., Kuo, C. J., Crabtree, G. R., and Blenis, J. (1992). Rapamycin-FKBP specifically blocks growth-dependent activation of and signalling by the 70 kd S6 protein kinases. *Cell* **69**, 1227–1236.

Corton, J. M., Gillespie, J. G., and Hardie, D. G. (1995). Role of the AMP-activated protein kinase in the cellular stress response. *Curr. Biol.* **4**, 315–324.

Cuenda, A., Rouse, J., Doza, Y. N., Meier, R., Cohen, P., Gallagher, T. F., Young, P. R., and Lee, J. C. (1995). SB 203580 is a specific inhibitor of a MAP kinase homologue which is stimulated by cellular stresses and interleukin-1. *FEBS Lett.*

Derijard, B., Raineaud, J., Barrett, T., Wu, I-H., Han, J., Ulevitch, R. J., and Davis, R. J. (1995). Independent human MAP-kinase signal transduction pathways defined by MEK and MKK isoforms. *Science* **267**, 682–685.

Doza, Y. N., Cuenda, A., Thomas, G. M., Cohen, P., and Nebreda, A. R. (1995). Activation of the MAP kinase homologue RK requires the phosphorylation of Thr-180 and Tyr-182 and both residues are phosphorylated in chemically stressed KB cells. *FEBS Lett.* **364**, 229–233.

Freshney, N. W., Rawlinson, L., Guesdon, F., Jones, E., Cowley, S., Hsuan, J., and Saklatvala, J. (1994). Interleukin-1 activates a novel protein kinase cascade that results in the phosphorylation of Hsp27. *Cell* **78**, 1039–1049.

Fry, D. W., Kraker, A. J., McMichael, A., Ambroso, L. A., Nelson, J. M., Leopold, W. R., Connors, R. W., and Bridges, A. J. (1994). A specific inhibitor of the epidermal growth factor receptor tyrosine kinase. *Science* **265**, 806–808.

Gupta, S., Campbell, D., Derijard, B., and Davis, R. J. (1994). Transcription factor ATF2 regulation by the JNK signal transduction pathway. *Science* **267**, 389–393.

Han, J., Lee, J. D., Bibbs, L., and Ulevitch, R. J. (1994). A MAP kinase targeted by endotoxin and hyperosmolarity in mammalian cells. *Science* **265**, 808–811.

Lavoie, J. N., Lambert, H., Hickey, E., Weber, L. A., and Landry, J. (1995). Modulation of cellular thermoresistance and actin filament stability accompanies phosphorylation-induced changes in the oligomeric structure of heat shock protein 27. *Mol. Cell. Biol.* **15**, 505–516.

Lee, J. C., Laydon, J. C., McDonnell, P. C., Gallagher, T. F., Kumar, S., Green, D., McNulty, D., Blumenthal, M. J., Heys, J. R., Landvatter, S. W., Strickler, L. E., McLaughlin, M. M., Siemens, I., Fisher, S., Livi, G. P., White, J. R., Adams, J. L., and Young, P. R. (1994a). A protein kinase involved in the regulation of inflammatory cytokine biosynthesis. *Nature* **372**, 739–746.

Lee, J. C., Laydon, J. T., and White, J. R. (1994b). Differential effects of the bicyclic imidazoles on cytokine synthesis in human monocytes and endothelial cells. *Agents Actions* **41**, C191–C192.

Maeda, T., Wurgler-Murphy, S. M., and Saito, H. (1994). A two-component system that regulates an osmosensing MAP kinase cascade in yeast. *Nature* **369**, 242–245.

Marshall, C. J. (1995). Specificity of receptor tyrosine kinase signalling: transient versus sustained extracellular signal-regulated kinase activation. *Cell* **80**, 179–185.

Milne, D. M., Campbell, L. E., Campbell, D. G., and Meek, D. M. (1995). p53 is phosphorylated *in vitro* and *in vivo* by an ultraviolet radiation-induced protein kinase characteristic of the c-Jun kinase, JNK1. *J. Biol. Chem.* **270**, 5511–5518.

Payne, D. M., Rossomando, A. J., Martino, P., Erikson, A. K., Her, J. H., Shabanowitz, J., Hunt, D. F., Weber, M. J., and Sturgill, T. W. (1991). Identification of the regulatory phosphorylation sites in pp42/mitogen-activated protein kinase (MAP kinase). *EMBO J.* **10**, 885–892.

Rouse, J., Cohen, P., Trigon, S., Morange, M., Alonso-Llamazares, A., Zamanillo, D., Hunt, T., and Nebreda, A. R. (1994). A novel kinase cascade triggered by chemical stress and heat shock which stimulates MAP kinase-activated protein kinase-2 and phosphorylation of the small heat shock proteins. *Cell* **78**, 1027–1037.

Saito, Y., Gomez, N., Campbell, D. G., Ashworth, A., Marshall, C. J., and Cohen, P. (1994). The threonine residues in MAP kinase kinase-1 phosphorylated *in vitro* by MAP kinase are also phosphorylated in nerve growth factor-stimulated rat phaeochromocytoma PC12 cells. *FEBS Lett.* **341**, 119–124.

Sanchez, I., Hughes, R. T., Mayer, B. J., Yee, K., Woodgett, J. R., Avruch, J., Kyriakis, J. M., and Zon, L. I. (1994). Role of SAPK/ERK kinase-1 in the stress-activated pathway regulating transcription factor c-Jun. *Nature* **372**, 794–798.

Stokoe, D., Campbell, D. G., Nakielny, S., Hidaka, H., Leevers, S., Marshall, C. J., and Cohen, P. (1992a). MAPKAP kinase-2; a novel protein kinase activated by mitogen-activated protein kinase. *EMBO J.* **11**, 3985–3992.

Stokoe, D., Engel, K., Campbell, D. G., Cohen, P., and Gaestel, M. (1992b). Identification of MAPKAP kinase-2 as a major enzyme responsible for the phosphorylation of the small mammalian heat shock proteins. *FEBS Lett.* **313**, 307–313.

Stokoe, D., Caudwell, F. B., Cohen, P. T. W., and Cohen, P. (1993). Specificity and structure of MAP kinase-activated protein kinase-2. *Biochem. J.* **296**, 842–849.

Sutherland, C., Campbell, D. G., and Cohen, P. (1993). Insulin-stimulated protein kinase-1 from rabbit skeletal muscle is the protein product of rsk^{mo}-2; identification of two threonines phosphorylated during activation by mitogen-activated protein kinase. *Eur. J. Biochem.* **212**, 581–588.

Yan, M., Dai, T., Deak, J. C., Kyriakis, J. M., Zon, L. I., Woodgett, J. R., and Templeton, D. J. (1994). Activation of stress-activated protein kinase by MEKK1 phosphorylation of its activator SEK1. *Nature* **372**, 798–800.

Fiona Burns
Allan Z. Zhao
Joseph A. Beavo

Department of Pharmacology
University of Washington
Seattle, Washington 98195

Cyclic Nucleotide Phosphodiesterases: Gene Complexity, Regulation by Phosphorylation, and Physiological Implications

I. Introduction

The importance of $3',5'$-cyclic nucleotides, cAMP and cGMP, has been well established in a wide variety of biological systems. It has been shown that cAMP and cGMP regulate gene transcription (Nagamine and Reich, 1985; Yamamoto *et al.*, 1988), cytoskeleton function (Greengard, 1987), postsynaptic potential (Sarvey *et al.*, 1989), neurotransmitter synthesis and release (Edelman *et al.*, 1981; Kandel and Schwartz, 1982; Nestler and Greengard, 1983; Schoffelmeer *et al.*, 1985), etc. The intracellular levels of cyclic nucleotides are regulated through synthesis by adenylyl and guanylyl cyclases and also degradation by phosphodiesterases (PDEs) (Bentley and Beavo, 1992). In many cases, PDEs have been shown, or at least implicated, to be important not only in maintaining a steady-state level of cAMP and cGMP, but also in modulating rapid oscillation of intracellular cyclic nucleotide level in response to extracellular stimuli (i.e., hormones, odorants, or light).

Our understanding of PDEs has been greatly expanded in the past several years thanks to a large body of molecular cloning, biochemical, and pharmacological studies. We now know that there are seven families of PDEs in mammals, all of which contain a core of 270 amino acids that are highly conserved and believed to form the catalytic domain of the enzyme. The molecular cloning data obtained so far clearly indicate that most of the PDE families consist of more than one gene product, and that many PDE genes are also expressed in multiple-splice variants. Furthermore, in some cases, different splice variant mRNAs are expressed in tissue or cell-type specific manners, and the activities of their encoded enzymes can be differentially regulated (Yan *et al.*, 1995; Sette *et al.*, 1994b; Sonnenburg *et al.*, 1993). It is likely that new members will be added to this PDE "superfamily" in the coming years.

In the following sections, we attempt to summarize much of the progress in the PDE field on the complexity of PDE genes and their expressions, the regulation by phosphorylation, and their physiological relevance. For this review, our attention will be mainly focused on those PDEs regulated by phosphorylation, which include PDE1 (Ca^{2+}/CaM activated), PDE3 (cGMP inhibited), PDE4 (cAMP specific), and PDE5 (cGMP binding). The PDE7 gene has not been well characterized and there has been no evidence that PDE2 (cGMP stimulated) is regulated by phosphorylation. In addition, the PDE6 (photoreceptor) family has been extensively reviewed elsewhere (Takemoto *et al.*, 1993). The nomenclature used in this chapter was adopted at the ASPET Colloquium on Multiple Phosphodiesterases, Newport Beach, California, April 22–23, 1994, and is described in a review of that meeting (Beavo *et al.*, 1994).

II. The Complexity and Expression of PDE Genes

Much of the progress in the PDE field has been centered around the characterization of cDNAs encoding different PDEs and their expression profiles. To better understand the structure and function of various PDEs, the regulation of their activities, and their physiological relevance, we believe it necessary to briefly summarize our current knowledge about the complexity of PDE genes and their expression.

A. PDEI Family (Ca^{2+}/CaM-Activated PDE)

Of all types of PDEs, Ca^{2+}/CaM-activated PDE is one of the most extensively studied and characterized. So far, there have been three different genes characterized within this family, i.e., PDE1A, PDE1B, and PDE1C

(Bentley and Beavo, 1992; Sonnenburg *et al.*, 1993; Yan *et al.*, 1995; Loughney *et al.*, 1996). Although PDE1A and PDE1B preferrentially hydrolyze cGMP, PDE1C has high affinity for both cAMP and cGMP (Yan *et al.*, 1995). The homology within the conserved catalytic domains of PDE1A, PDE1B, and PDE1C is greater than 80% at the amino acid level. There are at least two splice variants for both PDE1A and PDE1B, and at least five splice variants for PDE1C. Based on recent cDNA cloning and biochemical studies, we now know that the PDE1A gene encodes the previously characterized 58-kDa bovine lung isozyme, 59-kDa bovine heart isozyme, and 61-kDa bovine brain isozyme (Hansen and Beavo, 1982; Sharma *et al.*, 1984; Sharma and Wang, 1986), and that PDE1B encodes the so-called "63-kDa" CaM PDE originally purified from bovine brain (Bentley *et al.*, 1992). The recently identified PDE1C represents a group of complex splice variants with sequence differences at both 5' and 3' portions of the cDNAs. Although the calculated molecular weight of certain PDE1C splice variants is about 75 kDa, it remains to be determined whether these PDE1C splice variants actually encode the previously characterized 74-kDa CaM-activated PDE (Shenolikar *et al.*, 1985).

With various molecular techniques (such as *in situ* hybridization and RNase protection assay), we have begun to extensively map the expression profiles of all the PDE genes. One of the best examples of these studies is the recent *in situ* hybridization analysis of PDE1 gene expression in the mouse central nervous system (CNS) (Fig. 1) (Yan *et al.*, 1993; Yan and Beavo, 1994). It is apparent that different isozymes of the PDE1 families are expressed in distinct spatial domains of the CNS. Whereas the expression of PDE1A is restricted to the cerebral cortex, hippocampal pyramidal cell layer, and hippocampal CA1 and CA2 regions, PDE1B is abundantly expressed in the caudate–putamen, nucleus accumbens, olfactory tubercle, olfactory bulb, and granule layer of dendate gyrus. The distribution pattern of PDE1C is distinct from that of PDE1A and PDE1B with the highest level in neurons of the granule layer of the cerebellum, certain Purkinje cells, central amygdaloid nuclei, interpolar spinal trigem nucleus, glomerular, and external plexiform layer of olfactory bulb. The distinct spatial distribution profiles of these PDE1 genes clearly indicate their region-specific function in the CNS. For example, the expression pattern of PDE1B in the striatal region closely resembles that of dopamine receptors D_1 and D_2. One might expect that PDE1B would be involved in attenuating the duration and amplitude of the dopamine-regulated cAMP signal.

One of the interesting aspects of PDE1C expression is the recent realization that one of its splice variants, PDE1C2, is highly enriched in the olfactory epithelium sensory neurons (Yan *et al.*, 1995). Of several rat tissue RNAs examined by RNase protection assay, PDE1C2 was seen only in the olfactory epithelium, whereas other PDE1C splice variants were detected in cerebellum,

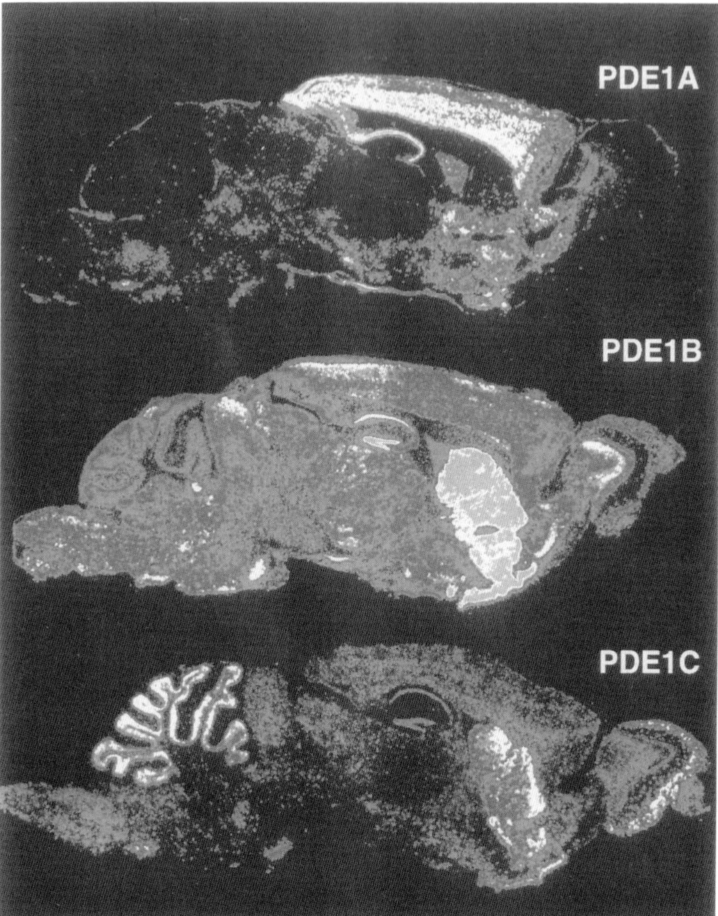

FIGURE 1   Spatial expression profiles of PDE1 family members in mouse central nervous system. These are darkfield digital images of *in situ* hybridization signals from different radiolabeled mouse PDE1 antisense probes in sagittal sections of mouse brain. The description of the expression profiles is given in the text. Data taken from Yan *et al.* (1993, 1995).

forebrain, testis, and heart. Unlike PDE1A and PDE1B, PDE1C2 displays very high affinity toward cAMP with K_m values about 20 or even 100 times lower than those of PDE1B or PDE1A, respectively (Yan *et al.*, 1995). In addition, PDE1A and PDE1B transcripts have been shown by *in situ* hybridization analysis to be absent in olfactory epithelium. Although further evidence is nceded, PDE1C2 most likely encodes the previously described CaM-stimulated PDE activity (with a high affinity for cAMP) present in olfactory cilia (Borisy *et al.*, 1992). Such a PDE activity has been shown to be important in modulating the subsecond cAMP signal in olfactory cilia (Jaworski *et al.*, 1995). An important feature of the olfactory transduction signal is its transient nature so that the receptor cells can be repeatedly stimulated. Following odorant stimulation, cAMP levels can decline nearly as fast (within ~200 msec) as they can rise (within ~50–100 msec) (Breer *et al.*, 1990). Therefore, concomitant with the shutoff of cAMP synthesis by phosphorylation of odorant receptors (Boekhoff and Breer, 1992; Boekhoff *et al.*, 1992; Dawson *et al.*, 1993; Schleicher *et al.*, 1993), there has to be one or more PDEs in olfactory epithelium responsible for the rapid removal of odorant-induced cAMP. The high level of PDE1C2 expression and its high affinity for cAMP suggest a major role of PDE1C2 in this scheme (Fig. 2). We emphasize that PDE1C2 is by no means the only PDE expressed in olfactory epithelium. PDE2, PDE3, and PDE4 have all been shown to be expressed in olfactory epithelium (D. Juilfs and J. Beavo, manuscript in preparation; A. Zhao and J. Beavo, unpublished observation).

What makes the PDE1C expression even more complex is its species variation. Based on RNase protection assay, although PDE1C is heavily expressed in mouse testis, its expression is barely detectable in rat testis (Yan *et al.*, 1995; A. Zhao, C. Yan, and J. Beavo, unpublished observation). This result is consistent with the previous biochemical observation from several laboratories that a CaM PDE with high affinity for cAMP can be easily detected in mouse testis, but not from rat testis of similar age (Geremia *et al.*, 1984; Rossi *et al.*, 1988). Endoh *et al.* (1995) have shown that although PDE3 is crucial in most mammalian cardiac muscle, hydrolysis of cAMP by cardiac PDE shows a wide range of species-dependent variations. More studies are needed to determine whether different species use different PDEs to achieve the same physiological consequence or use different PDEs to adapt to the physiological differences (of the same tissues) among species.

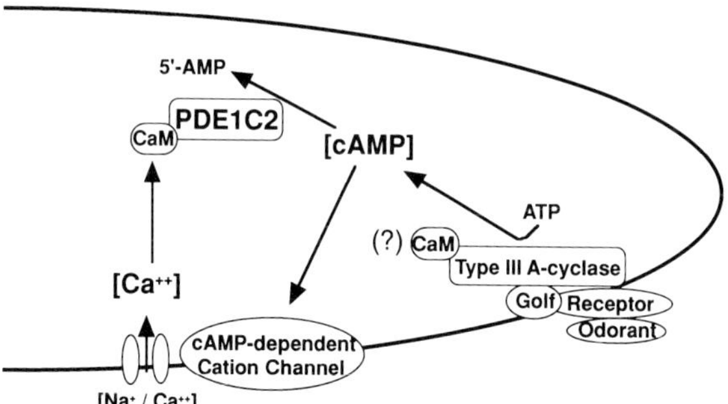

FIGURE 2 A model for the potential role of CaM PDE in odorant signal transduction pathways. Olfactory sensory pathway initially involves activation of odorant receptors. Activation of the receptor by the stimulus is transduced via a G protein (G_{olf}) to adenylyl cyclase. Activation of adenylyl cyclase results in an increase in the intracellular concentration of cAMP. Opening of cyclic nucleotide-dependent ion channels as a result of elevation of the cAMP level generates the electrical signal. The rapid termination of the response to the stimulus is proposed to be at least partially due to the stimulation of cAMP hydrolysis that is regulated by Ca^{2+} that enters through the channel and is mediated by the CaM PDE, PDE1C2.

B. PDE3 Family (cGMP-Inhibited PDE)

The PDE3 family (also known as the cGMP-inhibited PDE) consists of two members, PDE3A and PDE3B, both of which exhibit very similar kinetic properties (Manganiello *et al.*, 1992b). The full-length cDNAs encoding the human myocardial PDE3A and the rat adipocyte PDE3B have been isolated (Meacci *et al.*, 1991; Taira *et al.*, 1991). Consistent with previous biochemical data, the open reading frame of both these clones predicts a molecular weight of roughly 125 kDa (Beavo, 1988).

Earlier studies reported substantial variation in the molecular weight of the PDE3 isozymes, ranging from 63 kDa for the rat liver isoenzyme to 135 kDa for adipocyte PDE3 (Pyne *et al.*, 1987; Degerman *et al.*, 1990). This variation was very likely due to limited proteolysis of these PDE3 isoforms (Beavo, 1988). Based on assays of the purified PDE3 isozymes, these enzymes have a high affinity for both cAMP and cGMP as substrates, with K_ms ~1 μM. However, the V_{max} for cAMP hydrolysis is approximately 10-fold greater than the V_{max} for cGMP. Cyclic GMP potently inhibits cAMP hydrolyzing activity, which is thought to be due to competition at the catalytic site because the K_i is about the same as the K_m for cGMP hydrolysis (Degerman *et al.*, 1987; Harrison *et al.*, 1986). Given this, it is thought that the levels of cGMP within the cell are not substantially modulated by PDE3s, but that cGMP may regulate cAMP levels through the inhibition of PDE3. For example, one study indicates that cGMP will inhibit PDE3A activity and increase cAMP in intact platelets (Maurice and Haslam, 1990).

By RNA blot and RNase protection analysis, it is now know that PDE3A is mainly expressed in heart and platelets, whereas PDE3B is expressed mainly in adipose tissue and to a lesser extent in liver and brain (Meacci *et al.*, 1991; Taira *et al.*, 1991). Preliminary *in situ* hybridization analysis has mapped PDE3B expression to hippocampal and cerebellum regions (A. Zhao, C. Yan, and J. Beavo, unpublished observation). These results are consistent with those of recent biochemical and pharmacological studies using synthetic compounds, such as milrinone and cilostamide, that selectively and potently inhibit PDE3 activity (Pang, 1992). PDE3B has been purified from adipose tissue and rat liver, and PDE3A from cardiac tissue (Harrison *et al.*, 1986; Degerman *et al.*, 1987, 1988; Boyes and Loten, 1988; Pyne *et al.*, 1987). PDE3 isozymes in most cases are found in tissues in which cAMP and cGMP have similar functions such as platelets and vascular smooth muscle (Grant and Colman, 1984; Macphee *et al.*, 1986; Rascon *et al.*, 1992). In these tissues, it is likely that an increase in cGMP may augment the cAMP signal through inhibition of PDE3A activity.

The subcellular localization of the PDE3s differ in various cell types. Whereas most of the platelet PDE3A activity is cytosolic (Grant and Colman, 1984; Macphee *et al.*, 1986), in adipocytes and hepatocytes the activity appears to be associated with the membrane (Heyworth *et al.*, 1983; Manganiello and Vaughn, 1973; Zinman and Hollenberg, 1974). It has been suggested that the subcellular localization may be important in the functions of these enzymes. For example, although a sarcoplasmic reticulum-associated PDE3A accounts for only 1 or 2% of the total cAMP hydrolyzing activity in human hearts, several studies have shown that there is a correlation between the ability of cardiotonic agents to inhibit this enzyme and the increase of cardiac contractility (Kauffman *et al.*, 1987; Weishaar *et al.*, 1986).

C. PDE4 Family (cAMP-Specific PDEs)

The PDE4 isozymes, or cAMP-specific PDEs, preferentially hydrolyze cAMP with the K_m for cAMP hydrolysis occurring at ~1 μM in crude preparations (K_m cGMP >50 μM) (Conti *et al.*, 1992). Like the PDE1 family, the PDE4 gene family is also complex. To date, four distinct PDE4 genes, named PDE4A, PDE4B, PDE4C, and PDE4D, have been isolated by several groups. To add to this complexity, at least three of the four PDE4 genes express multiple proteins with heterologous N terminals (Davis *et al.*, 1989; Obernolte *et al.*, 1993; Swinnen *et al.*, 1989a,b). This is caused by either alternate splicing at the 5' end of the mRNA or alternative transcription start sites. Consequently, these mRNA variants generate the expression of PDE4 isoforms with different molecular weights and properties. It has been proposed that the differences in the 5' ends of these mRNAs may determine how the expressed proteins are regulated by hormones or subcellular location (see following sections).

The isolation of PDE4 genes has allowed researchers to determine the kinetic parameters of each PDE4 isozyme in a defined system (i.e., transfection assay or bacterial expression) free of other PDEs or other PDE4 family members (Obernolte *et al.*, 1993). Previous attempts at analyzing PDE4 kinetic parameters with tissue extracts have been hampered by the contamination of PDE3 due to the coexpression of PDE3 and PDE4 in many tissues. Obviously, the existence of multiple gene members and mRNA variants could have further complicated those efforts.

The complexity of the PDE4 gene family also partially explains why there has been substantial variation in the reported molecular mass and structure of these PDEs purified from crude tissues. Obviously, the possibility that some of the variation was due to limited proteolysis cannot be ruled out. For example, the molecular mass of the dog kidney (Epstein *et al.*, 1982; Thompson *et al.*, 1988) and the rat liver peripheral (Marchmont *et al.*, 1981) PDE4 isozymes have been reported as 71 and 52 kDa, respectively, whereas the molecular mass of native Sertoli cell PDE4 has been reported to be 135–150 kDa (Conti *et al.*, 1992).

Low levels of PDE4 activity have been reported in many tissues. PDE4 activity has been found in kidney, gonads, cardiac muscle, liver, tracheal smooth muscle, and in various regions of the brain (Epstein *et al.*, 1982; Lobban *et al.*, 1994; Marchmont *et al.*, 1981; McLaughlin *et al.*, 1993; Rossi *et al.*, 1985; Thompson *et al.*, 1988; Torphy *et al.*, 1991; Weishaar *et al.*, 1987). It also appears to be the major cAMP hydrolyzing activity in inflammatory cells (Dent *et al.*, 1991; Livi *et al.*, 1990; Obernolte *et al.*, 1993; Peachell *et al.*, 1992; Souness *et al.*, 1991; Turner *et al.*, 1993). Using a PCR approach, Conti's group has also found that some of the PDE4 mRNA variants are expressed in a tissue-specific manner (Monaco *et al.*, 1994).

PDE4 activities have been found in both cytosol and membrane depending on cell types. In endocrine cells the PDE4 activity is mainly cytosolic, whereas PDE4 activities have been associated with the membrane fraction in hepatocytes (Marchmont *et al.*, 1981). Two PDE4B isoforms have been found to be differentially expressed in the membrane (PDE4B2) and the cytosol (PDE4B1) of the brain (Houslay and Lobban, 1994; Lobban *et al.*, 1994; Obernolte *et al.*, 1993). These two isoforms differ in their N-terminal domains, indicating differences in this domain can determine subcellular localization of PDE4 isozymes (Shakur *et al.*, 1993).

D. PDE5 Family (or cGMP-Binding PDEs)

The PDE5, also termed the cGMP-binding PDE (cGB-PDE), preferentially hydrolyzes cGMP, with an estimated K_m of 5 μM (Francis *et al.*, 1990). Unlike the PDE1 isozymes that can also hydrolyze cGMP with high affinity, PDE5 activity is unaffected by Ca^{2+} and CaM or by pretreatment with trifluoperazine, a CaM antagonist. PDE5 activity can be inhibited by zaprinast (IC_{50} 0.5–1 μM) and dipyridimole (IC_{50} 1.5–10 μM) (Lugnier *et al.*, 1986). These drugs also inhibit PDE1 activity but the IC_{50} values are ~10-fold higher than those for the PDE5 enzyme (Ahn *et al.*, 1992).

cDNAs encoding the bovine and mouse PDE5 have been isolated (McAllister-Lucas *et al.*, 1994; F. Burns and J. Beavo, unpublished results). Alignment of the deduced amino acid sequences of the two cDNAs indicates that besides the conserved catalytic domain, both encode a conserved homologous region near the N terminus. This region contains two internal repeats, and studies have shown that these repeats encode two sites with distinct cGMP-binding properties (McAllister-Lucas *et al.*, 1994). Previous biochemical studies demonstrated that in addition to binding at the catalytic site, cGMP also binds to distinct nonhydrolytic sites on PDE5 (Francis *et al.*, 1980). It is believed that by interacting with the catalytic site, cGMP can induce a conformational change and increase cGMP binding to these nonhydrolytic sites (Francis *et al.*, 1980; Thomas *et al.*, 1990a). This hypothesis is supported by the observations that nonselective inhibitors, such as IBMX, that specifically interact with the catalytic site can increase cGMP binding 2- to 5-fold (Francis *et al.*, 1980; Hamet *et al.*, 1984). It is unclear whether cGMP binding to the noncatalytic sites can regulate cGMP hydrolysis. However, it has been suggested that by binding to these sites, cGMP stimulates cAMP-dependent protein kinase (cA-PK)- or cGMP-dependent protein kinase (cG-PK)-mediated phosphorylation of PDE5 (see the following for details). The binding of cGMP to nonhydrolytic sites has also been observed in two other PDE families, the PDE2s (cGMP-stimulated PDE) and the PDE6s (photoreceptor PDEs) (Gillespie and Beavo, 1989; Stroop and Beavo, 1991).

At least *in vitro*, PDE5 forms a homodimer (Francis *et al.*, 1990). The molecular mass of the native platelet and lung PDE5 has been reported to be ~180 kDa, with each monomer being ~93 kDa (Francis and Corbin, 1988; Hamet *et al.*, 1984; Thomas *et al.*, 1990a). This is consistent with the predicted molecular weight of 99 kDa from the ORF of the isolated PDE5 cDNA (McAllister *et al.*, 1993).

So far the tissue distribution of PDE5 has been studied mainly by biochemical and pharmacological methods. Inhibition of PDE5 activity by zaprinast has been used to determine the cellular location, tissue distribution, and physiological function of PDE5. As would be expected by its selectivity for cGMP hydrolysis, PDE5 has been found in tissues, such as smooth muscle, lung, platelets, and kidney, in which cGMP has been reported as an important physiological mediator (Ahn *et al.*, 1992; Hamet *et al.*, 1984; Malta, 1989; Wilkins *et al.*, 1990). PDE5 isozymes from lung and platelets have been purified to apparent homogeneity (Francis and Corbin, 1988; Hamet *et al.*, 1984; Thomas *et al.*, 1990a). The amounts of PDE5 expressed in the lung appear to be similar to those of cG-PK (Mumby *et al.*, 1982; Francis *et al.*, 1990). *In vivo* studies have shown that zaprinast can augment ANP-induced natriuresis and diuresis, suggesting that PDE5 is not only localized in the kidney but is important in regulating cGMP levels in this tissue (Wilkins *et al.*, 1990). Studies to date indicate that all PDE5s are cytosolic enzymes.

III. Regulation of PDE Activity by Phosphorylation

As discussed previously, various biological mediators, such as CaM and cGMP, have the ability to interact with and regulate the activity of specific PDE family members. In addition, regulation of cellular PDE activity via phosphorylation–dephosphorylation reactions has been widely reported. In this section, the current literature pertaining to the phosphorylation of isozymes within the PDE1, PDE3, PDE4, and PDE5 gene families is reviewed. Speculation on the physiological significance that phosphorylation may play in the modulation of cyclic nucleotide levels and on the possible effect of this mode of regulation on cellular function is presented.

A. Regulation of PDE1 Activity

Both the 59-kDa heart and 61-kDa brain isozymes (PDE1A1 and PDE1A2) have been shown to be phosphorylated *in vitro* by cA-PK (Florio *et al.*, 1994; Hashimoto *et al.*, 1989; Sharma and Wang, 1985). PDE1B, however, has been demonstrated to be phosphorylated *in vitro* by CaM-dependent kinase II (Sharma and Wang, 1986; Hashimoto *et al.*, 1989). Phosphorylation causes a decrease in binding affinity of CaM to both PDE1A and PDE1B, thereby decreasing enzyme activity. In the case of PDE1A1 and PDE1A2, site-directed mutagenesis has revealed that it is the phosphorylation on Ser120 that is responsible for the change in CaM affinity (Florio *et al.*, 1994).

The phosphorylation of CaM PDE has raised interesting issues regarding the possible physiological roles of these PDEs in the central nervous system. It has been demonstrated that CaM kinase II is involved in synaptic plasticity (Hanson and Schulman, 1992). Given that PDE1B is highly expressed in the dentate gyrus of the hippocampus, it seems possible that at least part of the effects of CaM kinase II could be mediated by phosphorylation of PDE1B, such that the decreased CaM PDE activity would help sustain the level of cAMP-mediated effects on synaptic strength and neuronal gene expression. Similarly, one could imagine that PDE1A might be involved in cAMP-mediated long-term potentiation (or depression). Thus, phosphorylation of PDE1A by cA-PK could help decrease its activity, therefore prolonging the amplitude and duration of the cAMP signal. This could allow a small signal initiated at a nerve ending to reach the cell body and nucleus so that longer-term transcriptional changes could trigger even more lasting changes in the synapse. Subsequent elevation of Ca^{2+} should eventually overcome the PDE1A inhibition and reset the whole system.

B. Regulation of PDE3 Activity

There is substantial *in vitro* and *in vivo* data supporting the idea that phosphorylation is important in regulating the cellular function of the PDE3 isozymes. Probably the most extensive phosphorylation studies have been performed on the particulate adipocyte PDE3B (Manganiello *et al.*, 1992a). In intact cells, both isoprenaline and insulin can stimulate the phosphorylation and activity of PDE3B (Degerman *et al.*, 1990; Smith *et al.*, 1991). Isoprenaline mediates its effects via the activation of cA-PK (Degerman *et al.*, 1990; Gettys *et al.*, 1988). *In vitro* data suggest that phosphorylation occurs at Ser-427 located within the N-terminal domain of PDE3B (Rascon *et al.*, 1994). Phosphorylation in response to insulin treatment appears to occur at a different serine residue(s) and is mediated via an unidentified serine/threonine kinase (Degerman *et al.*, 1990; Smith *et al.*, 1991). Although isoprenaline and insulin can activate PDE3B activity, stimulation in the presence of both these agents is greater than that with either agent alone (Smith *et al.*, 1991). The ability of these agents to synergistically activate adipocyte PDE3B appears contrary to what is known about the effects of these agents. Isoprenaline stimulates lipolysis via the elevation of cellular cAMP, whereas insulin mediates its antilipolytic action by lowering cellular cAMP. However, it has been observed that the metabolic effects of insulin are more pronounced after cAMP levels have already been elevated. Therefore, during lipolysis the synergistic action of insulin on catecholamine-activated PDE3B would presumably rapidly reduce cellular cAMP, cA-PK activation, and lipolysis, therefore facilitating the antilipolytic action of insulin (Fig. 3).

Cyclic AMP-PK can also phosphorylate and activate PDE3A isozymes from platelets, heart, and aortic smooth muscle (Grant *et al.*, 1988; Harrison *et al.*, 1986; Macphee *et al.*, 1988; Rascón *et al.*, 1992). In these cells, agonist-stimulated elevation in cAMP is usually transient. The rapid decrease in cellular cAMP levels is probably due in part to the activation of these PDE3A isoenzymes, suggesting that this is an important "negative feedback" mechanism and contributes to the functioning of the cell. For example, elevated cAMP levels inhibit platelet function. Therefore, a rapid decrease in cAMP could facilitate the prompt activation of platelets by agonists, an important regulatory mechanism in the blood coagulation response.

C. Regulation of PDE4 Activity

In many cell types, including Sertoli cells, lymphocytes, and monocytes, an increase in PDE4 activity has been associated with long-term stimulation by activators of adenylyl cyclase (Swinnen *et al.*, 1989b; Obernolte *et al.*, 1993; Torphy *et al.*, 1992). It was determined that this increase was not due to post-translational modification, but rather to transcriptional activation of PDE4. In addition, other PDE4 isozymes can be regulated in the short-term by post-translational modification. Incubation of thyroid FRTL-5 cells with

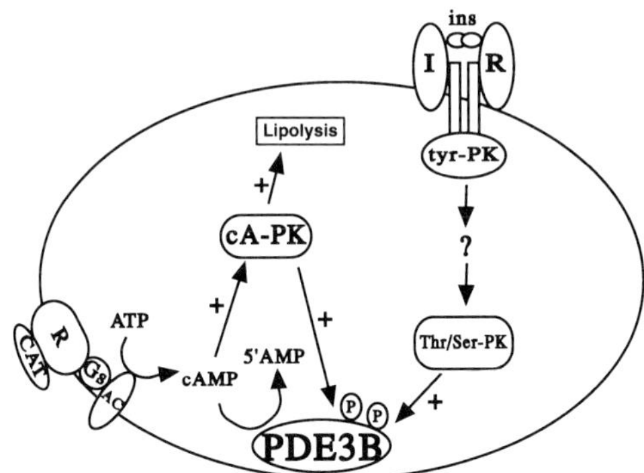

FIGURE 3   By binding to their cell surface receptors on adipocytes, catecholamines (CAT) can activate adenylyl cyclase. The resulting increase in cAMP can stimulate cA-PK, which phosphorylates hormone-sensitive lipase to mediate lipolysis. In addition, cA-PK can also phosphorylate and activate the membrane-bound PDE3. In addition, the PDE3 can be further phosphorylated and activated by an insulin-stimulated serine/threonine kinase.

thyroid-stimulating hormone (TSH) produces a biphasic increase in PDE4 activity. The initial increase occurs within 15 min and is associated with incorporation of $^{32}P_i$ into a 93-kDa polypeptide with identical electrophoretic properties to thyroid PDE4D, whereas the delayed increase was associated with an increase in PDE4D protein (Sette *et al.*, 1994a,c). The short-term effects were mimicked by forskolin and dibutyryl cAMP indicating that phosphorylation and activation were mediated by cA-PK. Phosphorylation and activation are not only observed with cA-PK, insulin is also known to stimulate a membrane-associated PDE4 activity in hepatocytes (Heyworth *et al.*, 1983). Phosphoamino analysis and antiphosphotyrosine blotting indicate that phosphorylation is on tyrosine residue(s) (Pyne *et al.*, 1989). This insulin-mediated phosphorylation and activation is thought to contribute to the antagonism of glucagon by insulin in the liver.

The temporal differences in the activation of PDE4 activity probably reflect physiological function. In the case of insulin, an immediate, rapid decrease in cellular cAMP levels is required to antagonize glycogenolysis and gluconeogenesis. The long- and short-term increases in PDE4 activity correlate with the desensitization of these cells to adenylyl cyclase activators. Thyroid cells become rapidly desensitized to TSH and other cAMP-elevating agents, whereas desensitization in Sertoli cells is slower. However, it could be speculated that as well as playing a role in tachyphylaxis, the short- and long-term regulation of PDE4 activity is important in other aspects of cell function. In contrast to other cell types, both proliferation and differentiation can be induced by TSH-mediated elevation in cAMP in thyroid cells (for review see Dumont *et al.*, 1992). Therefore, it could be speculated that in the thyroid, different "pools" of cAMP are involved in these processes and that the biphasic stimulation of PDE4 activity may be involved in regulating these different pools. Sertoli cells and germ cells are localized together in the seminiferous tubules. In response to FSH and the resulting increase in cAMP, Sertoli cells secrete peptides and other components into these tubules that are required for germ cell differentiation (Delmas and Sassone-Corsi, 1994). Therefore, the long-term induction the PDE4 activity may be involved in facilitating this process.

D. Regulation of PDE5 Activity

There are a few reports describing the phosphorylation of PDE5 isozymes. These have been mainly carried out by Corbin and co-workers on the purified bovine lung PDE, although other researchers have described the phosphorylation of a partially purified PDE5 from guinea pig lung. In addition, some recent data suggest that PDE5 can be phosphorylated *in vivo*. These reports are summarized as follows.

Thomas *et al.* (1990b) observed that purified bovine lung PDE5 is a substrate for both cA-PK- and cG-PK-catalyzed phosphorylation *in vitro*. Nevertheless, on phosphorylation, no change in the cGMP hydrolyzing activity of this PDE was observed. In the presence of cG-PK, PDE5 incorporated $^{32}P_i$ at a 10-fold faster rate than in the presence of equimolar concentrations of cA-PK, suggesting that the bovine lung PDE5 is a better substrate for cG-PK than for cA-PK. The phosphorylation site has now been identified as Ser-92 within the N-terminal domain of the protein (McAllister *et al.*, 1993). Phosphorylation of this isoenzyme was dependent on the presence of cGMP (Thomas *et al.*, 1990b). In the absence of cGMP, neither cA-PK nor cG-PK (activated by 8-bromo-cGMP, which has no effect on PDE5) catalyzed the phosphorylation of this PDE. Thomas *et al.* (1990b) observed that the effect of cGMP on phosphorylation could be mimicked by IBMX or nonhydrolyzable analogs. They therefore suggested that cGMP mediates its effects by binding to the noncatalytic site on the enzyme, inducing a conformational change that exposes the phosphorylation sites. These results suggest that the bovine lung PDE5 may be regulated by cG-PK in a cGMP-dependent manner.

In other studies, guinea pig lung PDE5 activity could be stimulated 11-fold in the presence of the catalytic subunit of cA-PK and ATP (Burns *et al.*, 1992). This activation could be reversed by addition of phosphatases suggesting a phosphorylation event was involved (Burns and Pyne, 1992). In agreement with the observations of Thomas *et al.* (1990b), activation of the PDE was enhanced in the presence of cGMP. The ability of the partially purified, but not the purified, PDE5 isozyme to be activated by cA-PK suggests that some other factor(s) may be involved in the activation process.

Wyatt *et al.* (1994) demonstrated that the PDE5 can be phosphorylated in intact bovine aortic smooth muscle cells. Smooth muscle cells incubated with 0.1 μM ANP for 1 min incorporated $[^{32}P]ATP$ into a phosphopeptide that could be immunoprecipitated with anti-PDE5 antiserum. In addition, this effect was only seen in early passage cells; late passage cells are reported to contain no cG-PK and no phosphorylation was observed.

These results therefore indicate that PDE5 isozymes can be phosphorylated by either cA-PK or cG-PK. It still remains to be seen if this phosphorylation leads to an activity change *in vivo*. However, if the changes are similar to those observed for guinea pig lung PDE5, phosphorylation may be an important negative feedback mechanism controlling cellular cGMP levels in response to cG-PK activation. In response to agents such as NO and ANP, the duration of cellular cGMP levels is usually transient. Therefore, activation of the PDE5 appears to be a likely mechanism by which cGMP can be rapidly reduced (Fig. 4). It appears that cA-PK-induced phosphorylation of PDE5 would only occur in situations in which both cAMP and cGMP levels are elevated suggesting that there is some degree of cross-talk between these two pathways in the regulation of cGMP levels.

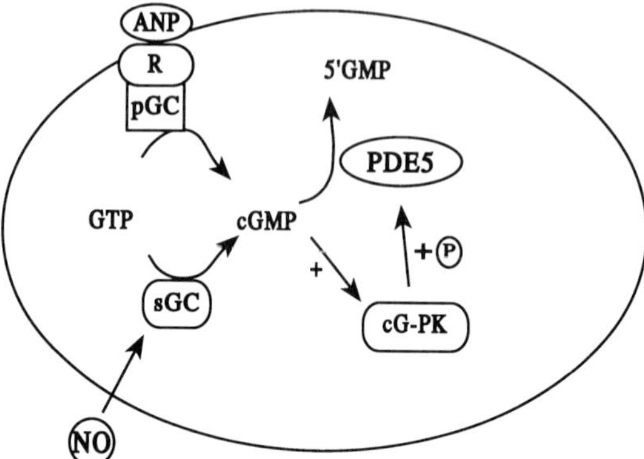

FIGURE 4   Both ANP and NO can elevate cGMP levels by stimulating either the particulate or soluble guanylyl cyclases, respectively. The resulting increase in cGMP activates cG-PK, which then mediates its physiological effects by catalyzing the phosphorylation of various proteins. One of the substrates for cG-PK is PDE5. It is speculated that phosphorylation stimulates PDE5 activity.

IV. Summary and Conclusions

Recent studies demonstrate that the regulation of PDEs by phosphorylation plays an important role in cellular function. Phosphorylation of PDEs appears not only to be an important autoregulatory mechanism that protects cells from overstimulation, but also is involved in physiological processes such as lipolysis and memory and learning. With the recent advances in molecular cloning of the PDEs, it has become apparent that many isoforms exist within single gene families and that these isoforms have different functions depending on tissue distribution and cellular localization. Determining the role that phosphorylation plays in regulating specific PDE isoforms in specific tissues and cell types will enable researchers to gain insight into the control of cyclic nucleotide signaling processes.

In addition, the complexity of PDE gene families and their expression patterns raises interesting issues for our future research in the PDE field and should also lead us to revisit some of the conclusions drawn in previous biochemical and pharmacological studies. The existence of multiple isozymes should explain in part why a wide range of kinetic constants has been reported for a given PDE. It is likely that many of the "pure" biochemical preparations used in many early kinetic studies are likely to have contained more than one isoform. The complexity of PDEs also poses significant challenges to future pharmacological studies on PDEs in terms of drug specificity. To make a PDE inhibitor more specific, different gene members of the same PDE family and different isozymes (encoded by various splice variants) of the same PDE gene should be considered. Because some PDE genes have been found to be expressed at high level in several tissues, in a subset of cells of a given tissue, or in a species-dependent manner, future studies of various PDE inhibitors will need to entail a complete understanding of the corresponding PDE expression profiles.

References

Ahn, H. S., Crim, W., Pitts, B., and Sybertz, E. J. (1992). Calcium-calmodulin-stimulated and cyclic-GMP-specific phosphodiesterases. Tissue distribution, drug sensitivity, and regulation of cyclic GMP levels. *Adv. Second Messenger Phosphoprotein Res.* **25**, 271–288.

Beavo, J. A. (1988). Multiple isozymes of cyclic nucleotide phosphodiesterase. *Adv. Second Messenger Phosphoprotein Res.* **22**, 1–38.

Beavo, J. A., Conti, M., and Heaslip, R. J. (1994). Multiple cyclic nucleotide phosphodiesterases. *Mol. Pharmacol.* **46**, 399–405.

Bentley, J. K., and Beavo, J. A. (1992). Regulation and function of cyclic nucleotides. *Curr. Opin. Cell. Biol.* **4**, 233–240.

Bentley, J. K., Kadlecek, A., Sherbert, C. H., Seger, D., Sonnenburg, W. H., Novack, J. P., and Beavo, J. A. (1992). Molecular cloning of cDNA encoding a "63"-kDa calmodulin-stimulated phosphodiesterase from bovine brain. *J. Biol. Chem.* **267**, 18676–18682.

Boekhoff, I., and Breer, H. (1992). Termination of second messenger signaling in olfaction. *Proc. Natl. Acad. Sci. U.S.A.* **89**, 471–474.

Boekhoff, I., Schleicher, S., Strotmann, J., and Breer, H. (1992). Odor-induced phosphorylation of olfactory cilia proteins. *Proc. Natl. Acad. Sci. U.S.A.* **89**, 11983–11987.

Borisy, F. F., Ronnett, G. V., Cunningham, A. M., Juilfs, D., Beavo, J., and Snyder, S. H. (1992). Calcium/calmodulin-activated phosphodiesterase expressed in olfactory receptor neurons. *J. Neurosci.* **12**, 915–923.

Boyes, S., and Loten, E. G. (1988). Purification of an insulin-sensitive cyclic AMP phosphodiesterase from rat liver. *Eur. J. Biochem.* **174**, 303–309.

Breer, H., Boekhoff, I., and Tareilus, E. (1990). Rapid kinetics of second messenger formation in olfactory transduction. *Nature* **345**, 65–68.

Burns, F., and Pyne, N. J. (1992). Interaction of the catalytic subunit of protein kinase A with the lung type V cyclic GMP phosphodiesterase: Modulation of non-catalytic binding sites. *Biochem. Biophys. Res. Commun.* **189**, 1389–1396.

Burns, F., Rodger, I. W., and Pyne, N. J. (1992). The catalytic subunit of protein kinase A triggers activation of the type V cyclic GMP-specific phosphodiesterase from guinea-pig lung. *Biochem. J.* **283**, 487–491.

Conti, M., Swinnen, J. V., Tsikalas, K. E., and Jin, S. L. (1992). Structure and regulation of the rat high-affinity cyclic AMP phosphodiesterases. A family of closely related enzymes. *Adv. Second Messenger Phosphoprotein Res.* **25**, 87–99.

Davis, R. L., Takayasu, H., Eberwine, M., and Myres, J. (1989). Cloning and characterization of mammalian homologs of the Drosophila dunce + gene. *Proc. Natl. Acad. Sci. U.S.A.* **86**, 3604–3608.

Dawson, T. M., Arriza, J. L., Jaworsky, D. E., Borisy, F. F., Attramadal, H., Lefkowitz, R. J., and Ronnett, G. V. (1993). Beta-adrenergic receptor kinase-2 and beta-arrestin-2 as mediators of odorant-induced desensitization. *Science* **259**, 825–829.

Degerman, E., Belfrage, P., Newman, A. H., Rice, K. C., and Manganiello, V. C. (1987). Purification of the putative hormone-sensitive cyclic AMP phosphodiesterase from rat adipose tissue using a derivative of cilostamide as a novel affinity ligand. *J. Biol. Chem.* **262**, 5797–5807.

Degerman, E., Manganiello, V. C., Newman, A. H., Rice, K. C., and Belfrage, P. (1988). Purification, properties and polyclonal antibodies for the particulate, low Km cAMP phosphodiesterase from bovine adipose tissue. *Second Messengers Phosphoproteins* **12**, 171–182.

Degerman, E., Smith, C. J., Tornqvist, H., Vasta, V., Belfrage, P., and Manganiello, V. C. (1990). Evidence that insulin and isoprenaline activate the cGMP-inhibited low-Km cAMP phosphodiesterase in rat fat cells by phosphorylation. *Proc. Natl. Acad. Sci. U.S.A.* **87**, 533–537.

Delmas, V., and Sassone-Corsi, P. (1994). The key role of CREM in the cAMP signaling pathway in the testis. *Mol. Cell. Endocrinol.* **100**, 121–124.

Dent, G., Giembycz, M. A., Rabe, K. F., and Barnes, P. J. (1991). Inhibition of eosinophil cyclic nucleotide PDE activity and opsonised zymosan-stimulated respiratory burst by 'type IV'-selective PDE inhibitors. *Br. J. Pharmacol.* **103**, 1339–1346.

Dumont, J. E., Lamy, F., Roger, P., and Maenhaut, C. (1992). Physiological and pathological regulation of thyroid cell proliferation and differentiation by thyrotropin and other factors. *Physiol. Rev.* **72**, 667–694.

Edelman, A. M., Raese, J. D., Lazar, M. A., and Barchas, J. D. (1981). Tyrosine hydroxylase: Studies on the phosphorylation of a purified preparation of the brain enzyme by the cyclic AMP-dependent protein kinase. *J. Pharmacol. Exp. Ther.* **216**, 647–653.

Epstein, P. M., Strada, S. J., Sarada, K., and Thompson, W. J. (1982). Catalytic and kinetic properties of purified high-affinity cyclic AMP phosphodiesterase from dog kidney. *Arch. Biochem. Biophys.* **218**, 119–133.

Endoh, M., Kawabata, Y., and Katano, Y. (1995). PDE isozyme specific regulation of cardiac cyclic AMP and contractility in mammalian myocardium: Clinical relevance. *Jpn. J. Pharmacol.* **67**(Suppl. I), 43.

Florio, V. A., Sonnenburg, W. K., Johnson, R., Kwak, K. S., Jensen, G. S., Walsh, K. A., and Beavo, J. A. (1994). Phosphorylation of the 61-kDa calmodulin-stimulated cyclic nucleotide phosphodiesterase at serine 120 reduces its affinity for calmodulin. *Biochemistry* **33**, 8948–8954.

Francis, S. H., and Corbin, J. D. (1988). Purification of cGMP-binding protein phosphodiesterase from rat lung. *Methods Enzymol.* **159**, 722–729.

Francis, S. H., Lincoln, T. M., and Corbin, J. D. (1980). Characterization of a novel cGMP binding protein from rat lung. *J. Biol. Chem.* **255**, 620–626.

Francis, S. H., Thomas, M. K., and Corbin, J. D. (1990). Cyclic GMP-binding cyclic GMP-specific phosphodiesterase from lung. *In* "Cyclic Nucleotide Phosphodiesterases: Structure, Regulation and Drug Action" (J. Beavo and M. D. Houslay, eds.), pp. 117–140. Wiley, Chichester, UK.

Geremia, R., Rossi, P., Mocini, D., Pezzotti, R., and Conti, M. (1984). Characterization of a calmodulin-dependent high-affinity cyclic AMP and cyclic GMP phosphodiesterase from male mouse germ cells. *Biochem. J.* **217**, 693–700.

Gettys, T. W., Vine, A. J., Simonds, M. F., and Corbin, J. D. (1988). Activation of the particulate low Km phosphodiesterase of adipocytes by addition of cAMP-dependent protein kinase. *J. Biol. Chem.* **263**, 10359–10363.

Gillespie, P. G., and Beavo, J. A. (1989). cGMP is tightly bound to bovine retinal rod phosphodiesterase. *Proc. Natl. Acad. Sci. U.S.A.* **86**, 4311–4315.

Grant, P. G., and Colman, R. W. (1984). Purification and characterization of a human platelet cyclic nucleotide phosphodiesterase. *Biochemistry* **23**, 1801–1807.

Grant, P. G., Mannarino, A. F., and Colman, R. W. (1988). cAMP-mediated phosphorylation of the low-Km cAMP phosphodiesterase markedly stimulates its catalytic activity. *Proc. Natl. Acad. Sci. U.S.A.* **85**, 9071–9075.

Greengard, P. (1987). Neuronal phosphoproteins. *Mol. Neurobiol.* **1**, 81–119.

Hamet, P., Coquil, J. F., Bousseau, L. S., Franks, D. J., and Tremblay, J. (1984). Cyclic GMP binding and phosphodiesterase: Implication for platelet function. *Adv. Cyclic Nucleotide Protein Phosphorylation Res.* **16**, 119–136.

Hansen, R. S., and Beavo, J. A. (1982). Purification of two calcium/calmodulin dependent forms of cyclic nucleotide phosphodiesterase by using conformation-specific monoclonal antibody chromatography. *Proc. Natl. Acad. Sci. U.S.A.* **79,** 2788–2792.

Hanson, P. I., and Schulman, H. (1992). Neuronal Ca^{2+}/calmodulin-dependent protein kinases. *Annu. Rev. Biochem.* **61,** 559–601.

Harrison, S. A., Reifsnyder, D. H., Gallis, B., Cadd, G. G., and Beavo, J. A. (1986). Isolation and characterization of bovine cardiac muscle cGMP-inhibited phosphodiesterase: A receptor for new cardiotonic drugs. *Mol. Pharmacol.* **29,** 506–514.

Hashimoto, Y., Sharma, R. K., and Soderling, T. R. (1989). Regulation of Ca^{2+}/calmodulin-dependent cyclic nucleotide phosphodiesterase by the autophosphorylated form of Ca^{2+}/calmodulin-dependent protein kinase II. *J. Biol. Chem.* **264,** 10884–10887.

Heyworth, C. M., Wallace, A. V., and Houslay, M. D. (1983). Insulin and glucagon regulate the activation of two distinct membrane-bound cyclic AMP phosphodiesterases in hepatocytes. *Biochem. J.* **214,** 99–110.

Houslay, M. D., and Lobban, M. (1994). Distinct brain region distribution patterns for the expression of the type-IV cyclic AMP phosphodiesterase (rPDE-IVB1), PDE-4 (rPDE-IVB2), RD1 (rPDE-IVA1) and the identification of a novel species "OLF-RD1" (rPDEIVA2). *FASEB J.* **8,** A369.

Jaworsky, D. E., Natsuzaki, O., Borisy, F. F., and Ronnett, G. V. (1995). Calcium modulates the rapid kinetics of the odorant-induced cyclic AMP signal in rat olfactory cilia. *J. Neurosci.* **15,** 310–318.

Kandel, E. R., and Schwartz, J. H. (1982). Molecular biology of learning: Modulation of transmitter release. *Science* **218,** 433–443.

Kauffman, R. F., Schenck, K. W., Utterback, B. G., Crowe, V. G., and Cohen, M. L. (1987). *In vitro* vascular relaxation by new inotropic agents: Relationship to phosphodiesterase inhibition and cyclic nucleotides. *J. Pharmacol. Exp. Ther.* **242,** 864–872.

Livi, G. P., Kmetz, P., McHale, M. M., Cieslinski, L. B., Sathe, G. M., Taylor, D. P., Davis, R. L., Torphy, T. J., and Balcarek, J. M. (1990). Cloning and expression of cDNA for a human low-Km, rolipram-sensitive cyclic AMP phosphodiesterase. *Mol. Cell. Biol.* **10,** 2678–2686.

Lobban, M., Shakur, Y., Beattie, J., and Houslay, M. D. (1994). Identification of two spliced variant forms of type-IVB cyclic AMP phosphodiesterase, DPD (rPDE-IVB1) and PDE-4 (rPDE-IVB2) in brain: Selective localization in membrane and cytosolic compartments and differential expression in various brain regions. *Biochem. J.* **304,** 399–406.

Loughney, K., Martins, T. J., Harris, E. A. S., Sadhu, K., Hicks, J. B., Sonnenburg, W. K., Beavo, J. A., and Ferguson, K. (1996). Isolation and characterization of cDNAs corresponding to two human calcium, calmodulin-regulated, 3′,5′-cyclic nucleotide phosphodiesterases. *J. Biol. Chem.* **271,** 796–806.

Lugnier, C., Schoeffter, P., Le, B. A., Strouthou, E., and Stoclet, J. C. (1986). Selective inhibition of cyclic nucleotide phosphodiesterases of human, bovine and rat aorta. *Biochem. Pharmacol.* **35,** 1743–1751.

Macphee, C. H., Harrison, S. H., and Beavo, J. A. (1986). Immunological identification of the major platelet low-Km cAMP phosphodiesterase: Probable target for anti-thrombotic agents. *Proc. Natl. Acad. Sci. U.S.A.* **83,** 6660–6663.

Macphee, C. H., Reifsnyder, D. H., Moore, T. A., Lerea, K. M., and Beavo, J. A. (1988). Phosphorylation results in activation of a cAMP phosphodiesterase in human platelets. *J. Biol. Chem.* **263,** 10353–10358.

Malta, E. (1989). Biphasic relaxant curves to glyceryl trinitrate in rat aortic rings. Evidence for two mechanisms of action. *Naunyn Schmiedebergs Arch. Pharmacol.* **339,** 236–243.

Manganiello, V., and Vaughn, M. (1973). An effect of insulin on cAMP phosphodiesterase activity in fat cells. *J. Biol. Chem.* **248,** 7164–7170.

Manganiello, V. C., Degerman, E., Smith, C. J., Vasta, V., Tornqvist, H., and Belfrage, P. (1992a). Mechanisms for activation of the rat adipocyte particulate cyclic-GMP-inhibited cyclic AMP phosphodiesterase and its importance in the antilipolytic action of insulin. *Adv. Second Messenger Phosphoprotein Res.* **25**, 147–164.

Manganiello, V. C., Degerman, E., Smith, C. J., Vasta, V., Tornqvist, H., and Belfrage, P. (1992b). Mechanisms for activation of the rat adipocyte particulate cyclic-GMP-inhibited cyclic AMP phosphodiesterase and its importance in the antilipolytic action of insulin. *Adv. Second Messenger Phosphoprotein Res.* **25**, 147–164.

Marchmont, R. J., Ayad, S. R., and Houslay, M. D. (1981). Purification and properties of the insulin-stimulated phosphodiesterase from rat liver plasma membranes. *Biochem. J.* **195**, 645–652.

Maurice, D. H., and Haslam, R. J. (1990). Molecular basis of the synergistic inhibition of platelet function by nitrovasodilators and activators of adenylate cyclase: Inhibition of cyclic AMP breakdown by cyclic GMP. *Mol. Pharmacol.* **37**, 671–681.

McAllister, L. L. M., Sonnenburg, W. K., Kadlecek, A., Seger, D., Trong, H. L., Colbran, J. L., Thomas, M. K., Walsh, K. A., Francis, S. H., Corbin, J. D., and Beavo, J. A. (1993). The structure of a bovine lung cGMP-binding, cGMP-specific phosphodiesterase deduced from a cDNA clone. *J. Biol. Chem.* **268**, 22863–22873.

McAllister-Lucas, L. M., Haik, T. L., Colbran, J. L., Sonnenburg, W. K., Seger, D., Beavo, J. A., Francis, S. H., and Corbin, J. D. (1994). Mutagenesis of a lung cGMP-specific phosphodiesterase provides evidence for two distinct sites for allosteric cGMP-binding with an essential aspartic acid at each site. *FASEB J.* **8**, A372.

McLaughlin, M. M., Cieslinski, L. B., Burman, M., Torphy, T. J., and Livi, G. P. (1993). A low-Km, rolipram-sensitive, cAMP-specific phosphodiesterase from human brain. Cloning and expression of cDNA, biochemical characterization of recombinant protein, and tissue distribution of mRNA. *J. Biol. Chem.* **268**, 6470–6476.

Meacci, E., Moos, M., Taira, M., Smith, C. J., Vasta, V., Degerman, E., Movsesian, M. A., Belfrage, P., and Manganiello, V. C. (1991). Cloning of cDNA for a human cardiac cGMP-inhibited low Km cAMP phosphodiesterase (CGI PDE). *FASEB J.* **5**, A454.

Monaco, L., Vicini, E., and Conti, M. (1994). Structure of 2 rat genes coding for closely related rolipram-sensitive cAMP phosphodiesterases—Multiple messenger RNA variants originate from alternative splicing and multiple start sites. *J. Biol. Chem.* **269**, 347–357.

Mumby, M. C., Martins, T. J., Chang, M. L., and Beavo, J. A. (1982). Identification of cGMP-stimulated cyclic nucleotide phosphodiesterase in lung tissue with monoclonal antibodies. *J. Biol. Chem.* **257**, 13283–13290.

Nagamine, Y., and Reich, E. (1985). Gene expression and cAMP. *Proc. Natl. Acad. Sci. U.S.A.* **82**, 4606–4610.

Nestler, E. J., and Greengard, P. (1983). Protein phosphorylation in the brain. *Nature* **305**, 583–588.

Obernolte, R., Bhakta, S., Alvarez, R., Bach, C., Zuppan, P., Mulkins, M., Jarnagin, K., and Shelton, E. R. (1993). The cDNA of a human lymphocyte cyclic-AMP phosphodiesterase (PDE IV) reveals a multigene family. *Gene* **129**, 239–247.

Pang, D. C. (1992). Tissue and species specificity of cardiac cAMP-phosphodiesterase inhibitors. *Adv. Second Messenger Phosphoprotein Res.* **25**, 307–320.

Peachell, P. T., Undem, B. J., Schleimer, R. P., MacGlashan, D. W. J., Lichtenstein, L. M., Cieslinski, L. B., and Torphy, T. J. (1992). Preliminary identification and role of phosphodiesterase isozymes in human basophils. *J. Immunol.* **148**, 2503–2510.

Pyne, N. J., Cooper, M. E., and Houslay, M. D. (1987). The insulin- and glucagon-stimulated 'dense-vesicle' high-affinity cyclic AMP phosphodiesterase from rat liver. Purification, characterization and inhibitor sensitivity. *Biochem. J.* **242**, 33–42.

Pyne, N. J., Cushley, W., Nimmo, H. G., and Houslay, M. D. (1989). Insulin stimulates the tyrosyl phosphorylation and activation of the 52 kDa peripheral plasma-membrane cyclic AMP phosphodiesterase in intact hepatocytes. *Biochem. J.* **261**, 897–904.

Rascón, A., Lindgren, S., Stavenow, L., Belfrage, P., Andersson, K. E., Manganiello, V. C., and Degerman, E. (1992). Purification and properties of the cGMP-inhibited cAMP phosphodiesterase from bovine aortic smooth muscle. *Biochim. Biophys. Acta* **1134**, 149–156.

Rasćon, A., Degerman, E., Taira, M., Meacci, E., Smith, C. J., Manganiello, V., Belfrage, P., and Tornqvist, H. (1994). Identification of the phosphorylation site *in vitro* for cAMP-dependent protein kinase on the rat adipocyte cGMP-inhibited cAMP phosphodiesterase. *J. Biol. Chem.* **269**, 11962–11966.

Rossi, P., Giorgi, M., Geremia, R., and Kincaid, R. L. (1988). Testis-specific calmodulin-dependent phosphodiesterase. A distinct high affinity cAMP isoenzyme immunologically related to brain calmodulin-dependent cGMP phosphodiesterase. *J. Biol. Chem.* **263**, 15521–15527.

Rossi, P., Pezzotti, R., Conti, M., and Geremia, R. (1985). Cyclic nucleotide phosphodiesterases in somatic and germ cells of mouse seminiferous tubules. *J. Reprod. Fertil.* **74**, 317–327.

Sarvey, J. M., Burgard, E. C., and Decker, G. (1989). Long-term potentiation: Studies in the hippocampal slice. *J. Neurosci. Methods* **28**, 109–124.

Schleicher, S., Boekhoff, I., Arriza, J., Lefkowitz, R. J., and Breer, H. (1993). A beta-adrenergic receptor kinase-like enzyme is involved in olfactory signal termination. *Proc. Natl. Acad. Sci. U.S.A.* **90**, 1420–1424.

Schoffelmeer, A. N., Wardeh, G., and Mulder, A. H. (1985). Cyclic AMP facilitates the electrically evoked release of radiolabelled noradrenaline, dopamine and 5-hydroxytryptamine from rat brain slices. *Naunyn Schmiedebergs Arch. Pharmacol.* **330**, 74–76.

Sette, C., Iona, S., and Conti, M. (1994a). The short-term activation of the rolipram-sensitive, cAMP-specific phosphodiesterase by thyroid-stimulating hormone in thyroid FRTL-5 cells is mediated by a cAMP-dependent phosphorylation. *J. Biol. Chem.* **269**, 9245–9252.

Sette, C., Vicini, E., and Conti, M. (1994b). Modulation of cellular responses by hormones—Role of cAMP specific, rolipram-sensitive phosphodiesterases. *Mol. Cell. Endocrinol.* **100**, 75–79.

Sette, C., Vicini, E., and Conti, M. (1994c). The ratPDE3/IVd phosphodiesterase gene codes for multiple proteins differentially activated by cAMP-dependent protein kinase. *J. Biol. Chem.* **269**, 18271–18274.

Shakur, Y., Pryde, J. G., and Houslay, M. D. (1993). Engineered deletion of the unique N-terminal domain of the cyclic AMP-specific phosphodiesterase RD1 prevents plasma membrane association and the attainment of enhanced thermostability without altering its sensitivity to inhibition by rolipram. *Biochem. J.* **292**, 677–686.

Sharma, R. K., and Wang, J. H. (1985). Differential regulation of bovine brain calmodulin-dependent cyclic nucleotide phosphodiesterase isoenzymes by cyclic AMP-dependent protein kinase and calmodulin-dependent phosphatase. *Proc. Natl. Acad. Sci. U.S.A.* **82**, 2603–2607.

Sharma, R. K., and Wang, J. H. (1986). Calmodulin and Ca^{2+}-dependent phosphorylation and dephosphorylation of the 63-kDa subunit-containing bovine brain calmodulin-stimulated cyclic nucleotide phosphodiesterase isozyme. *J. Biol. Chem.* **261**, 1322–1328.

Sharma, R. K., Adachi, A. M., Adachi, K., and Wang, J. H. (1984). Demonstration of bovine brain calmodulin-dependent cyclic nucleotide phosphodiesterase isozymes by monoclonal antibodies. *J. Biol. Chem.* **259**, 9248–9254.

Shenolikar, S., Thompson, W. J., and Strada, S. J. (1985). Characterization of a Ca^{2+}-calmodulin-stimulated cyclic GMP phosphodiesterase from bovine brain. *Biochemistry* **24**, 672–678.

Smith, C. J., Vasta, V., Degerman, E., Belfrage, P., and Manganiello, V. C. (1991). Hormone-sensitive cyclic GMP-inhibited cyclic AMP phosphodiesterase in rat adipocytes. Regulation of insulin- and cAMP-dependent activation by phosphorylation. *J. Biol. Chem.* **266**, 13385–13390.

Sonnenburg, W. K., Seger, D., and Beavo, J. A. (1993). Molecular cloning of a cDNA encoding the "61-kDa" calmodulin-stimulated cyclic nucleotide phosphodiesterase. Tissue-specific expression of structurally related isoforms. *J. Biol. Chem.* **268**, 645–652.

Souness, J. E., Carter, C. M., Diocee, B. K., Hassall, G. A., Wood, L. J., and Turner, N. C. (1991). Characterization of guinea-pig eosinophil phosphodiesterase activity. Assessment of its involvement in regulating superoxide generation. *Biochem. Pharmacol.* **42**, 937–945.

Stroop, S. D., and Beavo, J. A. (1991). Structure and function studies of the cGMP-stimulated phosphodiesterase. *J. Biol. Chem.* **266**, 23802–23809.

Swinnen, J. V., Joseph, D. R., and Conti, M. (1989a). Molecular cloning of rat homologues of the Drosophila melanogaster dunce cAMP phosphodiesterase: Evidence for a family of genes. *Proc. Natl. Acad. Sci. U.S.A.* **86**, 5325–5329.

Swinnen, J. V., Joseph, D. R., and Conti, M. (1989b). The mRNA encoding a high-affinity cAMP phosphodiesterase is regulated by hormones and cAMP. *Proc. Natl. Acad. Sci. U.S.A.* **86**, 8197–8201.

Taira, M., Meacci, E., and Manganiello, V. C. (1991). Two distinct cDNAs of cGMP-inhibited low Km cAMP phosphodiesterase (cGI-PDE) from rat adipose tissue and human heart. *Pharmacologist* **33**, 190.

Takemoto, D. J., Gonzalez, K., Undovichenko, J., and Cunnick, J. (1993). Cyclic GMP-regulated cyclic nucleotide phosphodiesterases. *Cell. Signalling* **5**, 549–553.

Thomas, M. K., Francis, S. H., and Corbin, J. D. (1990a). Characterization of a purified bovine lung cGMP-binding cGMP phosphodiesterase. *J. Biol. Chem.* **265**, 14964–14970.

Thomas, M. K., Francis, S. H., and Corbin, J. D. (1990b). Substrate- and kinase-directed regulation of phosphorylation of a cGMP-binding phosphodiesterase by cGMP. *J. Biol. Chem.* **265**, 14971–14978.

Thompson, W. J., Shen, C. C., and Strada, S. J. (1988). Preparation of dog kidney high-affinity cAMP phosphodiesterase. *Methods Enzymol.* **159**, 760–766.

Torphy, T. J., Zhou, H. L., Burman, M., and Huang, L. B. (1991). Role of cyclic nucleotide phosphodiesterase isozymes in intact canine trachealis. *Mol. Pharmacol.* **39**, 376–84.

Torphy, T. J., Zhou, H. L., and Cieslinski, L. B. (1992). Stimulation of beta adrenoceptors in a human monocyte cell line (U937) up-regulates cyclic AMP-specific phosphodiesterase activity. *J. Pharmacol. Exp. Ther.* **263**, 1195–1205.

Turner, N. C., Wood, L. J., Burns, F. M., Gueremy, T., and Souness, J. E. (1993). The effect of cyclic AMP and cyclic GMP phosphodiesterase inhibitors on the superoxide burst of guinea-pig peritoneal macrophages. *Br. J. Pharmacol.* **108**, 876–883.

Weishaar, R. E., Kobylarz, S. D., Quade, M. M., Steffen, R. P., and Kaplan, H. R. (1986). Multiple molecular forms of phosphodiesterase and the regulation of cardiac muscle contractility. *J. Cyclic Nucleotide Protein Phosphor. Res.* **11**, 513–527.

Weishaar, R. E., Kobylarz, S. D. C., and Kaplan, H. R. (1987). Subclasses of cyclic AMP phosphodiesterase in cardiac muscle. *J. Mol. Cell. Cardiol.* **19**, 1025–1036.

Wilkins, M. R., Settle, S. L., and Needleman, P. (1990). Augmentation of the natriuretic activity of exogenous and endogenous atriopeptin in rats by inhibition of guanosine $3',5'$-cyclic monophosphate degradation. *J. Clin. Invest.* **85**, 1274–1279.

Wyatt, T. A., Francis, S. H., McAllister-Lucas, L. M., Naftilan, A. J., and Corbin, J. D. (1994). Phosphorylation of cGMP-binding cGMP-specific phosphodiesterase in vascular smooth muscle. *FASEB J.* **8**, A2150.

Yamamoto, K. K., Gonzalez, G. A., Biggs, W. H., III, and Montminy, M. R. (1988). Phosphorylation-induced binding and transcriptional efficacy of nuclear factor CREB. *Nature* **334**, 494–498.

Yan, C., and Beavo, J. A. (1994). Molecular cloning and cellular distribution of the 75 kDa calmodulin-dependent phosphodiesterase. *FASEB J.* **8**, A81.

Yan, C., Bentley, J. K., Sonnenburg, W. K., and Beavo, J. A. (1993). Differential expression of the 61 kDa and 63 kDa calmodulin-dependent phosphodiesterases in the mouse brain. *J. Neurosci.* **14**, 973–984.

Yan, C., Zhao, A. Z., Bentley, J. K., Loughney, K., Ferguson, K., and Beavo, J. A. (1995). Molecular cloning and characterization of a novel calmodulin-dependent phosphodiesterase highly enriched in olfactory sensory neurons. *Proc. Natl. Acad. Sci. U.S.A.* **92**, 9677–9681.

Zinman, B., and Hollenberg, C. H. (1974). Effect of insulin and lipolytic agents on rat adipocyte low Km phosphodiesterase. *J. Biol. Chem.* **249**, 2182–2187.

Melanie H. Cobb*
Shuichan Xu*
Mangeng Cheng*
Doug Ebert*
David Robbins†
Elizabeth Goldsmith‡
Megan Robinson*

*The University of Texas Southwestern Medical Center
Department of Pharmacology
Dallas, Texas 75235

†University of California, San Francisco
Hooper Foundation
San Francisco, California 94143

‡The University of Texas Southwestern Medical Center
Department of Biochemistry
Dallas, Texas 75235

Structural Analysis of the MAP Kinase ERK2 and Studies of MAP Kinase Regulatory Pathways

I. Introduction

The MAP kinase pathway is stimulated by numerous hormones, growth factors, and oncogene products including Ras (Davis, 1993; Robbins *et al.*, 1994) and contributes to their spectrum of actions. The importance of intracellular processes believed to be regulated by the MAP kinases has focused attention on understanding the control of this pathway. Mechanisms regulating the MAP kinase pathway are complicated and inactivation methods plentiful (reviewed in Blenis, 1993; Davis, 1993; Campbell *et al.*, 1994; Cobb *et al.*, 1994; Hunter, 1995). A schematic representation of the MAP kinase pathway is shown in Figure 1. Receptor tyrosine kinases regulate the pathway through Ras; heterotrimeric G protein-coupled receptors also use Ras to activate the pathway, although there may also be Ras-independent mechanisms. Ras interacts with Raf bringing it to the membrane where it is activated by an unknown mechanism. Raf and perhaps three distinct types

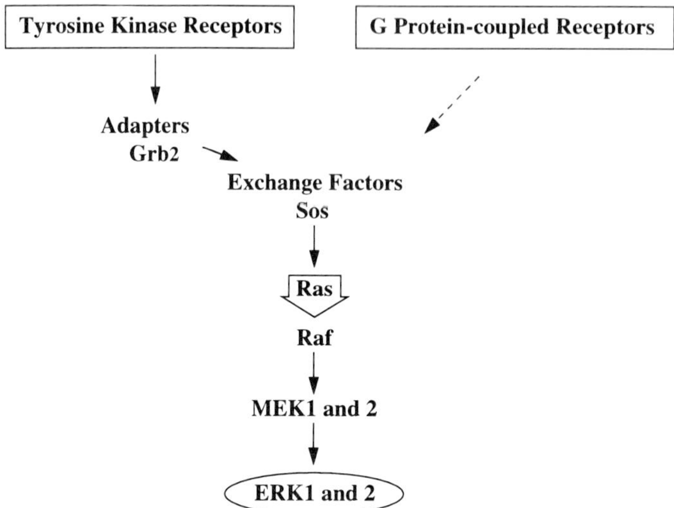

FIGURE I   Cascade regulating the MAP kinases ERK1 and ERK2.

of protein kinases, Mos, MEKK1, and the less well-characterized I-MEKK (Haystead *et al.*, 1994), phosphorylate and activate the MAP kinase kinases originally discovered by Ahn and Krebs (Ahn *et al.*, 1991). These enzymes, also known as MAP/ERK kinases or MEKs, are dual-specificity protein kinases with only the MAP kinases ERK1 and ERK2 as known substrates (Seger *et al.*, 1992). The MAP kinases, however, are pleiotropic, phosphorylating many substrates throughout the cell. This pathway has been repeated several times in yeast and mammalian cells, although mechanisms regulating the similar but parallel cascades are sketchier (Brewster *et al.*, 1993; Levin and Errede, 1993; Herskowitz, 1995).

This chapter focuses primarily on structure–function studies performed collaboratively between the Goldsmith and Cobb laboratories at the University of Texas Southwestern Medical Center in Dallas and includes experiments examining the relationship of MEKK1 to the MAP kinase pathway and potential feedback mechanisms in the pathway.

II. Regulation of MAP Kinases by Phosphorylation

Figure 2 summarizes work performed in many laboratories on regulation of ERK1 and ERK2. There is a single highly active form of either enzyme that is phosphorylated both on tyrosine and on threonine residues (T183 and Y185 in ERK2 indicated in Figure 3) by the dual-specificity protein kinases, MEK1 and MEK2 (Anderson *et al.*, 1990; Boulton and Cobb, 1991;

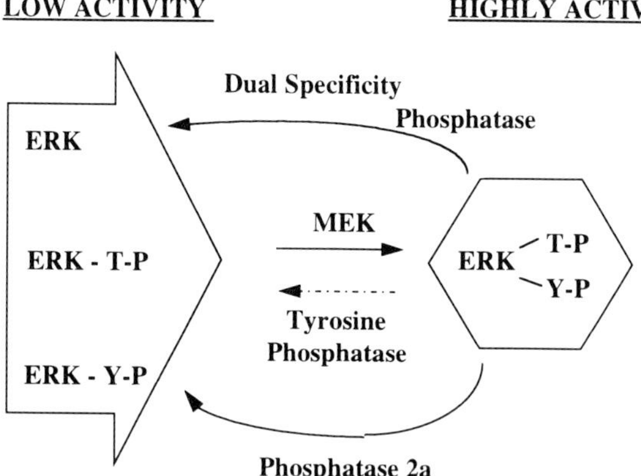

FIGURE 2 MAP kinases are regulated by phosphorylation. MEKs are dual-specificity kinases that phosphorylate threonine and tyrosine on ERK1 and ERK2. Three types of phosphatases may inactivate them by dephosphorylation of one or both residues.

Funasaka *et al.*, 1992; Zhang *et al.*, 1995). This results in an increase of ~1000-fold in activity compared to the unphosphorylated forms. There are three low-activity forms of each enzyme, the unphosphorylated protein and the two singly phosphorylated forms, that contain phosphate on only tyrosine or only threonine. These two singly phosphorylated ERKs have little more protein kinase activity than unphosphorylated proteins. Because tyrosine and threonine phosphorylations are required to activate the MAP kinases, there are three types of phosphatase that can inactivate them: dedicated dual-specificity phosphatases identified in several laboratories, phosphoserine/threonine phosphatases such as phosphoprotein phosphatase 2A, and phosphotyrosine phosphatases (Alessi *et al.*, 1993; Sontag *et al.*, 1993; Sun *et al.*, 1993; Zheng and Guan, 1993). Evidence for involvement of dual-specificity phosphatases and phosphoprotein phosphatase 2A in physiological regulation of MAP kinase is strong (Sontag *et al.*, 1993). Because ERK1 and ERK2 can autophosphorylate on tyrosine, the form containing only threonine phosphate may, in as yet undefined circumstances, be able to reactivate itself through autophosphorylation. *In vitro*, however, the autophosphorylation reaction proceeds only to a small extent with no more than a few percent of the protein becoming phosphorylated.

FIGURE 3 Phosphorylation lips in the related kinases ERK2 and ERK3. The primary sequence is shown with the residues exchanged in mutants underlined.

III. Structure–Function Studies of ERK2 ______________

We have undertaken a crystallographic analysis of the MAP kinases (Goldsmith and Cobb, 1994; Zhang *et al.*, 1994, 1995). The properties we wish to understand through these studies are (i) the basis of the unusual substrate specificity of the MAP kinases, (ii) the specificity of their recognition by MEKs, and (iii) the roles of two phosphorylations in controlling activity. In addition, we would like to create constitutively active mutants of these enzymes, because they would be invaluable research tools. Finally, we would like to determine what MAP kinase structures suggest about regulation and specificity of other protein kinases.

A. Description of the Structure

The analysis of the unphosphorylated form of ERK2 has been completed offering us new information about the low-activity state of the protein (Zhang *et al.*, 1994). The three-dimensional structure of ERK2 contains the two-domain organization found in all protein kinases whose structures have been determined thus far. The smaller N-terminal domain provides many contacts for ATP, and the larger C-terminal domain contains the major determinants for protein substrate interactions. The active site is formed at the interface of these two domains. ATP binds more deeply inside this cleft, whereas protein substrates bind more superficially. Compared to the active form of cyclic AMP-dependent protein kinase (Knighton *et al.*, 1991a,b), these two domains are rotated ~17° apart from each other, like an opened clam shell. In the active form, these two domains will be rotated toward each other to reduce the size of the cleft and allow the necessary interactions among opposing residues on each domain involved in substrate positioning and catalysis.

The two sites of regulatory phosphorylation, tyrosine 185 and threonine 183, are contained in a structure that lies at the mouth of the active site called the phosphorylation lip. The tyrosine side chain points in toward the active site. This probably accounts for the limited ability of these enzymes to autophosphorylate on tyrosine and is consistent with biochemical evidence that autophosphorylation is a unimolecular reaction (Robbins *et al.*, 1993). From the refinement of the ERK2 structure, an anion binding site has been identified on the surface of the C-terminal domain (Zhang *et al.*, 1995). Sulfate, acquired from the crystallization broth, is bound to this site in the crystal. This site lies near the phosphorylation lip and may be the binding site for phosphorylated Y185.

B. Substrate Specificity

The sites phosphorylated in a number of MAP kinase substrates have been identified. The minimal feature is a serine or threonine residue followed by proline (Davis, 1993). There are conformational features that determine which of the serine/threonine residues followed by proline are phosphorylated, because many are not recognized.

An overlay of the ERK2 backbone on top of the backbone of cAMP-dependent protein kinase in the substrate binding region suggests interactions that will contribute to substrate binding to ERK2. In cAMP-dependent protein kinase, the residues interacting with the amino acid following the phosphorylated site, the P+1 residue, form a pocket accommodating a large hydrophobic side chain, typical of the P+1 site in substrates for this kinase. In cAMP-dependent protein kinase, the residues forming this pocket are primarily two leucine residues and glutamic acid (Knighton *et al.*, 1991a). In ERK2, these residues themselves, an alanine and two arginine residues, contain large side chains that occupy the pocket present in cAMP-dependent protein kinase. Thus, in ERK2 there is no pocket to allow binding of a large side chain. In other proline-selective kinases, the residues in these positions are more like those in ERK2 than those in cAMP-dependent protein kinase.

A second characteristic of substrates of cAMP-dependent protein kinase is one or two basic residues preceding the phosphorylated residue in the P−2 or P−3 position. The basic side chains contribute a binding contact for negatively charged phosphate on ATP (Zhang *et al.*, 1994; Zhang *et al.*, 1995). In ERK2, this binding contact is provided by the enzyme itself and, therefore, is not a requirement in the substrate. Thus, this may account for the absence of basic residues in MAP kinase substrates.

C. Studies of MEK Specificity

One of our strategies to analyze the specificity of MEK–ERK interactions takes advantage of the related protein kinase ERK3. An ERK3 cDNA was isolated in the original cloning effort that identified ERK1 and ERK2 (Boulton *et al.*, 1991). ERK3 is 62 kDa and has a catalytic core nearly 50% identical to ERK1 and a C terminus that extends ~180 amino acids beyond the catalytic core. Little is known about its substrates or function. Comparison of sequences within the phosphorylation lip (Fig. 3) indicates that the ERK3 lip is of the same length and similar in sequence to ERK2. However, it does not retain both phosphorylation sites. The tyrosine phosphorylation site in the MAP kinases is replaced by glycine in ERK3. Mutants were created by exchanging the residues indicated in Figure 3 in the ERK2 and ERK3 lips. The abilities of purified MEK1 and MEK2 to phosphorylate ERK2, and T183S, Y185G ERK2 were compared. MEK1 and MEK2 phosphorylate ERK2; they do not phosphorylate T183S or Y185G ERK2, indicating that Y185 is a critical determinant for recognition of MAP kinase by MEK. This is consistent with the finding that Y185F ERK2 is a MEK substrate (M. Cheng *et al.*, unpublished data). However, because peptides that contain the lip sequence are not phosphorylated by MEK, other features of ERKs are also important for their recognition.

D. Predicted Roles of the Two Phosphorylated Residues in the Active Structure

To examine the contributions of the phosphorylation of each site for activity, ERK2 mutants were constructed in which one or both sites were replaced with an uncharged amino acid (L'Allemain *et al.*, 1992; Robbins *et al.*, 1993). Table I presents a qualitative description of the basal and active forms of some of these mutants. Earlier studies *in vitro* and in intact cells indicated that tyrosine is the first site phosphorylated, although threonine is phosphorylated if tyrosine is absent, as in Y185F ERK2. All the phosphorylation site mutants are approximately equal in activity to the unphosphorylated wild-type enzyme. Upon phosphorylation by MEK, none of these mutants is activated more than 10-fold. We also used two strategies in an attempt to create constitutively active ERK2 proteins. First, we substituted negatively charged residues for the phosphorylated residues. Glutamate could partially replace phosphothreonine but could not substitute for phosphotyrosine (Robbins *et al.*, 1993). The T183E ERK2 mutant was activated nearly 100-fold by phosphorylation on tyrosine only. Second, we noted that T183S, Y185G ERK2 was activated to some extent by phosphorylation

TABLE I   Mutations to Test the Effects of Phosphorylation on ERK2 Activity

	Residue		Myelin basic protein kinase activity	
	183	*185*	*−MEK*	*+MEK*
ERK2	T	Y	Low	Very high
	Y	T	Low	?
	T	F	Low	Low
	A	Y	Low	Low
	A	F	Low	[a]
	E	Y	Low	High
	T	E	Low	Low
	E	E	Low	[a]
	S	G	Low	Low
	E	G	Low	[a]

Note. The phosphorylated residues in wild-type ERK2 are indicated in the top line. Protein kinase activities of various mutants with and without prior phosphorylation by MEK are indicated as low or high.

[a] Proteins are not MEK substrates because they lack the T and Y phosphorylation sites.

using partially purified cell extracts. Therefore, we made the double-mutant T183E + Y185G ERK2 to combine these two mutations; however, this mutant also lacked elevated catalytic activity.

Four of these mutants have been crystallized and their three-dimensional structures determined. From these mutant structures, we have a much better idea about how these protein kinases are regulated (Zhang *et al.*, 1995). The three-dimensional structures of the mutants suggest that the phosphorylation lip of ERK2 is structurally unstable. The phosphorylation lip is disordered to varying extents in the mutant enzymes. The least disorder is apparent in T183E ERK2 in which the major difference in structure is the replaced glutamate side chain itself. The most surprising result is that the point mutant Y185E ERK2 is disordered throughout much of the lip. The disorder in the Y185 mutants indicates that a critical function of the tyrosine side chain is to tether the lip structure in the low-activity state. Point mutations, especially of tyrosine to phenylalanine, are usually made with the expectation that their effects on structure will be negligible. For this residue of ERK2, this is not the case; this point mutant causes significant structural perturbations. In the double-mutant T183E, Y185E ERK2, most of this phosphorylation lip is disordered and several of the contacting structures are also disordered.

What is the explanation for this disorder? Because relatively small inputs of energy appear to change the conformation of this part of the molecule, we conclude that an internal loop in this protein is poised to undergo conformational transitions with only a small amount of energy provided, for example, by binding to MEK or upon phosphorylation. Thus, the flexibility of this surface loop is a major aspect of its regulation by phosphorylation.

These studies provide a compelling picture of the roles of each phosphorylated residue in the activated structure. Interaction of phosphorylated Y185 with a surface anion binding site on the C-terminal domain may help stabilize a lip conformation compatible with high catalytic activity. Phosphorylated T183 probably interacts with basic residues, one of which is in the N-terminal domain, to stabilize a domain rotation that closes the active site, a global conformation change that is also necessary for high catalytic activity.

E. The Phosphorylation Lip Is an Important Regulatory Element of the Protein Kinase Family

The phosphorylation lip is one of the most variable regions among the protein kinases (Hanks *et al.*, 1988). Differences in both sequence and length are common even within kinase subfamilies. This internal loop is apparently used by the protein kinases to service a wide variety of regulatory mechanisms (Goldsmith and Cobb, 1994).

F. Gain-of-Function MAP Kinase Mutants

There have been several genetic screens to identify activated forms of MAP kinases. Gain-of-function mutations have been reported in two MAP kinases, FUS3 from the pheromone response pathway of *S. cerevisiae,* and the *Drosophila rolled* gene product (Brill *et al.,* 1994; Brunner *et al.,* 1994). Although they have greater activity according to various genetic assays, none of the mutations confers high protein kinase activity on MAP kinase *in vitro.* The mutations were mapped onto the structure of ERK2 to analyze possible mechanisms leading to their gain of function. Interestingly, all of the dominant mutations are on the surface of ERK2, whereas the recessive mutations are largely buried in the N-terminal domain. One possible explanation for the recessive mutations is that they influence the structure of the N-terminal domain. The dominant mutations could all involve interactions with other molecules. Particularly suggestive is residue 230, which is likely to influence interactions with substrates or regulators.

IV. Mammalian MEK Kinase Activities

A. Role of MEKK1 in the MAP Kinase Pathway

In yeast, several distinct MAP kinase pathways called modules (Marsh *et al.,* 1991; Kurjan, 1992; Levin and Errede, 1993; Herskowitz, 1995) have been identified; the best studied of these functions in the pheromone response pathway. In this signaling cascade, STE11 is the MAP kinase kinase kinase that activates the MEK homolog STE7. STE7 then phosphorylates and activates the MAP kinase homolog FUS3. Lange-Carter *et al.* (1993) have cloned a mammalian homolog of STE11, which they named MEK kinase (MEKK1). Because of its similarity to STE11, MEKK1 was expected to be upstream of MAP kinase. At least three distinct MAP kinase modules have now been identified in mammals (Davis, 1994). Several laboratories have implicated MEKK1 in the regulation of these other MAP kinase modules in addition to the MAP kinase pathway (Minden *et al.,* 1994; Xu *et al.,* 1995; Yan *et al.,* 1994); these modules control the Jun–N-terminal kinase/ stress-activated protein kinases (JNK/SAPK) that activate the transcriptional regulator c-Jun, and p38/HOG, an enzyme that activates expression of tumor necrosis factor by lipopolysaccharide (Han *et al.,* 1994; Lee *et al.,* 1994). An important issue is to define in which of these mammalian MAP kinase modules MEKK1 participates and how its activity and its specificity are regulated.

MEK1 and MEK2 are activated in cells transfected with various catalytic fragments of rat MEKK1 (Xu *et al.,* 1995). This was the case in 293 cells, PC12 cells, and Jurkat cells. Both cotransfected and endogenous MEK activities were stimulated by expression of MEKK1. MEKK1 was also a powerful activator of the MAP kinase homolog, JNK/SAPK. Cotransfection of JNK/ SAPK with MEKK1 from rat or mouse profoundly increased JNK/SAPK activity as measured by phosphorylation of a glutathione-S-transferase–Jun fusion protein. This supports the idea that MEKK1 has the ability to regulate MEK activities from two distinct MAP kinase modules, the ERK and the JNK/SAPK pathways. However, despite activation of MEK1 and MEK2, ERK2 activity was increased very little in cells expressing MEKK1. Thus, MEKK1 had a limited capacity to activate MAP kinase. Therefore, MEKK1 appears to be coupled to activation of JNK/SAPK; however, this is not because of the selective phosphorylation of activators of JNK/SAPK, because MEK1 and MEK2 are also MEKK1 targets. The mechanisms underlying this apparent uncoupling of MEKs from MAP kinase activation remain unclear.

In mammalian cells, the protein kinases Raf (Dent *et al.*, 1992; Howe *et al.*, 1992; Kyriakis *et al.*, 1992) and Mos (Posada *et al.*, 1993; Resing *et al.*, 1994), in addition to MEKK1 (Yan and Templeton, 1994), have been reported to phosphorylate and activate MEKs. These enzymes have been isolated by immunoprecipitation or partial purification and have not been tested in purified form, free from contaminating eukaryotic protein kinases. Interestingly, MEKK1 immunoprecipitated from transfected mammalian cells phosphorylates the catalytic subunit of phosphatase 2A nearly as well as it does MEK1 (S. Xu, unpublished data); however, when isolated from bacterial expression systems, it retains the ability to phosphorylate MEK1 (see below) but no longer phosphorylates phosphatase 2A. This suggests that a distinct endogenous protein kinase is also tightly associated with MEKK1 immunoprecipitated from eukaryotic cells or that MEKK1 is modified to broaden its specificity in mammalian cells. Thus, we confirmed that MEKK1 activated MEKs not only in intact cells, but also in purified form *in vitro*.

A constitutively active fragment of rat MEKK1 containing its catalytic domain (MEKK-C) and lacking most of its N-terminal sequence was expressed in bacteria to examine the capacity of the purified enzyme to activate MEKs *in vitro*. MEKK-C activated both MEK1 and MEK2 by over 100-fold, yielding MEKs that quantitatively activated ERK1 and ERK2 (Xu *et al.*, 1995). The sites phosphorylated on MEK1 by MEKK-C were the same as those phosphorylated by c-Mos (Resing *et al.*, 1994) and Raf-1 (Alessi *et al.*, 1994). *In vitro* partially purified MEKK and Raf-1 phosphorylated comigrating phosphopeptides of MEK1; these phosphopeptides were also found following EGF stimulation of cells (Gardner *et al.*, 1994). MEKK-C incorporated ~2 mol of phosphate per mole MEK1 and MEK2, activating them as measured by the enhanced ability of each MEK to phosphorylate catalytically defective (K52R) ERK2. The most highly phosphorylated MEK proteins were largely shifted in electrophoretic mobility and had specific activites of 105 and 113 nmol/min/mg, in reasonable agreement with the specific activity reported for enzyme purified from A431 cells (Seger *et al.*, 1992).

MEKK-C is phosphorylated in mammalian cells whether it is expressed alone or coexpressed with MEK1 or MEK2 as determined by incorporation of ^{32}P into the protein (Fig. 4A). Treatment of the protein with phosphoprotein phosphatase 2A results in loss of 50–70% of its activity (Fig. 4B). The same is true for longer forms of the protein. Okadaic acid blocks the ability of phosphatase 2A to reduce the kinase activity of MEKK1, confirming that loss of activity is due to dephosphorylation of the kinase.

The high specific activity of MEKK-C isolated from bacteria in the absence of any modification supports the concept that the catalytic domain of this type of protein kinase will be constitutively active in the absence of

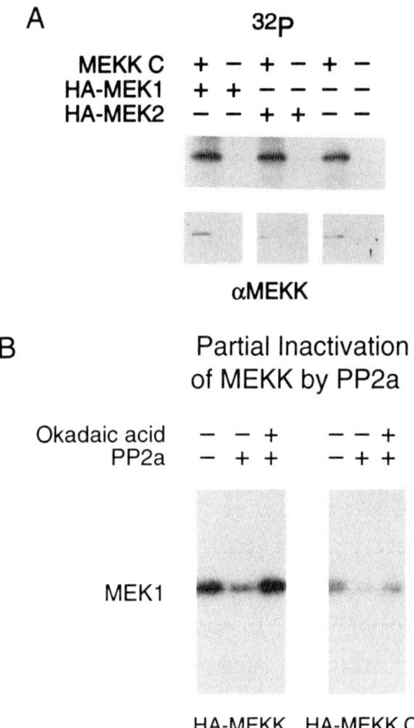

FIGURE 4   MEKK1 and MEKK-C are phosphorylated in mammalian cells. (A) ^{32}P-labeling of HA-MEKK-C. (Top) ^{32}P content of MEKKC; (bottom) immunoblot showing immunoreactive MEKK-C band. (B) HA-MEKK1 or HA-MEKK-C was incubated with (+) or without (−) phosphatase 2A and okadaic acid as described in text. Activity was measured using MEK1 (20 μg/ml) as substrate. The immunoprecipitated HA-MEKK and HA-MEKK-C were incubated in buffer (20 mM Tris–Cl, pH 7.4, 2 mM EGTA, 2 mM DTT, 2 mM, MgCl$_2$, 500 μg/ml BSA, 0.05% Brij-35, 5μg/ml aprotinin, 5 μg/ml leupeptin, and 100 μg/ml PMSF) alone, buffer plus 40 μg/ml PP2A, or buffer plus 40 μg/ml PP2A and 2 μM okadaic acid at 30°C for 60 min. Reactions were terminated by washing the pellets with lysis buffer and then 0.25 M Tris–Cl, pH 7.6, plus 0.1 M NaCl. The kinase activity of the MEKK was assayed using 50 μg/ml recombinant MEK1 as substrate in the presence of 0.2 μM okadaic acid. The cDNAs of mouse MEK1 (from Anne Gardner and Gary Johnson, National Jewish Center, Denver, CO) and human MEK2 (from Kun Liang Guan, University of Michigan) were subcloned into pCEP4HA. Histidine-tagged MEK1, MEK2, and the catalytic domain of MEKK (MEKK-C) were purified on Ni^{2+}–NTA–agarose (Xu *et al.,* 1995). Antibodies: Antisera raised against a C-terminal peptide of MEKK (Lange-Carter *et al.,* 1993) were described previously. A monoclonal antibody to the HA epitope (12CA5) was purchased from BabCo (Berkeley, CA).

its N-terminal domain. Because this appears to be the case, it is likely that regulatory mechanisms for MEKK1 will be contained in its N-terminal domain. Possible regulatory mechanisms attributable to this N-terminal region have not yet been elucidated. As suggested previously, apparently the activity of the catalytic domain itself is modestly altered by phosphorylation.

B. Endogenous MEK Kinase Activities of Cell Lysates

When PC12 cells are stimulated with NGF, both ERK and MEK activities are dramatically increased (Gómez and Cohen, 1991; Ahn *et al.*, 1992). Presumably, the activities of the bona fide MEK activators are also enhanced. Therefore, the composition of MEK kinases that can phosphorylate MEK in extracts from NGF-stimulated PC12 cells was examined using recombinant MEK1 as a substrate. In lysates from NGF-stimulated cells fractionated on MonoQ, two major peaks of kinase activity phosphorylating MEK were found. These activities comigrated with ERK1 and ERK2, as detected by Western blotting using an ERK-specific antiserum. To characterize these activities further, fresh lysates were chromatographed on Q-Sepharose and the activity phosphorylating MEK was pooled, concentrated, and subjected to gel filtration to resolve the MAP kinases from other possible MEK kinases of larger size that should appear in the void volume. One major peak of MEK kinase activity was identified (Fig. 5, top). The chromatographic

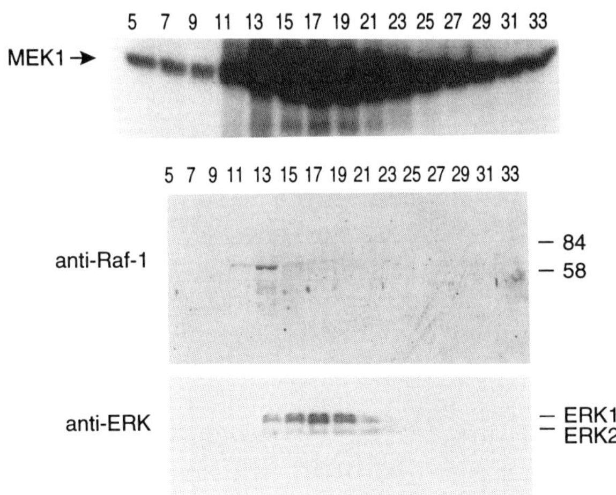

FIGURE 5   MEK kinase activity in cell lysates. Lysates from NGF-stimulated PC12 cells were chromatographed on Q-Sepharose and then on Ultrogel AcA54. The phosphorylation of MEK1 by the gel filtration fractions was monitored (top). Immunoreactive Raf-1 (middle), and ERK1 and ERK2 (bottom) were detected with antibodies to Raf-1 and to ERK1 and ERK2. Antibodies: Antiserum raised against subdomain XI of ERK1 (Boulton and Cobb, 1991) was described previously. A polyclonal antibody recognizing Raf-1 was obtained from Santa Cruz Biotechnology.

behavior of the activity was consistent with that of MAP kinase. This activity, which was present at highest amounts in fractions 17 and 19, comigrated with myelin basic protein kinase activity and with ERK1 and ERK2 immunoreactivity (Fig. 5, bottom). Immunoblotting of fractions from the gel filtration column with antibodies to Rsk and Raf-1 (Fig. 5, middle) showed that these proteins eluted primarily in the void volume, fractions 11 to 13, indicative of their larger sizes. MEKK1 and B-Raf also would be expected to elute in these fractions, whereas c-Mos has not been reported in PC12 cells. These results suggest that ERKs constitute a major portion of the detectable MEK kinase activity able to phosphorylate MEK in PC12 cells and are consistent with several other reports that MAP kinases phosphorylate MEKs (Matsuda *et al.*, 1993; Gardner *et al.*, 1994). This finding raises the possibility that ERKs may exhibit feedback regulatory activity toward MEKs.

C. Effects of Phosphorylation by ERK1 on MEK1 and MEK2

The effect of MAP kinase phosphorylation on MEK1 activity was assessed by measuring MEK activity toward K52R ERK2 following phosphorylation by purified ERK1. MEK1 was phosphorylated by ERK1 to stoichiometries of greater than 1.4 mol/mol; this phosphorylation did not stimulate the ability of MEK1 to phosphorylate K52R ERK2 (Fig. 6, top). To determine if phosphorylation by ERK1 impaired the activation of MEK1, MEK1 first was phosphorylated with purified ERK1 and then activated by MEKK-C. Prephosphorylation of MEK1 by ERK1 had no effect on its activation by MEKK-C (Fig. 6B, middle).

Next, we determined if phosphorylation by ERK1 impaired the activity of MEK1 that had already been activated. To ascertain this, MEK1 was phosphorylated by MEKK-C and then effects of ERK1 phosphorylation on its activity were examined (Fig. 6, bottom). Following phosphorylation by ERK1, MEK activity was reduced by no more than 20–30%, leading to the conclusion that phosphorylation of MEK1 by ERK1 does not significantly alter its catalytic activity *in vitro* measured under these conditions.

The two known sites of phosphorylation on MEK1 by MAP kinase lie within a MEK-specific insert (T292) and near the C terminus (T386) (Gardner *et al.*, 1994; Mansour *et al.*, 1994). The double phosphorylation site mutant T292A/T386A MEK1 obtained from Natalie Ahn and Sam Mansour (University of Colorado, Boulder) was compared to the wild-type protein as an ERK1 substrate. Although both the mutant and wild-type proteins were phosphorylated and activated by MEKK-C, only the wild-type protein was phosphorylated by ERK1 as previously reported (Gardner *et al.*, 1994; Mansour *et al.*, 1994). Mutation in these sites resulted in no change in the catalytic activity of MEK1.

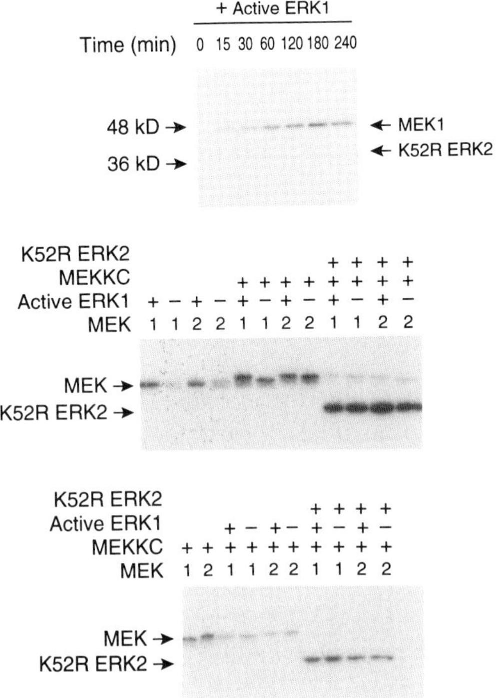

FIGURE 6   Effects of active ERK1 on MEK1 and MEK2 activaton by MEKK-C. (Top) Time course of MEK1 phosphorylation by active ERK1. MEK1 (10 μg/ml) was incubated with active ERK1 and [γ-^{32}P]ATP at 30°C. At indicated times, an aliquot was transferred to another tube to measure its ability to phosphorylate K52R ERK2 in a 5-min assay. The autoradiogram is shown. (Middle) MEK1 (10 μg/ml) and MEK2 (10 μg/ml) were first autophosphorylated or phosphorylated with ERK1 at 30°C overnight as indicated. MEKK-C was added to each tube and incubated for another 4 hr. The activities of the MEKs were then determined using K52R ERK2 as a substrate. The autoradiogram is shown. (Bottom) MEK1 (10 μg/ml) and MEK2 (10 μg/ml) were first phosphorylated by MEKK-C for 4 hr, then autophosphorylated or phosphorylated by active ERK1 as indicated for 4 hr. The activities of MEKs were measured as described previously and the autoradiograph is shown.

MEK1 and MEK2 are nearly 85% identical; the major differences in sequence lie at the N terminus and in the poorly conserved MEK-specific insert in the C-terminal domain of the kinase. Several MAP kinase consensus phosphorylation sites exist in both proteins. The C-terminal site is conserved in MEK1 and MEK2 (T386 and T394, respectively), but the site in the insert (T292 of MEK1) is not. However, based on the known substrate specificity of the MAP kinases, another possible MAP kinase site is present in the MEK2 insert (S295). Thus, we tested the possibility that phosphorylation of MEK2 by ERK1 may differentially influence its enzymatic activity. ERK1 incorporated approximately 0.95 mol phosphate/mol MEK2 but had no effect on MEK2 activity either before or after activation by MEKK-

C. Phosphoamino acid analysis indicates that both MEK1 and MEK2 are phosphorylated by ERK1 primarily on Thr, suggesting that S295 is not a site of ERK1 phosphorylation on MEK2.

V. Conclusions

In the best-defined mammalian MAP kinase pathway, there are two ERKs and two MEKs that have very similar properties and appear to be interchangeable in biochemical assays, leading to some suggestions that these isoforms may be redundant. Thus, it will be important to identify the unique features, if any, of these enzyme pairs. It is interesting that phosphorylation by MAP kinase had so little effect on MEK activity, because phosphorylation of MEK1 by cdc2 at a site (S286) close to the MAP kinase site (T292) in the C-terminal insert greatly inhibits MEK1 activity (Rossomando *et al.*, 1994). The poorly conserved insert in the C-terminal domain of MEK1 is rich in proline and contains sequences that meet a minimum consensus for binding to SH3 domains. Although we have been unable to determine any effects of MAP kinase phosphorylation on MEK1 or MEK2, Weber's group (Jelinek *et al.*, 1994) has found that mutation of this MAP kinase phosphorylation site in MEK1 abolishes its interaction with Ras/Raf. Thus, association with other proteins may be controlled by phosphorylation of MEKs by MAP kinase.

MAP kinase pathways are more numerous and individually complicated than could have been predicted even 2 years ago. From future studies of the interactions among components of MAP kinase pathways and identification of factors conferring specificity on distinct MAP kinase modules, we hope to understand the properties that lead to what we now recognize as sometimes subtle differences in targeting, specificity, and selectivity in MAP kinase modules and the biological impact of these differences.

Acknowledgments

We thank Jeff Frost, Jessie Hepler, Meg Phillips, Tom Geppert, and members of the Cobb lab (University of Texas Southwestern) for valuable discussions and critical reading of the manuscript, Clark Garcia for preparation of some of the bacterial proteins, Colleen Vanderbilt for assistance with isolation of the rat MEKK1 cDNA and immunoblotting, and Jo Hicks for preparation of the manuscript. We also thank Natalie Ahn and Sam Mansour (University of Colorado, Boulder) for the MEK1 mutants used in this study. This work was supported by Grants I1243 to M.H.C. and I1128 to E.G. from the Welch Foundation, National Institutes of Health research Grants DK34128 (M.H.C.) and DK46993 (E.G.), a Council for Tobacco Research research grant, and a fellowship from the Juvenile Diabetes Association (to S.X.).

References

Ahn, N. G., Robbins, D. J., Haycock, J. W., Seger, R., Cobb, M. H., and Krebs, E. G. (1992). Identification of an activator of the microtubule-associated protein 2 kinases ERK1 and

ERK2 in PC12 cells stimulated with nerve growth factor or bradykinin. *J. Neurochem.* **59**, 147–156.

Ahn, N. G., Seger, R., Bratlien, R. L., Diltz, C. D., Tonks, N. K., and Krebs, E. G. (1991). Multiple components in an epidermal growth factor-stimulated protein kinase cascade. *In vitro* activation of a myelin basic protein/microtubule-associated protein 2 kinase. *J. Biol. Chem.* **266**, 4220–4227.

Alessi, D. R., Saito, Y., Campbell, D. G., Cohen, P., Sithanandam, G., Rapp, U., Ashworth, A., Marshall, C. J., and Cowley, S. (1994). Identification of the sites in MAP kinase kinase-1 phosphorylated by p74^{raf-1}. *EMBO J.* **13**, 1610–1619.

Alessi, D. R., Smythe, C., and Keyse, S. M. (1993). The human CL100 gene encodes a Tyr/Thr-protein phosphatase which potently and specifically inactivates MAP kinase and suppresses its activation by oncogenic ras in Xenopus oocyte extracts. *Oncogene* **8**, 2015–2020.

Anderson, N. G., Maller, J. L., Tonks, N. K., and Sturgill, T. W. (1990). Requirement for integration of signals from two distinct phosphorylation pathways for activation of MAP kinase. *Nature* **343**, 651–653.

Blenis, J. (1993). Signal transduction via the MAP kinases: Proceed at your own RSK. *Proc. Natl. Acad. Sci. U.S.A.* **90**, 5889–5892.

Boulton, T. G., and Cobb, M. H. (1991). Identification of multiple extracellular signal-regulated kinases (ERKs) with antipeptide antibodies. *Cell Regul.* **2**, 357–371.

Boulton, T. G., Nye, S. H., Robbins, D. J., Ip, N. Y., Radziejewska, E., Morgenbesser, S. D., DePinho, R. A., Panayotatos, N., Cobb, M. H., and Yancopoulos, G. D. (1991). ERKs: A family of protein-serine/threonine kinases that are activated and tyrosine phosphorylated in response to insulin and NGF. *Cell* **65**, 663–675.

Brewster, J. L., de Valoir, T., Dwyer, N. D., Winter, E., and Gustin, M. C. (1993). An osmosensing signal transduction pathway in yeast. *Science* **259**, 1760–1763.

Brill, J. A., Elion, E. A., and Fink, G. R. (1994). A role for autophosphorylation revealed by activated alleles of *FUS3*, the yeast MAP kinase homolog. *Mol. Cell. Biol.* **5**, 297–312.

Brunner, D., Oellers, N., Szabad, J., Biggs, W. H., III, Zipursky, S. L., and Hafen, E. (1994). A gain-of-function mutation in Drosophila MAP kinase activates multiple receptor tyrosine kinase signaling pathways. *Cell* **76**, 875–888.

Campbell, J. S., Seger, R., Graves, J. D., Graves, L. M., Jensen, A. M., and Krebs, E. G. (1994). The MAP kinase cascade. *Recent Prog. Horm. Res.* **50**, 131–159.

Cobb, M. H., Hepler, J. E., Cheng, M., and Robbins, D. (1994). The mitogen-activated protein kinases, ERK1 and ERK2. *Sem. Cancer Biol.* **5**, 261–268.

Davis, R. (1994). MAPKS: New JNK expands the group. *TIBS* **19**, 470–473.

Davis, R. J. (1993). The mitogen-activated protein kinase signal transduction pathway. *J. Biol. Chem.* **268**, 14553–14556.

Dent, P., Haser, W., Haystead, T. A. J., Vincent, L. A., Roberts, T. M., and Sturgill, T. W. (1992). Activation of mitogen-activated protein kinase kinase by v-Raf in NIH 3T3 cells and in vitro. *Science* **257**, 1404–1407.

Funasaka, Y., Boulton, T., Cobb, M., Yarden, Y., Fan, B., Lyman, S. D., Williams, D. E., Anderson, D. M., Mishima, Y., and Halaban, R. (1992). c-Kit-kinase rapidly stimulates tyrosine phosphorylation and activates MAP-2 kinase in normal melanocytes, but is not functional in melanomas. *Mol. Biol. Cell* **3**, 197–209.

Gardner, A. M., Vaillancourt, R. R., Lange-Carter, C. A., and Johnson, G. L. (1994). MEK-1 phosphorylation by MEK kinase, Raf, and mitogen-activated protein kinase. Analysis of phosphopeptides and regulation of activity. *Mol. Biol. Cell* **5**, 193–201.

Goldsmith, E., and Cobb, M. H. (1994). Protein kinases. *Curr. Opin. Struct. Biol.* **4**, 833–840.

Gómez, N., and Cohen, P. (1991). Dissection of the protein kinase cascade by which nerve growth factor activates MAP kinases. *Nature* **353**, 170–173.

Han, J., Lee, J.-D., Bibbs, L., and Ulevitch, R. J. (1994). A MAP kinase targeted by endotoxin and hyperosmolarity in mammalian cells. *Science* **265**, 808–811.

Hanks, S. K., Quinn, A. M., and Hunter, T. (1988). The protein kinase family: Conserved features and deduced phylogeny of the catalytic domains. *Science* **241**, 42–52.

Haystead, C. M. M., Gregory, P., Shirazi, A., Fadden, P., Mosse, C., Dent, P., and Haystead, T. A. J. (1994). Insulin activates a novel adipocyte mitogen-activated protein kinase kinase kinase that shows rapid phasic kinetics and is distinct from c-Raf. *J. Biol. Chem.* **269**, 12804–12808.

Herskowitz, I. (1995). MAP kinase pathways in yeast: For mating and more. *Cell* **80**, 187–197.

Howe, L. R., Leevers, S. J., Gómez, N., Nakielny, S., Cohen, P., and Marshall, C. J. (1992). Activation of the MAP kinase pathway by the protein kinase raf. *Cell* **71**, 335–342.

Hunter, T. (1995). Protein kinases and phosphatases: The yin and yang of protein phosphorylation and signaling. *Cell* **80**, 225–236.

Jelinek, T., Catling, A. D., Reuter, C. W. M., Moodie, S. A., Wolfman, A., and Weber, M. J. (1994). RAS and RAF-1 form a signalling complex with MEK-1 but not MEK-2. *Mol. Cell. Biol.* **14**, 8212–8218.

Knighton, D. R., Zheng, J., Ten Eyck, L. F., Ashford, V. A., Xuong, N.-H., Taylor, S. S., and Sowadski, J. M. (1991a). Crystal structure of the catalytic subunit of cyclic adenosine monophosphate-dependent protein kinase. *Science* **253**, 407–413.

Knighton, D. R., Zheng, J., Ten Eyck, L. F., Xuong, N.-H., Taylor, S. S., and Sowadski, J. M. (1991b). Structure of a peptide inhibitor bound to the catalytic subunit of cyclic adenosine monophosphate-dependent protein kinase. *Science* **253**, 414–429.

Kurjan, J. (1992). Pheromone response in yeast. *Annu. Rev. Biochem.* **61**, 1097–1129.

Kyriakis, J. M., App, H., Zhang, X.-F., Banerjee, P., Brautigan, D. L., Rapp, U. R., and Avruch, J. (1992). Raf-1 activates MAP kinase-kinase. *Nature* **358**, 417–421.

L'Allemain, G. L., Her, J.-H., Wu, J., Sturgill, T. W., and Weber, M. J. (1992). Growth factor-induced activation of a kinase activity which causes regulatory phosphorylation of p42/microtubule-associated protein kinase. *Mol. Cell. Biol.* **12**, 2222–2229.

Lange-Carter, C. A., Pleiman, C. M., Gardner, A. M., Blumer, K. J., and Johnson, G. L. (1993). A divergence in the MAP kinase regulatory network defined by MEK kinase and Raf. *Science* **260**, 315–319.

Lee, J. C., Laydon, J. T., Mcdonnell, P. C., Gallagher, T. F., Kumary, S., Green, D., McNulty, D., Blumenthal, M. J., Heys, J. R., Landvatter, S. W., Strickler, J. E., McLaughlin, M. M., Siemens, I. R., Fisher, S. M., Livi, G. P., White, J. R., Adams, J. L., and Young, P. R. (1994). A protein kinase involved in the regulation of inflammatory cytokine biosynthesis. *Nature* **372**, 739–746.

Levin, D. E., and Errede, B. (1993). A multitude of MAP kinase activation pathways. *J. NIH Res.* **5**, 49–52.

Mansour, S. J., Matten, W. T., Hermann, A. S., Candia, J. M., Rong, S., Fukasawa, K., Vande Woude, G. F., and Ahn, N. G. (1994). Transformation of mammalian cells by constitutively active MAP kinase kinase. *Science* **265**, 966–970.

Marsh, L., Neiman, A. M., and Herskowitz, I. (1991). Signal transduction during pheromone response in yeast. *Annu. Rev. Cell Biol.* **7**, 699–728.

Matsuda, S., Gotoh, Y., and Nishida, E. (1993). Phosphorylation of *Xenopus* mitogen-activated protein (MAP) kinase kinase by MAP kinase kinase and MAP kinase. *J. Biol. Chem.* **268**, 3277–3281.

Minden, A., Lin, A., McMahon, M., Lange-Carter, C., Dérijard, B., Davis, R. J., Johnson, G. L., and Karin, M. (1994). Differential activation of ERK and JNK mitogen-activated protein kinases by Raf-1 and MEKK. *Science* **266**, 1719–1723.

Posada, J., Yew, N., Ahn, N. G., Vande Woude, G. F., and Cooper, J. A. (1993). Mos stimulates MAP kinase in *Xenopus* oocytes and activates a MAP kinase kinase in vitro. *Mol. Cell. Biol.* **13**, 2546–2553.

Resing, K. A., Mansour, S. J., Hermann, A. S., Johnson, R. S., Candia, J. M., Fukasawa, K., Vande Woude, G. F., and Ahn, N. G. (1994). Unpublished manuscript.

Robbins, D. J., Zhen, E., Cheng, M., Xu, S., Ebert, D., and Cobb, M. H. (1994). MAP kinases ERK1 and ERK2: Pleiotropic enzymes in a ubiquitous signaling network. *Adv. Cancer Res.* **63**, 93–116.

Robbins, D. J., Zhen, E., Okami, H., Vanderbilt, C., Ebert, D., Geppert, T. D., and Cobb, M. H. (1993). Regulation and properties of extracellular signal-regulated protein kinases 1 and 2 *in vitro. J. Biol. Chem.* **268**, 5097–5106.

Rossomando, A. J., Dent, P., Sturgill, T. W., and Marshak, D. R. (1994). Mitogen-activated protein kinase kinase 1 (MKK1) is negatively regulated by threonine phosphorylation. *Mol. Cell. Biol.* **14**, 1594–1602.

Seger, R., Ahn, N. G., Posada, J., Munar, E. S., Jensen, A. M., Cooper, J. A., Cobb, M. H., and Krebs, E. G. (1992). Purification and characterization of mitogen-activated protein kinase activator(s) from epidermal growth factor-stimulated A431 cells. *J. Biol. Chem.* **267**, 14373–14381.

Sontag, E., Federov, S., Robbins, D., Cobb, M., and Mumby, M. (1993). The interaction of SV40 small tumor antigen with protein phosphatase 2A stimulates the MAP kinase pathway and induces cell proliferation. *Cell* **75**, 887–897.

Sun, H., Charles, C. H., Lau, L. F., and Tonks, N. K. (1993). MKP-1 (3CH134), an immediate early gene product, is a dual specificity phosphatase that dephosphorylates MAP kinase in vivo. *Cell* **75**, 487–493.

Xu, S., Robbins, D., Frost, J., Dang, A., Lange-Carter, C., and Cobb, M. H. (1995). MEKK1 phosphorylates MEK1 and MEK2 but does not cause activation of MAP kinase. *Proc. Natl. Acad. Sci. U.S.A.* **92**, 6808–6812.

Yan, M., Dal, T., Deak, J. C., Kyriakis, J. M., Zon, L. I., Woodgett, J. R., and Templeton, D. J. (1994). Activation of stress-activated protein kinase by MEKK1 phosphorylation of its activator SEK1. *Nature* **372**, 798–800.

Yan, M., and Templeton, D. J. (1994). Identification of 2 serine residues of MEK-1 that are differentially phosphorylated during activation by raf and MEK kinase. *J. Biol. Chem.* **269**, 19067–19073.

Zhang, F., Strand, A., Robbins, D., Cobb, M. H., and Goldsmith, E. J. (1994). Atomic structure of the MAP kinase ERK2 at 2.3 Å resolution. *Nature* **367**, 704–710.

Zhang, J., Zhang, F., Ebert, D., Cobb, M. H., and Goldsmith, E. J. (1995). Activity of the MAP kinase ERK2 is controlled by a flexible surface loop. *Nature Structure* **3**, 299–307.

Zheng, C.-F., and Guan, K.-L. (1993). Dephosphorylation and inactivation of the mitogen-activated protein kinase by a mitogen-induced Thr/Tyr protein phosphatase. *J. Biol. Chem.* **268**, 16116–16119.

Patricia T. W. Cohen
Mao Xiang Chen
Christopher G. Armstrong
Medical Research Council Protein Phosphorylation Unit
Department of Biochemistry
The University
Dundee DD1 4HN, Scotland, United Kingdom

Novel Protein Phosphatases That May Participate in Cell Signaling

Phosphorylation and dephosphorylation of serine and threonine residues in proteins is a key mechanism for the regulation of cellular functions and is particularly important for the transmission of cellular signals. The major family of protein phosphatases that dephosphorylate serine and threonine residues includes PP1, PP2A, and the calcium/calmodulin-regulated protein phosphatase (PP2B, calcineurin-CNA, CMP), which were identified as enzyme activities in the cell cytosol (Cohen, 1989; Shenolikar, 1994). However, low-stringency cDNA cloning, mutant analysis, and polymerase chain reactions using oligonucleotide primers constructed to regions conserved in these protein phosphatases have identified many other phosphatases in this family (Cohen *et al.*, 1990; Chen *et al.*, 1992). On further analysis, these novel phosphatases have been shown to carry out distinct cellular functions. A phylogenetic tree of the fully sequenced phosphatase catalytic subunits belonging to this family that are present in the human species and in the yeast *Saccharomyces cerevisiae* is shown in Figure 1. Three isoforms of human PP1 have been identified, whereas there is only one in *S. cerevisiae*. Two isoforms of PP2A are known in both the human species and *S. cerevisiae*. Three isoforms of human calcium/calmodulin-regulated protein phosphatase have

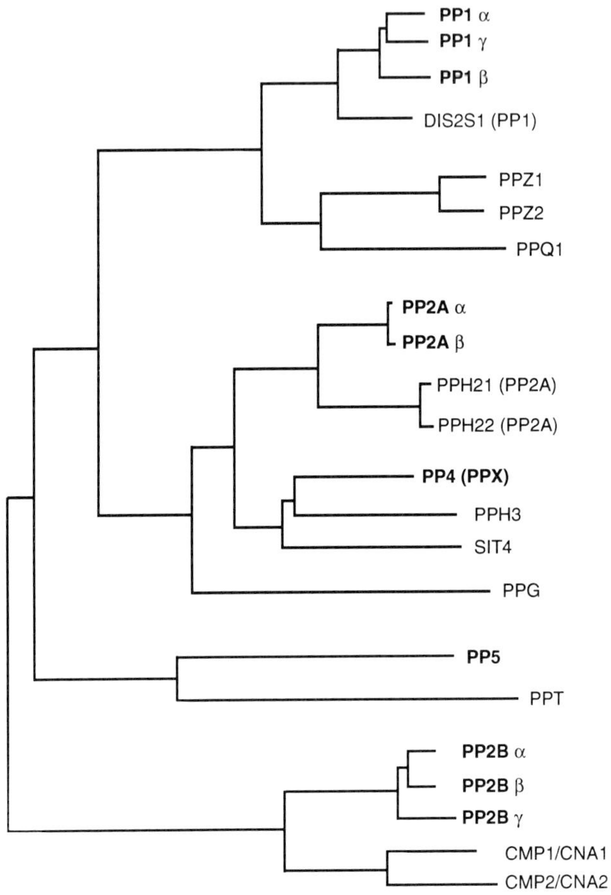

FIGURE I   Phylogenetic tree depicting the relationship between human (bold type) and *S. cerevisiae* (plain type) protein serine/threonine phosphatases. The tree is derived from a distance matrix after progressive multiple alignment (Feng and Doolittle, 1990). The sequences used are human PP1α (Barker *et al.*, 1990), PP1β (Barker *et al.*, 1994), PP1γ (Barker *et al.*, 1993), PP2Aα and PP2Aβ (Arino *et al.*, 1988; Hemmings *et al.*, 1988), PP4 (Brewis *et al.*, 1993), PP5 (this chapter), PP2Bα1 (Muramatsu and Kincaid, 1993), PP2Bβ3 (McPartlin, 1991), PP2Bγ (Muramatsu and Kincaid, 1992) and *S. cerevisiae* PP1 (DIS2S1) (Ohkura *et al.*, 1989), PPH21 and PPH22 (Sneddon *et al.*, 1989), SIT4 (Arndt *et al.*, 1989), PPH3 (Ronne *et al.*, 1991), PPG (Posas, Clotet, *et al.*, 1993), CMP1 and CMP2, also termed CNA1 and CNA2 (Cyert *et al.*, 1991; Liu *et al.*, 1991), PPZ1 and PPZ2 (Hughes *et al.*, 1993), PPQ (Chen *et al.*, 1993), and PPT1 (Chen *et al.*, 1994). Only the phosphatase domains (starting 60 amino acids prior to the conserved GDXHG motif and terminating 50 amino acids following the conserved SAPNY motif) were employed to prevent spurious alignment of the additional dissimilar domains. The average amino acid identity between human PP1 and PP2B isoforms is 34%.

been identified—one being a testis-specific form—whereas yeast contains only two isoforms of this phosphatase. In addition to these proteins with recognized enzyme activities, there are three novel protein phosphatases, PPZ1, PPZ2 (Hughes *et al.*, 1993; Lee *et al.*, 1993; Posas, Casamayor, *et al.*, 1993), and PPQ (Chen *et al.*, 1993), which show about 60% identity to PP1 in the catalytic domain, have long N-terminal domains, and have been shown to possess distinct cellular functions from PP1. Three novel protein phosphatases, termed human PP4 (originally designated PPX) (da Cruz e Silva *et al.*, 1988; Brewis *et al.*, 1993), *S. cerevisiae* SIT4 (Arndt *et al.*, 1989), and PPG (Posas *et al.*, 1993b), show about 60% identity to PP2A and perform different cellular roles to PP2A. Yeast PPH3 (Ronne *et al.*, 1991) is structurally most similar to PP4, but it is not clear whether it is the functional homolog of PP4. However, the novel protein phosphatases, human PP5 and *S. cerevisiae* PPT (Chen *et al.*, 1994), are not closely related to any of the known protein phosphatases and therefore these enzymes form a distinct new subfamily.

I. Protein Phosphatase 5

Human PP5 and yeast PPT comprise a catalytic domain preceded by a long N-terminal domain that contains four 34-amino-acid or tetratricopeptide repeat (TPR) motifs (Fig. 2A). The overall identity between human PP5 and yeast PPT is 42%; the identity is higher (46%) in the phosphatase domain than in the N-terminal domain containing the TPR motifs (36%). From this level of sequence identity, it is unclear whether PP5 and PPT are functional homologs or whether the PP5/PPT subfamily will contain other members. The four TPR motifs are degenerate, showing considerable sequence divergence, but nevertheless a consensus sequence can be deduced that shows appreciable sequence similarity to the TPR motifs identified in other proteins (Fig. 3). Hydrophobic residues occur at several positions—4, 11, 17, 24, and 28, amino acids with small side chains are at positions 8 and 20, and a proline residue is often present at position 32. By comparison with the TPR structures deduced for other proteins, the schematic structure of PP5 is thought to be as shown in Figure 2B. Subdomains A and B (Fig. 3) may form amphipathic α-helical regions in each repeat and the proline residue may allow the protein chain to fold back on itself. The four repeats thus are believed to form a compact structure with the bulky hydrophobic residues of one repeat locking into the pocket formed by the small side chains of other residues. The TPR motif in PP5 is followed by 50-amino-acid residues that link it to the conserved region of the phosphatase catalytic domain, which is presumed to take up a globular structure. TPR domains in other proteins are believed to be sites of protein–protein interactions

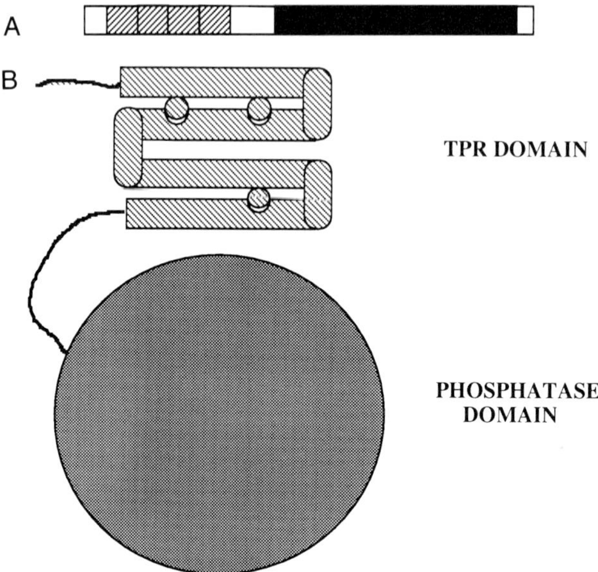

FIGURE 2   (A) Domain structure of human PP5 (497 amino acids) and *S. cerevisiae* PPT1 (513 amino acids). The four TPR motifs are denoted by hatched boxes and the conserved region of the catalytic domain by a black box. (B) Schematic model of PP5. The four TPR motifs form a compact structure with the large hydrophobic amino acids of one repeat interlocking with the pocket formed by the small side chains of an adjacent repeat. Proline residues allow the polypeptide chain to fold back on itself between TPR motifs. A short section of the polypeptide links the TPR domain to the phosphatase catalytic domain and it is thought that the TPR domain may target the phosphatase domain to its site of action.

and therefore it seems plausible that the PP5 TPR domain may target the phosphatase domain close to its substrate(s).

Full-length PP5 expressed as a soluble enzyme in *Escherichia coli* will dephosphorylate phosphorylase and histone H1 when these substrates are phosphorylated by protein kinase A and cdc2, respectively. Although PP5 is three times more active against histone than phosphorylase, the specific activity of PP5 against both substrates is substantially lower than that of PP1 and PP2A catalytic subunits and therefore these substrates are unlikely to be dephosphorylated by PP5 *in vivo*. Using phosphorylase as substrate PP5 was found to be very sensitive to the tumor promoters microcystin and okadaic acid, the IC_{50} for both toxins being in the nanomolar range and that for okadaic acid being less than that of PP1 (Chen *et al.*, 1994).

Analysis of poly(A) RNA by blotting shows the presence of a 2.3-kb PP5 mRNA in all tissues examined including brain, liver, lung, heart, placenta, skeletal muscle, kidney, and pancreas. The level does not vary greatly from tissue to tissue, although the PP5/PP1β ratio is highest in liver. The subcellular localization of PP5 was examined using two different affinity-purified

<u>DOMAIN A</u>   <u>DOMAIN B</u>

```
Human  PP5      AEELKTQANDYFKAKDYENAIKFYSQAIELNPSN        55
S.cer  PPT      ALERKNEGNVFVKEKHFLKAIEKYTEAIDLDSTQ        45

Human  PP5      AIYYGNRSLAYLRTECYGYALGDATRAIELDKKY        89
S.cer  PPT      SIYFSNRAFAHFKVDNFQSALNDCDEAIKLDPKN        79

Human  PP5      IKGYYRRAASNMALGKFRAALRDYETVVKVKPHD  KD  125
S.cer  PPT      IKAYHRRALSCMALLEFKKARKDLNVLLKAKPND  PA  115

Human  PP5      AKMKYQECNKIVKQKAFERAIAGDEHKRSVVDSL       159
S.cer  PPT      ATKALLTCDRFIREERFRKAIGGAENEAKISLCQ       149


PP5+PPT  consensus   A..Ø..RA..ØØK...Ø..AØ.DØ..AØKØ.P..
                                R


TPR  consensus  for
other  proteins      ...Ø..ØG.ØØ.....Ø...A...Ø..AØ...P..
                        A              S          S
                        S                         G
```

FIGURE 3 Comparison of the N-terminal TPR domains of human PP5 and *S. cerevisiae* PPT1 to the TPR consensus sequence derived from other TPR-containing proteins. The single-letter code is used for amino acids. ϕ Indicates any amino acid with a large hydrophobic side chain. Domain A and domain B are thought to form amphipathic α-helices, which interact at their hydrophobic faces. Residues double underlined in PP5 and PPT1 conform to the PP5/PPT1 consensus sequence. The TPR consensus for other proteins is derived from all TPR motifs in the following sequences: nuc2[+], CDC23, CDC16, crn, SSN6, SK13, TFIIIC, PRP6, STI1, MAS70, PAS10, p58, and FKBP59.

polyclonal antibodies, one raised against the phosphatase domain and the other against the TPR domain. The primary antibodies were coupled to fluorescein-labeled secondary antibodies and detection was by immuno-fluorescent confocal microscopy. Anti-PP5 antibodies stained the nuclei of HeLa cells strongly, whereas only a weak stain was observed in the cyto-plasm when a minimal fixation method was used (2% paraformaldehyde for 5 min) (Figs. 4A and 4C). The nuclear staining was uneven and the nucleolus was unstained. Both antibodies gave essentially the same results and immunodepletion with excess antigen revealed very little nonspecific staining (Fig. 4H). In dividing cells, after nuclear membrane breakdown, the PP5 stain was distributed throughout the cytoplasm and excluded from the condensed DNA (Figs. 4D–4G). When the nuclear membrane was re-formed in late telophase, the anti-PP5 immunofluorescence was again in the nucleus although the intensity of the stain was much weaker than that in the neighboring interphase cells. The distribution of PP5 was similar in A431 cells and normal human fibroblasts. Serum deprivation for 4 days reduced the PP5 staining of HeLa cells to about the half the intensity (Chen *et al.*, 1994).

Table I lists the known proteins that contain TPR motifs. The majority are located in the nucleus, although several have been located to mitochon-dria or peroxisomes or are likely to be cytoplasmic. The nuclear TPR-

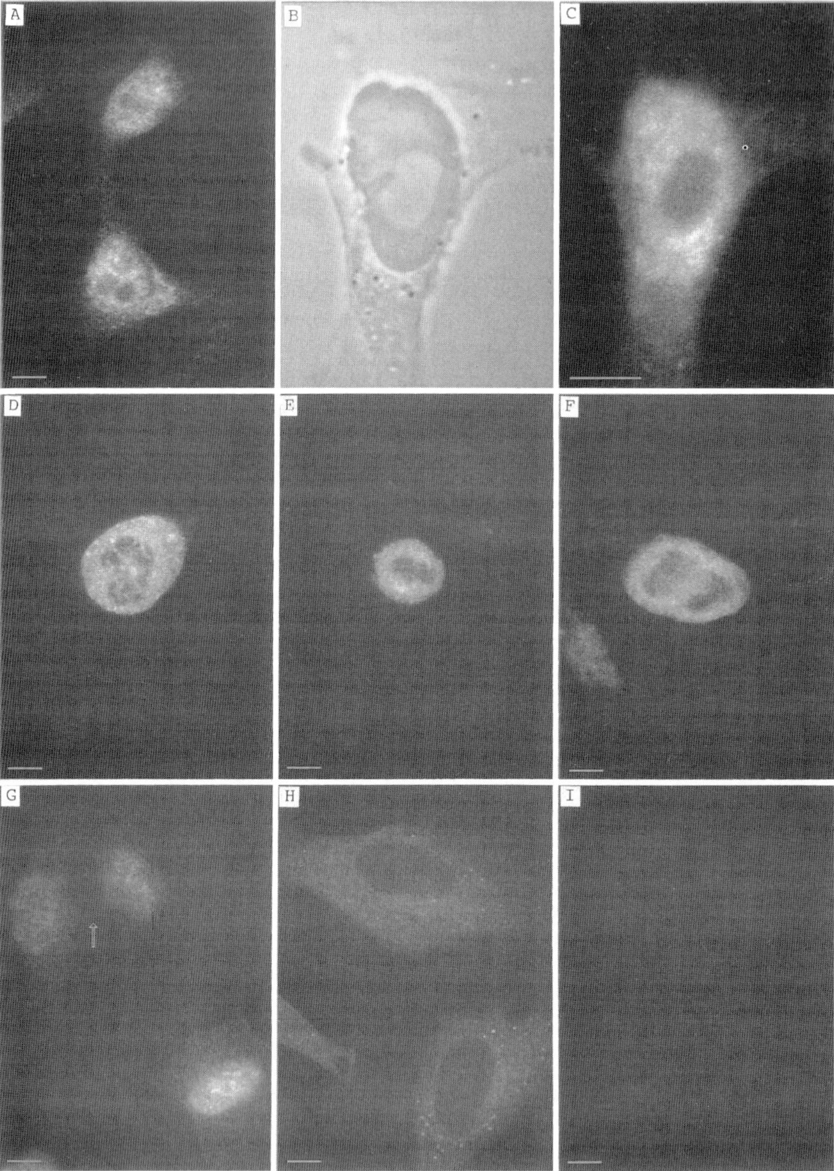

FIGURE 4   Immunofluorescent localization of PP5 in HeLa cells fixed with *p*-formaldehyde. PP5 anti-TPR domain antibodies were detected with fluorescein-labeled donkey anti-rabbit IgG antibodies (white stain). Fluorescent micrographs were taken with a confocal microscope. (A and C) Staining of the nucleus with PP5 anti-TPR domain antibodies; (B) phase image of nuclei from same cell as in C; (D–G) mitotic cells stained with PP5 anti-TPR domain antibodies, (D) prophase, (E) metaphase, (F) anaphase, and (G) late telophase, where the arrow shows the position of the midbody between two daughter cells; the cell at the lower right-hand corner is in interphase. (H) Cells stained with PP5 anti-TPR, domain antibodies that had been incubated with excess TPR domain polypeptide for 3 hr; (I) cells stained with fluorescein-labeled donkey anti-rabbit IgG antibodies. The scale bars denote 10 μm.

TABLE I   Tetratricopeptide Repeat (TPR)-Containing Proteins

Protein	Number of repeats	Localization	Function	Reference
CDC27	10	ND	Mitosis	Sikorski *et al.* (1991)
nuc2[+]	10	Nucleus	Mitosis	Hirano *et al.* (1990)
bimA[+]	10	Nucleus	Mitosis	Mirabito and Morris (1993)
CDC16	10	Nucleus	Mitosis	Icho and Wickner (1987)
cut9[+]	10	Nucleus	Mitosis	Samejima and Yanagida (1994)
CDC23	9	Nucleus	Mitosis	Sikorski *et al.* (1993)
crn	16	ND	Proliferation of neuroblasts	Zhang *et al.* (1991)
SSN6	10	Nucleus	Transcription	Schultz *et al.* (1990)
SKI13	8	Nucleus	Transcription	Rhee *et al.* (1989)
SIN3	4	Nucleus	Transcription	Wang and Stillman (1993)
TFIIIC	11	Nucleus	Transcription	Marck *et al.* (1993); Rameau *et al.* (1994)
PRP6	9	Nucleus	Splicing	Legrain and Choulika (1990)
STI1	8	ND	Stress response	Nicolet and Craig (1989)
IEP SSP 3352	8	Nucleus	Stress response	Honoré *et al.* (1992)
PP5	4	Nucleus	Phosphatase	Chen *et al.* (1994)
PPT	4	ND	Phosphatase	Chen *et al.* (1994)
MAS70	7	Mitochondria	Protein import	Hase *et al.* (1983)
MOM72	7	Mitochondria	Protein import	Sikorski *et al.* (1991)
PAS8	7	Peroxisomes	Protein import	McCollum *et al.* (1993)
PAS10	8	Peroxisomes	Protein import	van der Leij *et al.* (1993)
p58	9	ND	Inhibitor of PKR	Lee *et al.* (1994)
FKBP59	3	ND	Cyclophilin	Radanyi *et al.* (1994)
FKBP52	4	ND	Cyclophilin	Peattie *et al.* (1992)
Cyp40	3	ND	Cyclophilin	Kieffer *et al.* (1993)
ERBC	3	ND	Estrogen receptor–binding protein	Ratajczak *et al.* (1993)

Note. TPR-containing proteins that are or are likely to be in the nucleus are listed above the space. ND, not determined. IEP SSP 3352, PP5, p58, FKBP59, FKBP52, Cyp40, and ERBC are mammalian proteins, crn is from *Drosophila,* and all other proteins are of fungal origin. CDC27, nuc2[+], and bimA[+] are homologous proteins in different yeasts, as are CDC16 and cut9[+].

containing proteins can be divided into two groups, those involved in cell division cycle processes and those involved in RNA biogenesis. Mutagenesis has shown that TPR motifs are essential to the proteins containing them (Schultz *et al.,* 1990; Sikorski *et al.,* 1993), and there is evidence that TPR domains may form complexes with the same or different TPR domains. The proteins CDC27 and CDC23, which are essential for passage through the G_2-M stage of the cell cycle, have been shown to form homo- and heterodimers *in vitro* interacting via their TRP motifs. Using the yeast two-hybrid system, they have also been shown to interact *in vivo* (Sikorski *et al.*, 1991). Genetic and biochemical studies indicate that TPR domain-containing proteins may also interact functionally with proteins containing repeats of 43 amino acids, similar to those found in the β-subunit of G proteins such β-transducin. Examples of such functional interactions include cell cycle proteins CDC23/CDC20 and CDC16/20, and proteins in RNA biogenesis, SSN6/TUP1, PRP6/PRP4, SKI3/MAK11 (Goebl and Yanagida, 1991). If PP5 is functionally linked to another TPR-containing protein or a $G\beta$-related protein, it could form part of a signaling pathway that might influence nuclear functions such as RNA biogenesis and/or cell division. By comparison with most other TPR-containing proteins, PP5 has a low number of repeats (only four). Consequently, its interactions with other proteins may be weak, consistent with a transient regulation by dephosphorylation.

To ascertain whether PP5 might be involved in the regulation of cell division or RNA biogenesis, we tested the effects of the DNA synthesis inhibitors, hydroxyurea and aphidicholin, and the RNA synthesis inhibitors, α-amanitin and actinomycin D, on cultured HeLa cells. Of these four inhibitors, α-amanitin was the only one to show an effect on the subcellular location of PP5. Although control cells showed the presence of PP5 predominantly in the nucleoplasm and not in the nucleolus as expected (Fig. 5B), incubation of HeLa cells with 50 μg/ml of α-amanitin for 5 hr caused PP5 to move to the nucleolus so that anti-PP5 antibodies now stained the nucleolus more than the nucleoplasm (Fig. 5D). When RNA synthesis is inhibited by α-amanitin, the nucleolus breaks down into smaller fragments and it is these more numerous smaller fragments of the nucleolus that are stained with anti-PP5. Both anti-TPR and anti-phosphatase domain antibodies gave the same result. α-Amanitin inhibits both RNA polymerases II and III but not RNA polymerase I. Because the IC_{50} for RNA polymerase II is 25 ng/ml and that for RNA polymerase III is 20 μg/ml, we tested the effect of different concentrations of α-amanitin on HeLa cells. Although 2 μg/ml had no effect on the subcellular location of PP5, 50 μg/ml α-amanitin caused movement of PP5 into the nucleolus, implicating RNA polymerase III inhibition as a key factor in subcellular relocation of PP5. The results suggest the model depicted in Figure 6. PP5 may bind to or be located near RNA polymerase III in the nucleoplasm. When α-amanitin causes the inhibition of RNA polymerase III, this releases PP5 and causes it to migrate to the nucleolus where it may interact with and presumably inhibit RNA polymerase I. Coordinate regulation of RNA polymerases I and III is to be expected because RNA polymerase I synthesizes the large ribosomal RNA species, whereas RNA polymerase III is responsible for the

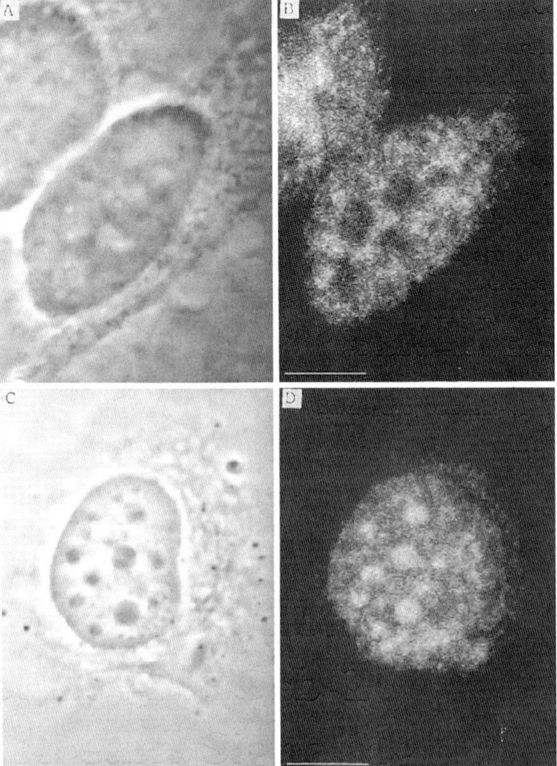

FIGURE 5   Immunofluorescent localization of PP5 in HeLa cells incubated with 50 μg/ml α-amanitin for 2 hr. PP5 anti-TPR domain antibodies were detected with fluorescein-labeled donkey anti-rabbit IgG antibodies (white stain). Fluorescent micrographs were taken with a confocal microscope. (A) Phase image of nuclei from same cells as in B that is stained with anti-TPR domain antibodies in the absence of α-amanitin; (C) phase image of nuclei from same cells as in D that is stained with anti-TPR domain antibodies in the presence of 50 μg/ ml α-amanitin.

synthesis of the 5S ribosomal RNA. In normal circumstances when cellular growth and ribosomal RNA synthesis are to be switched off, it is possible that PP5 could receive an incoming signal that will inhibit RNA polymerase III, causing the movement of PP5 into the nucleolus and the subsequent inhibition of RNA polymerase I. The role of PP5 as a signal molecule between RNA polymerases III and I is supported by a higher requirement for PP5 during cell growth. As described previously, interphase cells in logarithmic growth have a higher level of PP5 than cells in late telophase or serum-deprived cells.

The rat homolog of the human PP5 molecule described here has been cloned (Chinkers, 1994; Becker *et al.*, 1994). The amino acid sequences of the rat and human protein differ at only 10 positions (2% of the sequence).

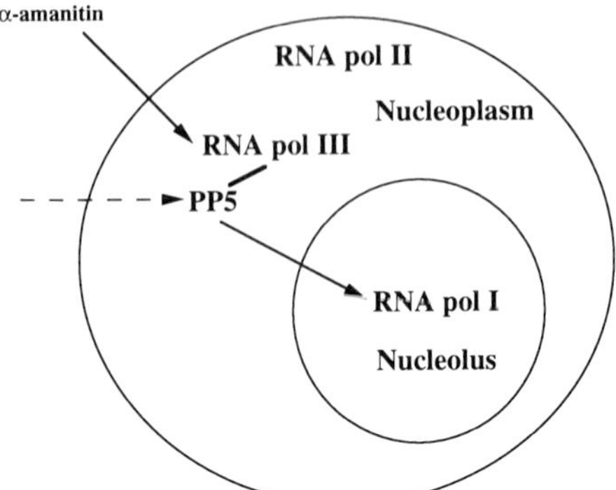

FIGURE 6   Model for the role of PP5 as a signal molecule between RNA polymerases III and I. During normal cellular growth, PP5 is in the nucleoplasm and is postulated to interact with RNA polymerase III. When α-amanitin inhibits RNA polymerase III, PP5 migrates to the nucleolus where it may interact with and inhibit RNA polymerase I. This could effect coordinate inhibition of the synthesis of the 5S ribosomal RNA by RNA polymerase III and large ribosomal RNA species by RNA polymerase I. In normal circumstances when cellular growth and ribosomal RNA synthesis are to be switched off, it is possible that PP5 could receive a incoming signal that will inhibit RNA polymerase III, causing the movement of PP5 into the nucleolus and the subsequent inhibition of RNA polymerase I.

Chinkers (1994) identified the rat PP5 using the protein kinase domain of the guanylyl cyclase receptor in the yeast two-hybrid system, suggesting that PP5 may interact directly with the guanylate cyclase receptor. Because PP5 is present at low levels in the cytoplasm, it is possible that this phosphatase could be involved in signaling from the receptor to the nucleus. However, we have been unable to detect any subcellular movement of PP5 in response to atrial natriuretic factor in dog kidney MDCK cells (M. X. Chen and P. T. W. Cohen, data not shown). In addition, the level of PP5 is higher than that of the guanylyl cyclase receptor and the occurrence of PP5 in all cell types compared to the specialized distribution of guanylyl cyclase receptors indicates a more universal role for PP5. A role in RNA biosynthesis is more likely, but it is of course possible that, like a number of other protein serine/threonine phosphatases, PP5 may perform more than one cellular function.

II. Protein Phosphatase Y

PPY was identified in *Drosophila melanogaster* (Dombrádi *et al.*, 1989a) and because no homologs in mammals or yeast have as yet been identified, it is necessary to study *Drosophila* to investigate the function of this protein phosphatase. Figure 7 shows a phylogenetic tree of *Drosophila* protein phosphatases. The tree possesses the same overall structure as that found

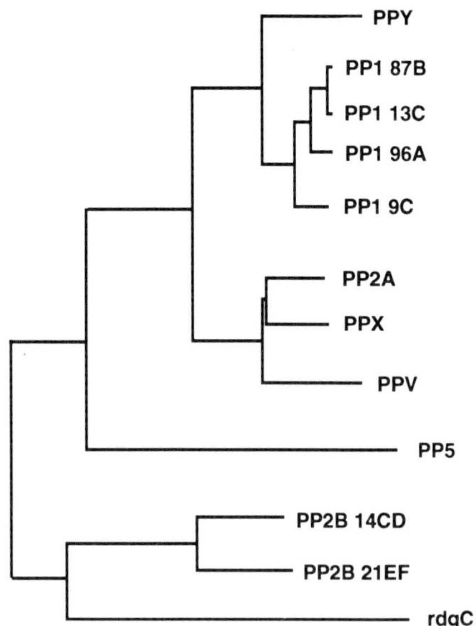

FIGURE 7 Phylogenetic tree of *Drosophila* protein phosphatases. The tree is derived from a distance matrix after progressive multiple alignment (Feng and Doolittle, 1990). The sequences used are *Drosophila* PP1 87B (Dombrádi *et al.*, 1989b), PP1 9C, PP1 96A (Dombrádi *et al.*, 1990b), PP1 13C (Dombrádi *et al.*, 1993), PP2A (Orgad *et al.*, 1990), PP5 (L. B. Brown and P. T. W. Cohen, unpublished data), PP2B 21EF (Guerini *et al.*, 1992), PP2B 14CD (Brown *et al.*, 1994), and rdgC (Steele *et al.*, 1992). The average amino acid identity between the PP1 and PP2B isoforms is 35%.

for the human species and *S. cerevisiae*. Four isoforms of PP1, one isoform of PP2A, and two isoforms of PP2B have been sequenced although at least three PP2B isoforms are known to exist. *Drosophila* also contains homologs of novel phosphatases in the PP2A subfamily, namely PP4 (N. D. Brewis and P. T. W. Cohen, unpublished data) and PPV, which is the functional homolog of yeast SIT4 (Mann *et al.*, 1993). RdgC contains a long C-terminal Ca^{2+}-binding domain and is in the PP2B subfamily (Steele *et al.*, 1992), whereas PP5 identifies the fourth subfamily (L. Brown and P. Cohen, unpublished data). PPY belongs to the PP1 subfamily and is 64% identical in amino acid sequence to *Drosophila* PP1 isoforms termed PP1 9C, PP1 13C, PP1 87B, and PP1 96A. However, this degree of sequence similarity is much lower than the >80% sequence similarity shared by the PP1 isoforms, suggesting that PPY may not simply be an isoform of PP1 but may perform a distinct cellular function.

Two main criteria have been used to distinguish PP1 activity from that of other protein serine/threonine phosphatases; namely, its inhibition by the small thermostable proteins, inhibitor 1 and inhibitor 2, and its ability to dephosphorylate the β-regulatory subunit of phosphorylase kinase faster than the α-subunit. Therefore, to test whether PPY is a serine/threonine phosphatase with activity distinct from that of PP1, we expressed PPY in the pCW vector in *E. coli* and examined its substrate specificity and the effect of inhibitors. Bacterially expressed PPY was able to dephosphorylate phosphorylase that had been phosphorylated by protein kinase A, although the specific activity of PPY was 40-fold lower than that of PP1. Using phosphorylase as substrate, the effect of both inhibitor 2 and phosphorylated inhibitor 1 was found to be markedly different on the two phosphatases, with PPY being unaffected by either inhibitor and PP1 being completely inhibited by nanomolar concentrations of both (Fig. 8). However, inhibitors with a broader specificity, such as the tumor promoter okadaic acid, inhibited

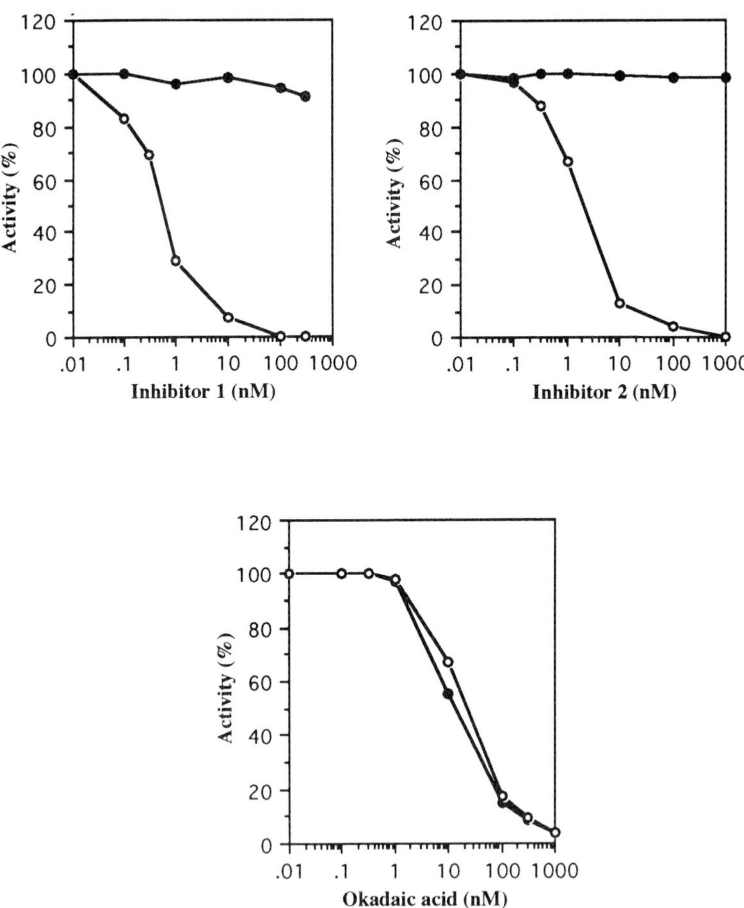

FIGURE 8   Effect of protein phosphatase inhibitors, inhibitor 1, inhibitor 2, and okadaic acid, on PPY and PP1. PPY (solid circles) and PP1 (open circles) were assayed at 0.17 mU/ml using ^{32}P-labeled phosphorylase as a substrate.

PPY at the same concentrations as observed for PP1. Figure 9 shows that the dephosphorylation of phosphorylase kinase regulatory subunits by PPY also differs from that of PP1; PPY dephosphorylates both α- and the β-subunits at approximately equal rates whereas PP1 shows a preference for dephosphorylation of the β-subunit. Thus, the enzymatic activity of PPY is quite distinct from that of PP1, and PPY must be regarded as a novel enzyme rather than an isoform of PP1 (Armstrong *et al.*, 1995).

Expression of PPY mRNA during the *Drosophila* life cycle is shown in Figure 10. In embryos, no PPY mRNA is detectable; larvae possess a very low level of a single 1.4-kb PPY transcript. In pupae, the level of this transcript is much higher, whereas in adults the PPY transcript is expressed at the same level in males but is not present at all in females. This result indicates that PPY is a male-specific protein and was a little surprising at first because PPY cDNA was isolated from a *Drosophila* head library! However, analysis of RNA from gonads showed that a PPY-specific transcript was abundantly expressed in testis but not present in ovary. No PPY transcript was detected in head RNA, but it was difficult to obtain very good RNA preparations from heads. Confirmation that PPY was a testis-specific phosphatase was obtained by immunoblotting with affinity-purified anti-PPY antibodies, which detected a 36-kDa PPY protein in testis extracts but not in ovary extracts, in male or female heads, abdomens, and thoraxes, or whole adult flies (Fig. 11). The lack of detection of PPY protein in male

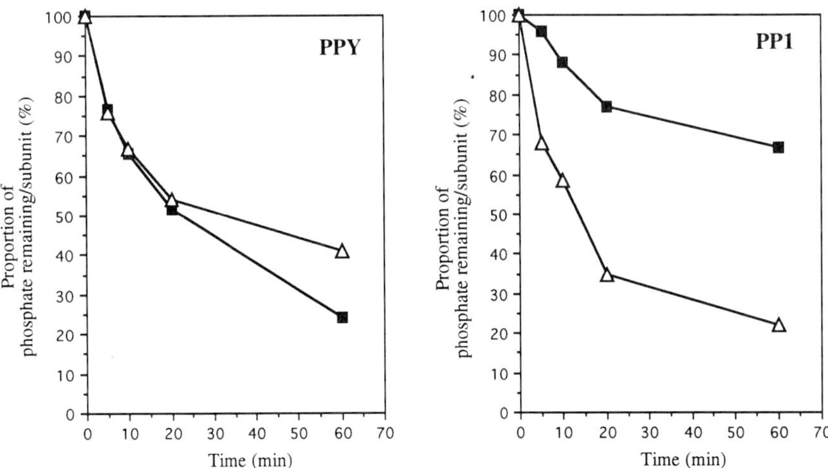

FIGURE 9 Dephosphorylation of the α- and β-subunits of phosphorylase kinase by bacterially expressed PPY and PP1. The dephosphorylation of the α-subunit is shown by squares and the dephosphorylation of the β-subunit by triangles. The ^{32}P-labeled phosphorylase kinase contained 0.8 mol of phosphate in the α-subunit and 0.85 mol in the β-subunit. Phosphate release from the α- and β-subunits was quantified as in Stewart *et al.* (1981). PP1 was used at 500 mU/ml and PPY at 2 mU/ml.

A

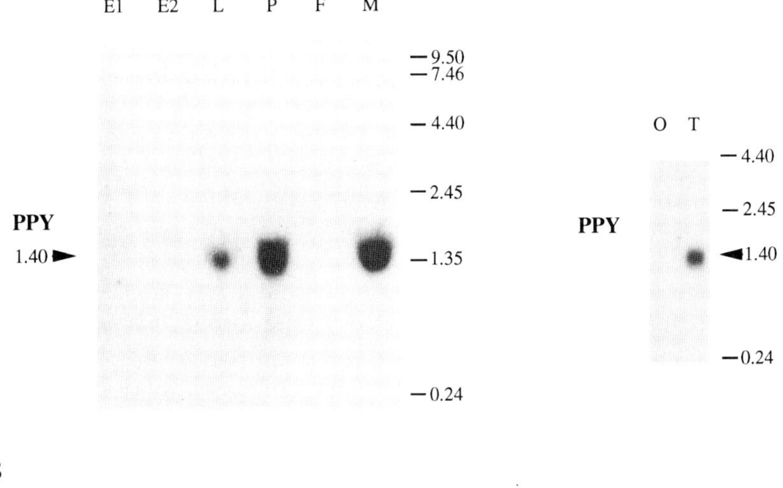

B

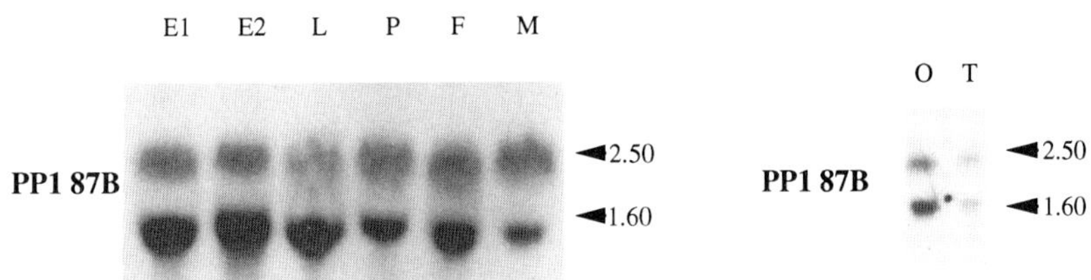

FIGURE 10   PPY mRNA levels during the *Drosophila* life cycle and in *Drosophila* tissues. (A) Total RNA from each developmental stage, ovary, and testis was fractionated on a denaturing agarose gel, blotted onto nylon membrane, and hybridized with the full-length PPY cDNA. (B) The blot was stripped and rehybridized with PP1 87B cDNA to control for loading variation [PP1 87B transcripts are expressed at a constant level during development (Dombrádi *et al.*, 1989(b)]. E1, 0- to 4-hr old embryos; E2, 4- to 24-hr-old embryos; L, larvae; P, pupae; F, female adult flies; M, male adult flies. Transcript sizes are indicated by arrows and calculated from their mobility relative to that of RNA size markers.

abdomens and whole adult male flies is most likely due to the low level of the PPY protein, which was estimated by immunoblotting to be 0.03% of the total protein in testis and only 0.0002% of the total protein in adult males. Overall, the results indicate that PPY is a testis-specific protein serine/threonine phosphatase and that its original identification in a *Drosophila* head library was likely to be due to contamination of the starting material for the preparation of the library with *Drosophila* abdomens.

The cellular and subcellular localization of PPY within the testis was examined initially by light microscopy using anti-PPY antibodies coupled via secondary antibodies to peroxidase staining. *Drosophila* testis is a coiled

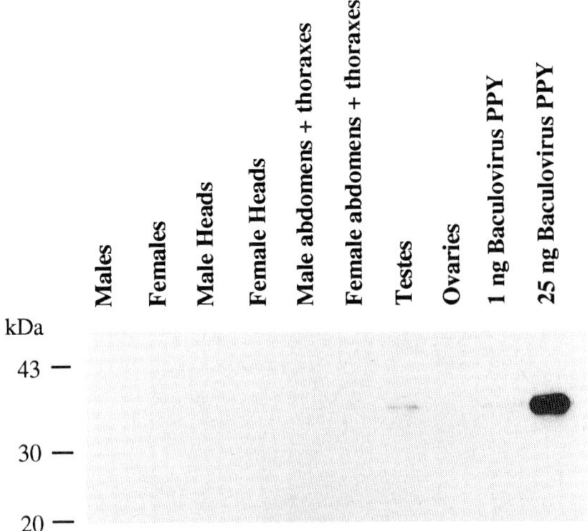

FIGURE 11   Immunoblot analysis of PPY protein in *Drosophila*. Soluble protein (10 μg) from *Drosophila* extracts was separated by SDS–PAGE, blotted onto nitrocellulose membrane, and probed with anti-PPY antibodies. The right-hand lanes contain pure PPY protein expressed from a baculovirus vector in insect cells.

tubular structure throughout which spermatogenesis occurs. During progression from the apical tip to terminal region, the germ cells undergo mitosis, meiosis, and differentiation with mature spermatozoa being released from the terminal region. Anti-PPY staining is absent or very low in the apical tip but present in increasing amounts during progression to the terminal region (Fig. 12). The staining appears punctate, occurring only in somatic cyst cells that encase the germ cells and being absent from the germ cells themselves. At the subcellular level, PPY is predominantly present in the cyst cell nuclei. Examination of anti-PPY staining reveals two distinct morphologies—a round form and a long thin form (Fig. 12)—corresponding to the nuclei of the two types of cyst cells that surround each developing bundle of germ cells. The nuclei of head cyst cells that encase the spermatocyte heads have a round morphology, whereas the nuclei of the tail cyst cells encasing the spermatocyte tails are long and thin. Antibodies against a specific C-terminal peptide sequence in PPY gave the same results as antibodies raised against the whole PPY protein expressed from a baculovirus vector in an insect cell system. To confirm the localization of PPY in cyst cell nuclei, we examined a *Drosophila*-enhancer trap line in which β-galactosidase, modified to include a nuclear localization signal, is expressed in cyst cells (Gönczy *et al.*, 1992). In this strain, colocalization of PPY and β-galactosidase was observed using immunofluorescent confocal microscopy. A weak anti-PPY staining was also observed in the cyst cell cytoplasm, although the predominant stain was in the nuclei (Armstrong *et al.*, 1995).

To consider the role of PPY in *Drosophila* testis, it is necessary to know about the function of cyst cells. There is evidence that cyst cell function is

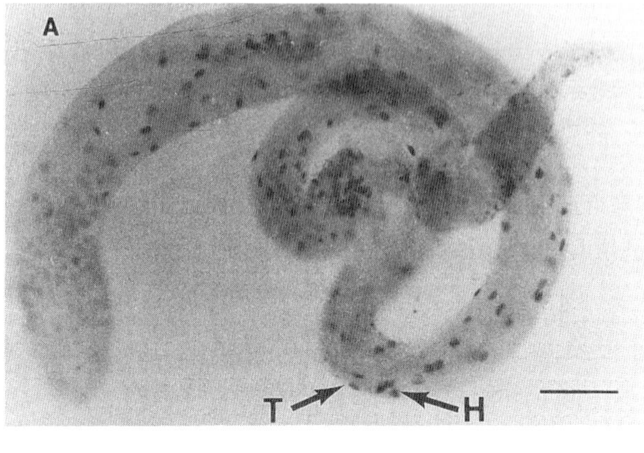

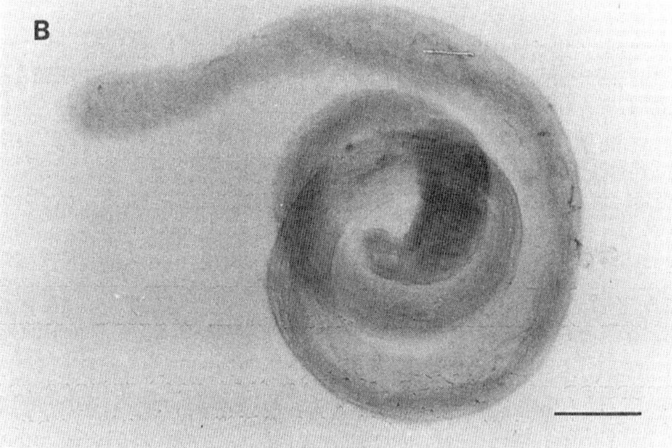

FIGURE 12   Immunolocalization of PPY in *Drosophila* testis. Testes were dissected from young adult males, fixed in 4% paraformaldehyde, and permeabilized in 0.1% Triton X-100 + 0.1% sodium deoxycholate. Affinity-purified anti-PPY antibodies were detected using horseradish peroxidase-conjugated anti-rabbit antibodies. (A) Testis stained with anti-PPY antibodies. (B) Testis stained following immunodepletion of anti-PPY antibodies on PPY– Sepharose. Head (H) cyst cell nuclei (round) and tail (T) cyst cell nuclei (elongated) are indicated by arrows. The scale bars represent 100 μm.

essential for normal spermatogenesis to occur. β3-Tubulin is a specialized isoform of tubulin that in testis is found only in cyst cells (Kimble *et al.*, 1989). Mutation in the gene encoding it results in a male sterile phenotype, demonstrating that β3-tubulin and by implication cyst cell function are necessary for normal spermatogenesis (Kimble *et al.*, 1990). It is likely that cyst cells play a major supporting role for the developing germ cells, perhaps supplying nutrients to them. A possible function of PPY in the nucleus might be to increase transcription of genes required for a continuous supply of nutrients.

A striking characteristic of cyst cells is that once formed they do not undergo further mitosis. Two cyst progenitor cells are anchored at the apical tip of the testis and these divide asymmetrically to yield two cyst cells that encase a single germ cell. The cyst cells remain associated with the developing germ cells throughout germ cell mitosis, meiosis, and differentiation to mature spermatocytes, but do not themselves undergo further cell division. They do, however, differentiate into head and tail cyst cells (Fuller, 1993). The absence of PPY from the apical region and its presence in more terminal regions is consistent with a role for PPY in arresting cell division. In addition, PPY, although unaffected by inhibitor 1 and inhibitor 2, is structurally related to PP1 87B, which is essential for exit from mitosis (Axton *et al.*, 1990; Dombrádi *et al.*, 1990a).

Another interesting function of cyst cells is that they provide a signal for spermatocyte differentiation. This was shown by studies employing germ cell transplantation techniques. When XX pole cells (precursor of germ cells) from wild-type embryos are transplanted into XY *oskar301* embryos, which fail to produce any viable pole cells, the transplanted pole cells undergo spermatogenesis, indicating that signals from the somatic environment of the germ cells determine their sexual identity. When XY germ cells are transplanted to XX *oskar301* embryos, primary spermatocytes are produced, indicating that although the XY germ cells possess autonomous information concerning their sexual identity, they require some inductive signal from the testis soma to enable completion of spermatogenesis from the primary spermatocyte stage (Steinman-Zwicky *et al.*, 1989). Because cyst cells are intimately associated with the germ cells, they are ideal somatic candidates for provision of such a signal (Fuller, 1993). The temporal expression of PPY in the cyst cells is consistent with the timing of a putative somatic signal leading to spermatocyte differentiation. Such a signal could be direct or through a change in transcription of certain genes.

In conclusion, PPY is a protein serine/threonine phosphatase specific to somatic cyst cells of testis. The distribution of PPY, coupled with its unique biochemical properties, suggests that PPY may be required to prevent cyst cell division, increase transcription of specific genes, and/or provide a signal for spermatocyte differentiation. The closest counterpart to cyst cells in mammals is the Sertoli cells, and it will be interesting to see if these cells possess a homolog of *Drosophila* PPY.

III. Summary of the Functions of Novel Protein Phosphatases

The functions of several novel protein phosphatases in the major family of protein serine/threonine phosphatases are beginning to be delineated and the information is summarized in Table II. PPZ1, PPZ2, PPQ, and PPY

TABLE II Functions of Novel Protein Phosphatases

Protein phosphatase	Occurrence	Function	Inhibition by okadaic acid and microcystin	Reference
PPZ1, PPZ2	Yeast	Regulation of cell integrity via a kinase cascade	ND	Hughes *et al.* (1993); Lee *et al.* (1993); Posas, Casamayor, *et al.* (1993)
PPQ	Yeast	Regulation of protein synthesis	ND	Chen *et al.* (1994)
PPY	*Drosophila* testis	Spermatocyte differentiation?	Yes	Armstrong *et al.* (1995)
PP4 (PPX)	Mammals, *Drosophila*	Microtubule nucleation?	Yes	Brewis *et al.* (1993)
PPV/SIT4	Mammals, *Drosophila,* yeast	Regulation of transcription and G_1/S stage of the cell cycle	ND	Arndt *et al.* (1989); Fernandez-Sarabia *et al.* (1992); Mann *et al.* (1993)
PPG	Yeast	Regulation of glycogen metabolism	ND	Posas, Clotet, *et al.* (1993)
PP5/PPT	Mammals, *Drosophila,* yeast	Regulation of RNA biosynthesis?	Yes	Chen *et al.* (1994)
rgdC	*Drosophila* eye	Visual signal transduction	ND	Steele *et al.* (1992); Byk *et al.* (1993)

Note. ND, not determined.

belong to the PP1 subfamily. PPZ1 and PPZ2 participate in the maintenance of cellular integrity and osmotic stability (Hughes *et al.*, 1993) and may be at least partially under the regulation of a protein kinase cascade initiated by protein kinase C (Lee *et al.*, 1993). PPQ is involved in the regulation of protein synthesis (Chen *et al.*, 1993) and may influence translational accuracy (Vincent *et al.*, 1994). These three enzymes have so far only been identified in the yeast *S. cerevisiae,* but, at least in the case of PPQ, it seems likely that a homolog will exist in higher organisms. PPY, as discussed previously, is likely to play a role in the somatic cellular signals required for spermatocyte differentiation and one might expect it to have a counterpart in mammals.

PP4 (originally termed PPX), PPV/SIT4, and PPG belong to the PP2A subfamily. PP4 has been shown to be highly concentrated at centrosomes, indicating that it may play a role in nucleation of microtubules (Brewis *et al.*, 1993). SIT4 in *S. cerevisiae* has been shown to be involved in the regulation of transcription of a number of genes including G_1 cyclins (Fernandez-Sarabia *et al.*, 1992). It is essential for passage though the G_1 to S stage of the cell cycle. PPV in *Drosophila* has been shown to be the functional homolog SIT4 (Mann *et al.*, 1993). No human homolog is known but a possible homolog has been identified in rats (Becker *et al.*, 1994). PPG participates in the regulation of glycogen synthase levels in *S. cerevisiae* (Posas, Clotet, *et al.*, 1993), but it is unclear whether a homolog in higher organisms will exist. The PP5/PPT subfamily is known in yeast, *Drosophila,* rat, and human species, and as discussed previously may be involved in signals regulating the biosynthesis of RNA. RdgC in the PP2B subfamily has only been identified in *Drosophila* and is essential to prevent light-induced photoreceptor degeneration (Steele *et al.*, 1992). It participates in visual signal transduction catalyzing the dephosphorylation of rhodopsin (Byk *et al.*, 1993).

Only a few of the novel protein phosphatases have been tested for inhibition by the tumor promoters okadaic acid and microcystin. These, like PP1 and PP2A, are inhibited by naturally occurring toxins in the nanomolar range (Table II). Although a number of novel protein phosphatases, such as PP4 and PP5, are found in all tissues, some are tissue specific. RdgC is found only in the eye, whereas PPY is located only in the testis. It appears likely that many more protein phosphatases, particularly tissue-specific forms, will be uncovered in the future.

Acknowledgments

This work was supported by the Medical Research Council, London UK. C.G.A. is a recipient of a Medical Research Council Studentship.

References

Arino, J., Woon, C. W., Brautigan, D., Miller, T. B., and Johnson, G. L. (1988). Human liver phosphatase 2A: cDNA and amino acid sequence of two catalytic subunit isotypes. *Proc. Natl. Acad. Sci. U.S.A.* **85**, 4252–4256.

Armstrong, C. G., Mann, D. J., Berndt, N., and Cohen, P. T. W. (1995). *Drosophila* PPY, a novel male specific protein serine/threonine phosphatase localised in somatic cells of the testis. *J. Cell Sci.* **108**, 3367–3375.

Arndt, K. T., Styles, C. A., and Fink, G. R. (1989). A suppressor of *HIS4* transcription defect encodes a protein with homology to the catalytic subunit of protein phosphatases. *Cell* **56**, 527–537.

Axton, J. M., Dombrádi, V., Cohen, P. T. W., and Glover, D. M. (1990). One of the protein phosphatase 1 isoenzymes in *Drosophila* is essential for mitosis. *Cell* **63**, 33–46.

Barker, H. M., Brewis, N. D., Street, A. J., Spurr, N. K., and Cohen, P. T. W. (1994). Three genes for protein phosphatase 1 map to different human chromosomes: Sequence, expression and gene localisation of protein serine/threonine phosphatase 1 beta (PPP1CB). *Biochim. Biophys. Acta* **1220**, 212–218.

Barker, H. M., Craig, S. P., Spurr, N. K., and Cohen, P. T. W. (1993). Sequence of human protein serine/threonine phosphatase 1 gamma and localization of the gene (PPP1CC) encoding it to chromosome bands 12q24.1–q24.2. *Biochim. Biophys. Acta* **1178**, 228–233.

Barker, H. M., Jones, T. A., da Cruz e Silva, E. F., Spurr, N. K., Sheer, D., and Cohen, P. T. W. (1990). Localization of the gene encoding a type I protein phosphatase catalytic subunit to human chromosome band 11q13. *Genomics* **7**, 159–166.

Becker, W., Kentrup, H., Klumpp, S., Schultz, J. E., and Joost, H. G. (1994). Molecular cloning of a protein serine/threonine phosphatase containing a putative regulatory tetratricopeptide repeat domain. *J. Biol. Chem.* **269**, 22586–22592.

Brewis, N. D., Street, A. J., Prescott, A. R., and Cohen, P. T. W. (1993). PPX, a novel protein serine/threonine phosphatase localized to centrosomes. *EMBO J.* **12**, 987–996.

Brown, L., Chen, M. X., and Cohen, P. T. W. (1994). Identification of cDNA encoding a *Drosophila* calcium/calmodulin regulated protein phosphatase, which has its most abundant expression in the early embryo. *FEBS Lett.* **339**, 124–128.

Byk, T., Bar-Yaacov, M., Doza, Y. N., Minke, B., and Selinger, Z. (1993). Regulatory arrestin cycle secures the fidelity and maintenance of the fly photoreceptor cell. *Proc. Natl. Acad. Sci. U.S.A.* **90**, 1907–1911.

Chen, M. X., Chen, Y. H., and Cohen, P. T. W. (1992). Polymerase chain reactions using *Saccharomyces, Drosophila* and human DNA predict a large family of protein serine/ threonine phosphatases. *FEBS Lett.* **306**, 54–58.

Chen, M. X., Chen, Y. H., and Cohen, P. T. W. (1993). PPQ, a novel protein phosphatase containing a Ser + Asn-rich amino terminal domain, is involved in the regulation of protein synthesis. *Eur. J. Biochem.* **218**, 689–699.

Chen, M. X., McPartlin, A. E., Brown, L., Chen, Y. H., Barker, H. M., and Cohen, P. T. W. (1994). A novel human protein serine/threonine which possesses four tetratricopeptide repeat (TPR) motifs and localizes to the nucleus. *EMBO J.* **13**, 4278–4290.

Chinkers, M. (1994). Targeting of a distinctive protein-serine phosphatase to the protein kinase- like domain of the atrial natriuretic peptide receptor. *Proc. Natl. Acad. Sci. U.S.A.* **91**, 11075–11079.

Cohen, P. (1989). The structure and regulation of protein phosphatases. *Annu. Rev. Biochem.* **58**, 453–508.

Cohen, P. T. W., Brewis, N. D., Hughes, V., and Mann, D. J. (1990). Protein serine/threonine phosphatases; An expanding family. *FEBS Lett.* **268**, 355–359.

Cyert, M. S., Kunisawa, R., Kaim, D., and Thorner, J. (1991). Yeast has homologs (CNA1 and CNA2 gene products) of mammalian calcineurin, a calmodulin-regulated phosphoprotein phosphatase. *Proc. Natl. Acad. Sci. U.S.A.* **88**, 7376–7380.

da Cruz e Silva, O. B., da Cruz e Silva, E. F., and Cohen, P. T. W. (1988). Identification of a novel protein phosphatase catalytic subunit by cDNA cloning. *FEBS Lett.* **242,** 106–110.

Dombrádi, V., Axton, J. M., Barker, H. M., and Cohen, P. T. W. (1990a). Protein phosphatase 1 activity in *Drosophila* mutants with abnormalities in mitosis and chromosome condensation. *FEBS Lett.* **275,** 39–43.

Dombrádi, V., Axton, J. M., Brewis, N. D., da Cruz e Silva, E. F., Alphey, L., and Cohen, P. T. W. (1990b). *Drosophila* contains three genes that encode distinct isoforms of protein phosphatase 1. *Eur. J. Biochem.* **194,** 739–745.

Dombrádi, V., Axton, J. M., Glover, D. M., and Cohen, P. T. W. (1989a). Molecular cloning and chromosomal localization of a novel *Drosophila* protein phosphatase. *FEBS Lett.* **247,** 391–395.

Dombrádi, V., Axton, J. M., Glover, D. M., and Cohen, P. T. W. (1989b). Cloning and chromosomal localization of *Drosophila* cDNA encoding the catalytic subunit of protein phosphatase 1α. High conservation between mammalian and insect sequences. *Eur. J. Biochem.* **183,** 603–610.

Dombrádi, V., Mann, D. J., Saunders, D. C., and Cohen, P. T. W. (1993). Cloning of the fourth functional gene for protein phosphatase 1 in *Drosophila melanogaster* from its chromosomal localisation. *Eur. J. Biochem.* **212,** 177–183.

Feng, D. F., and Doolittle, R. F. (1990). Progressive alignment and phylogenetic free construction of protein sequences. *Methods Enzymol.* **183,** 375–387.

Fernandez-Sarabia, M. J., Sutton, A., Zhong, T., and Arndt, K. T. (1992). SIT4 protein phosphatase is required for the normal accumulation of *SWI4, CLN1, CLN2* and *HSC26* RNAs during late G1. *Genes Dev.* **6,** 2417–2428.

Fuller, M. T. (1993). Spermatogenesis in *Drosophila*. *In* "The Development of *Drosophila melanogaster*," pp. 71–147. Cold Spring Harbor Laboratory Press, Cold Spring Harbor, NY.

Goebl, M., and Yanagida, M. (1991). The TPR snap helix: A novel protein repeat motif from mitosis to transcription. *Trends Biochem. Sci.* **16,** 173–177.

Gönczy, P., Viswanathan, S., and DiNardo, S. (1992). Probing spermatogenesis in *Drosophila* with P-element enhancer detectors. *Development* **114,** 89–98.

Guerini, D., Montell, C., and Klee, C. B. (1992). Molecular cloning and characterisation of the genes encoding the two subunits of *Drosophila melanogater* calcineurin. *J. Biol. Chem.* **267,** 2254–2259.

Hase, T., Reizman, H., Suda, K., and Schatz, G. (1983). Import of proteins into mitochondria: Nucleotide sequence of the gene for a 70-kd protein of the yeast mitochondrial outer membrane. *EMBO J.* **2,** 2169–2172.

Hemmings, B. A., Wernet, W., Mayer, R., Maurer, F., Hofsteenge, J., and Stone, S. R. (1988). The nucleotide sequence of the cDNA encoding the human lung protein phosphatase 2A beta catalytic subunit. *Nucleic Acids Res.* **16,** 11366.

Hirano, T., Kinoshita, N., Morikawa, K., and Tanagida, M. (1990). Snap helix with knob and hole: Essential repeats in *S. pombe* nuclear protein *nuc2*+. *Cell* **60,** 319–328.

Honoré, B., Leffers, H., Madsen, P., Rasmussen, H. H., Vandekerckhove, J., and Celis, J. E. (1992). Molecular cloning and expression of a transformation-sensitive human protein containing the TPR motif and sharing identity to the stress inducible yeast protein *STI1*. *J. Biol. Chem.* **267,** 8485–8491.

Hughes, V., Müller, A., Stark, M. J. R., and Cohen, P. T. W. (1993). Both isoforms of protein phosphatase Z are essential for the maintenance of cell size and integrity in *Saccharomyces cerevisiae* in response to osmotic stress. *Eur. J. Biochem.* **216,** 269–279.

Icho, T., and Wickner, R. B. (1987). Metal binding, nucleic acid-binding finger sequences in the *CDC16* gene of *Saccharomyces cerevisiae*. *Nucleic Acids Res.* **15,** 8439–8450.

Kieffer, L. J., Seng, T. W., Li, W., D. G., O., Handshumacher, R. E., and Bayney, R. M. (1993). Cyclophilin-40, a protein with homology to the p59 component of the steroid receptor complex. *J. Biol. Chem.* **268,** 12303–12310.

Kimble, M., Dettman, R., and Raff, E. C. (1990). The β3-tubulin gene of *Drosophila melanogaster* is essential for viability and fertility. *Genetics* **126,** 991–1006.

Kimble, M., Incardona, J. P., and Raff, E. C. (1989). A variant β-tubulin isoform of *Drosophila melanogaster* (β3) is expressed primarily in tissues of mesodermal origin in embryos and pupae, and is utilized in populations of transient microtubules. *Dev. Biol.* **131**, 415–429.

Lee, K. S., Hines, L. K., and Levin, D. E. (1993). A pair of functionally redundant genes (PPZ1 and PPZ2) encoding type 1-related protein phosphatases function within the PKC1-mediated pathway. *Mol. Cell. Biol.* **13**, 5843–5853.

Lee, T. G., Tang, N., Tompson, S., Miller, J., and Katze, M. G. (1994). The 58,000-Dalton cellular inhibitor of the interferon-induced double-stranded RNA-activated protein kinase (PKR) is a member of the tetratricopeptide repeat family of proteins. *Mol. Cell. Biol.* **14**, 2331–2342.

Legrain, P., and Choulika, A. (1990). The molecular characterisation of PRP6 and PRP9 yeast genes reveals a new cysteine/histidine motif common to several splicing factors. *EMBO J.* **9**, 2775–2781.

Liu, Y., Ishii, S., Tokai, M., Tsutsumi, H., Ohki, O., Akada, R., Tanaka, K., Tsuchiya, E., Fukui, S., and Miyakawa, T. (1991). The *Saccharomyces cerevisiae* genes (*CMP1* and *CMP2*) encoding calmodulin-binding proteins homologous to the catalytic subunit of mammalian protein phosphatase 2B. *Mol. Gen. Genet.* **227**, 52–59.

Mann, D. J., Dombrádi, V., and Cohen, P. T. W. (1993). *Drosophila* protein phosphatase V functionally complements a *SIT4* mutant in *Saccharomyces cerevisiae* and its amino terminal region can confer this complementation to a heterologous phosphatase catalytic domain. *EMBO J.* **12**, 4833–4822.

Marck, C., Lefebvre, O., Carles, C., Riva, M., Chaussivert, N., Ruet, A., and Sentenec, A. (1993). The TFIIIB-assembling subunit of yeast transcription factor TFIIIC has both tetratricopeptide repeats and basic helix–loop–helix motifs. *Proc. Natl. Acad. Sci. U.S.A.* **90**, 4027–4031.

McCollum, D., Monosov, E., and Subramani, S. (1993). The *pas8* mutant of *Pichia pastoris* exhibits the peroxisomal protein import deficiencies of Zellweger syndrome cells—The PAS8 protein binds to the COOH-terminal tripeptide peroxisomal targeting signal and is a member of the TPR protein family. *J. Cell Biol.* **121**, 761–774.

McPartlin, A. E., Barker, H. M., and Cohen, P. T. W. (1991). Identification of a third alternatively spliced cDNA encoding the catalytic subunit of protein phosphatase 2B beta. *Biochim. Biophys. Acta* **1088**, 308–310.

Mirabito, P. M., and Morris, N. R. (1993). BIMA, a TPR-containing protein required for mitosis, localises to the spindle pole body in *Aspergillus nidulans*. *J. Cell Biol.* **120**, 959–968.

Muramatsu, T., and Kincaid, R. L. (1992). Molecular cloning and chromosomal mapping of the human gene for the testis specific catalytic subunit of calmodulin-dependent protein phosphatase (calcineurin A). *Biochem. Biophys. Res. Commun.* **188**, 265–271.

Muramatsu, T., and Kincaid, R. L. (1993). Molecular cloning of a full-length cDNA encoding the catalytic subunit of human calmodulin-dependent protein phosphatase (Calcineurin Aα). *Biochim. Biophys. Acta* **1178**, 117–120.

Nicolet, C. M., and Craig, E. A. (1989). Isolation and characterization of *STI1*, a stress-inducible gene from *Saccharomyces cerevisiae*. *Mol. Cell. Biol.* **9**, 3638–3646.

Ohkura, H., Kinoshita, N., Miyatani, S., Toda, T., and Yanagida, M. (1989). The fission yeast dis2$^+$ gene required for chromosome disjoining encodes one of two putative type 1 protein phosphatases. *Cell* **57**, 997–1007.

Orgad, S., Brewis, N. D., Alphey, L., Axton, J. M., Dudai, Y., and Cohen, P. T. W. (1990). The structure of protein phosphatase 2A is as highly conserved as that of protein phosphatase 1. *FEBS Lett.* **275**, 44–48.

Peattie, D. A., Harding, M. W., Fleming, M. A., DeCenzo, M. T., Lippke, J. A., Livingstone, D. J., and Benasutti, M. (1992). Expression and characterization of human FKBP52, an immunophilin that associates with the 90 kDa heat shock protein and is a component of steroid receptor complexes. *Proc. Natl. Acad. Sci. U.S.A.* **89**, 10974–10978.

Posas, F., Casamayor, A., and Ariño, J. (1993). The PPZ protein phosphatases are involved in the maintenance of osmotic stability of yeast cells. *FEBS Lett.* **318**, 282–286.

Posas, F., Clotet, J., Muns, M. T., Corominas, J. A., Casamayor, A., and Ariño, J. (1993). The gene PPG encodes a novel yeast protein phosphatase involved in glycogen accumulation. *J. Biol. Chem.* **268**, 1349–1354.

Radanyi, C., Chambraud, B., and Baulieu, E. E. (1994). The ability of the immunophilin FKBP59-HBI to interact with the 90-kDa heat shock protein is encoded by its tetratricopeptide repeat domain. *Proc. Natl. Acad. Sci. U.S.A.* **91**, 11197–11201.

Rameau, G., Puglia, K., Crowe, A., Sethy, I., and Willis, I. (1994). A mutation in the second largest subunit of TFIIIC increases the rate-limiting step in transcription by RNA polymerase III. *Mol. Cell. Biol.* **14**, 822–830.

Ratajczak, T., Carrello, A., Mark, P. J., Warner, B. J., Simpson, R. J., Moritz, R. L., and House, A. K. (1993). The cyclophilin component of the unactivated estrogen receptor contains a tetratricopeptide repeat domain and shares identity with p59 (FKBP59). *J. Biol. Chem.* **268**, 13187–13192.

Rhee, S.-K., Icho, T., and Wickner, R. B. (1989). Structure and nuclear localization signal of the SKI3 antiviral protein of *Saccharomyces cerevisiae. Yeast* **5**, 149–158.

Ronne, H., Carlberg, M., Hu, G.-Z., and Nehlin, J. O. (1991). Protein phosphatase 2A in *Saccharomyces cerevisiae:* Effects on cell growth and bud morphogenesis. *Mol. Cell. Biol.* **11**, 4876–4884.

Samejima, I., and Yanagida, M. (1994). Bypassing anaphase by fission yeast cut9 mutation: Requirement of cut9 + to initiate anaphase. *J. Cell Biol.* **127**, 1655–1670.

Schultz, J., Marshall-Carlson, L., and Carlson, M. (1990). The N-terminal TPR region is the functional domain of *SSN6*, a nuclear phosphoprotein of *Saccharomyces cerevisiae. Mol. Cell. Biol.* **10**, 4744–4756.

Shenolikar, S. (1994). Protein serine/threonine phosphatases—New avenues for cell regulation. *Annu. Rev. Cell Biol.* **10**, 55–86.

Sikorski, R. S., Michaud, W. A., and Hieter, P. (1993). p62cdc23 of *Saccharomyces cerevisiae:* A nuclear tetratricopeptide repeat protein with two mutable domains. *Mol. Cell. Biol.* **13**, 1212–1221.

Sikorski, R. S., Michaud, W. A., Wootton, J. C., Boguski, M. S., Connelly, C., and Hieter, P. (1991). TPR proteins as essential components of the yeast cell cycle. *In* "Cold Spring Harbor Symposiumon Quantitative Biology," pp. 663–673. Cold Spring Harbor Laboratory Press, Cold Spring Harbor, NY.

Sneddon, A. A., Cohen, P. T., and Stark, M. J. (1990). *Saccharomyces cerevisiae* protein phosphatase 2A performs an essential cellular function and is encoded by two genes. *EMBO J.* **9**, 4339–4346.

Steele, F. R., Washburn, T., Rieger, R., and O'Tousa, J. E. (1992). *Drosophila retinal degeneration* C (*rdgC*) encodes a novel serine/threonine protein phosphatase. *Cell* **69**, 669–676.

Steinman-Zwicky, M., Schmid, H., and Nothiger, R. (1989). Cell-autonomous and inductive signals can determine the sex of the germ line of *Drosophila* by regulating the gene *Sxl. Cell* **57**, 157–166.

Stewart, A., Hemmings, B. A., Cohen, P., Goris, J., and Merlevede, W. (1981). The ATP-Mg dependent protein phosphatase and protein phosphatase-1 have identical substrate specificities. *Eur. J. Biochem.* **115**, 197–205.

van der Leij, I., Franse, M. M., Elgersma, Y., Distel, B., and Tabak, H. F. (1993). PAS10 is a tetratricopeptide-repeat protein that is essential for the import of most matrix proteins into peroxisomes of *Saccharomyces cerevisiae. Proc. Natl. Acad. Sci. U.S.A.* **90**, 11782–11786.

Vincent, A., Newman, G., and Liebman, S. (1994). Yeast translational allosuppressor, *SAL6:* A new member of the PP1-like phosphatase family with a long serine-rich N-terminal extension. *Genetics* **138**, 597–607.

Wang, H., and Stillman, D. J. (1993). Transcriptional repression in *Saccharomyces cerevisiae* by a SIN3-LexA fusion protein. *Mol. Cell. Biol.* **13**, 1805–1814.

Zhang, K., Smouse, D., and Perrimon, N. (1991). The *crooked neck* gene of *Drosophila* contains a motif found in a family of yeast cell cycle genes. *Genes Dev.* **5**, 1080–1091.

N. K. Tonks

Cold Spring Harbor Laboratory
Cold Spring Harbor, New York 11724-2208

Protein Tyrosine Phosphatases and the Control of Cellular Signaling Responses

I. Introduction

Studies of protein tyrosine kinases (PTKs) have established the critical importance of tyrosine phosphorylation in the signaling events that underlie the control of such fundamental physiological processes as cell growth and differentiation, cell cycle, and cytoskeletal function. However, it is important to realize that *in vivo* protein tyrosine phosphorylation is a reversible, dynamic process in which the net level of phosphate in a target substrate reflects not only the activity of the PTKs that phosphorylate it, but also the competing action of the protein tyrosine phosphatases (PTPs) that catalyze the dephosphorylation reaction. Thus, it is also important to look at the physiological roles of tyrosine phosphorylation from the perspective of the PTPs.

My introduction to the field came in 1985 as a post doc in Eddy Fischer's lab at the University of Washington, Seattle. At that time, although progress

Advances in Pharmacology, Volume 36

had been made in characterizing the PTKs, it was clear that PTPs must exist but their number and properties were poorly defined. One of the major technical difficulties that held back the study of PTPs was the requirement for a suitably phosphorylated purified protein substrate with which to measure enzyme activity. This was a particularly acute problem for the PTPs because at this time their physiological substrates were either not yet clearly identified or unavailable in the quantity required for routine assays. Therefore, those in the field turned to artificial substrates, including proteins that were only phosphorylated by PTKs after denaturation and chemical modification to expose their tyrosyl side chains. However, many such substrates were plagued by poor solubility properties and low stoichiometries of phosphory-lation. We spent a considerable period of time developing reduced carbox-amidomethylated and maleylated (RCM) lysozyme as our routine substrate. It displayed excellent solubility properties and was phosphorylated by the insulin receptor PTK to high stoichiometry at predominantly a single site. The development of this substrate gave us the means to assay PTP activity over a broad range of substrate concentrations. It also enabled us to exploit the ability of the insulin receptor PTK to utilize ATPγS as a phosphate donor and thus to generate a highly thiophosphorylated derivative of RCM lysozyme. This was very important because the PTPs retain the ability to bind thiophosphorylated substrates but they tend not to dephosphorylate them to a significant extent. Therefore, by preparing a column of thiophos-phorylated RCM lysozyme bound to Sepharose, we had generated a power-ful affinity chromatography approach to the isolation of PTPs. This proved to be the key step in allowing us to purify to apparent homogeneity the major PTP from human placenta. This enzyme was termed PTP1B (Tonks *et al.,* 1988a,b).

In collaboration with Harry Charbonneau and Ken Walsh, also of the University of Washington, we determined the complete primary sequence of PTP1B (Charbonneau *et al.,* 1988, 1989). This illustrated two important points. First, the sequence of PTP1B did not show significant similarity to the sequences that were available for Ser/Thr, acid, or alkaline phosphatases. Therefore, unlike the kinases that all appear to be derived from a common ancestor, the phosphatases have evolved in separate families. However, we were fortunate that there was already a substantial body of literature on CD45, also known as the leukocyte common antigen, which is a transmem-brane glycoprotein that is a surface marker of hematopoietic cells and that had been cloned by several groups. When we compared the sequence of PTP1B with those in the various databases, we noted a high level of identity between PTP1B and each of the two tandemly repeated intracellular domains of CD45 (Charbonneau *et al.,* 1988). What made this observation particu-larly provocative was the predicted structure of CD45. In addition to the tandem array of two PTP domains in the intracellular segment, CD45 pos-sesses a single transmembrane domain and a highly glycosylated, cysteine-rich extracellular segment that displays all the hallmarks of a ligand-binding motif. Thus, the exciting possibility was raised that CD45 may represent a prototype for a family of receptor-linked PTPs with the capacity to play a direct role in modulating cellular signaling responses through ligand-regulated dephosphorylation of tyrosyl residues in proteins at the plasma membrane. This proposal was further strengthened by our demonstration of intrinsic PTP activity in CD45 (Tonks, Charbonneau, *et al.*, 1988).

These data were published in 1988. What followed has been an explosion of interest in the PTPs and now a wide variety of these enzymes have been identified in sources as diverse as mammals, *Drosophila, C. elegans, Dictyostelium,* yeast, and viruses (Tonks, 1993). PTPs have even been detected in prokaryotes (Potts *et al.*, 1993). In fact, it has been shown that an essential virulence determinant of many strains of the pathogenic bacterium *Yersinia,* the causative agent of the plague or black death, is a PTP termed Yop, the activity of which is essential for virulence (Guan and Dixon, 1993).

Currently, some 50 PTPs have been identified and it is predicted, based on current data from genome sequencing efforts, that there will be ~500 PTPs in humans. Thus, the PTPs will rival the PTKs in their structural diversity and complexity. Each PTP contains at least one conserved segment of ~240 residues that is assumed to delineate an independently folding catalytic domain. Within this domain is the sequence motif [I/V]HCXAGX-XR[S/T]G that uniquely defines the PTP family of enzymes. The Cys residue, which is absolutely conserved, is involved in forming a thiophosphate intermediate as part of the catalytic mechanism (Guan and Dixon, 1991). Although the PTPs display a high degree of similarity in their catalytic domains, they can be readily differentiated on the basis of the structure of their noncatalytic segments. Most strikingly, like the PTKs, the PTPs can be subdivided into receptor-like and nontransmembrane, cytoplasmic forms. In addition, a subfamily of dual-specificity phosphatases, which also contain this signature motif, has been described recently (Fig. 1). In this chapter, I discuss various aspects of the structure, regulation, and function of members of each of the above categories. Due to space limitations, I focus primarily on contributions from my own lab. The following reviews provide a more comprehensive description of the recent advances in the field: Stone and Dixon (1994), Sun and Tonks (1994), Tonks (1993), and Walton and Dixon (1993).

II. Receptor-like PTPs

Generally, the structure of receptor PTPs can be described in terms of an intracellular segment containing two phosphatase domains (although

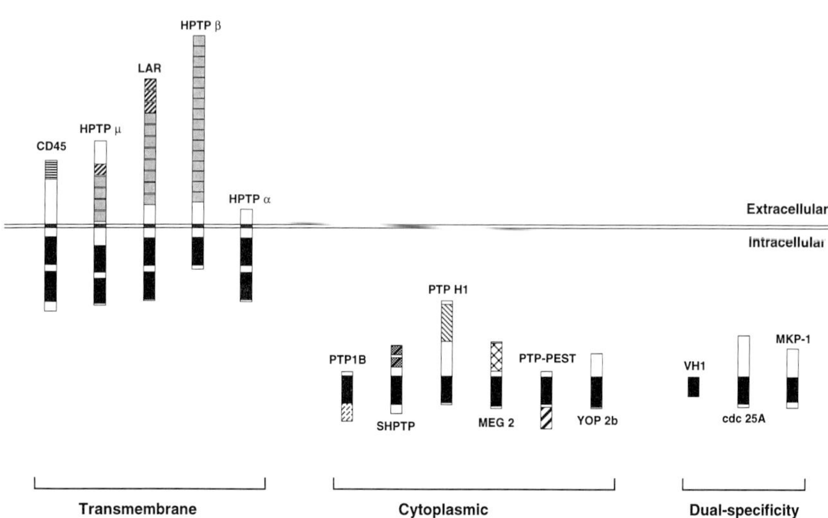

FIGURE I   Structural organization of representatives of the PTP family of enzymes. The PTPs can be classified as (i) transmembrane, receptor-like; (ii) nontransmembrane, cytoplasmic; or (iii) dual-specificity enzymes. The conserved catalytic domains are shown in black. The noncatalytic, regulatory segments fused to either the N or C terminus of the catalytic domains are described in the text.

there are now several examples with only a single catalytic domain), a single transmembrane domain and a variable extracellular segment. The significance of the two-domain arrangement that characterizes the intracellular segment of most receptor-like PTPs is unclear. Although there is the obvious potential for cooperative interactions between domains, it remains a point of controversy as to whether in fact both domains, in particular domain II, are active. In some cases, it has been proposed that the function of domain II is to control the specificity of domain I (Streuli *et al.*, 1990). Nevertheless, it seems clear that for PTPα domains I and II both possess intrinsic activity but differ in their substrate specificity (Wang and Pallen, 1991). Somewhat more controversial is the suggestion that both domains in CD45 are also active but subjected to different modes of regulation, in particular domain II, which requires phosphorylation to manifest its activity (Stover and Walsh, 1994). Presumably, as the physiological substrates of these enzymes are identified such controversies will be clarified.

In contrast to the similarity in the arrangement of their intracellular segments, these receptor-like PTPs can be distinguished on the basis of the structure of their extracellular segments, which presumably reflects an equivalent diversity in the variety of ligands to which they may respond. However, currently there is a distinct paucity of information with respect to the identity of such ligands. It was originally suggested that CD22, a B cell surface protein, interacted with the smallest, 180-kDa form of CD45

(Stamenkovic *et al.*, 1991). Subsequently, it has been shown that CD22 is a lectin that binds generally to sialoglycoproteins (Sgroi *et al.*, 1993). Its interaction with CD45 is not specific and the significance of such an interaction with respect to cell signaling remains unclear. An alternatively spliced form of the extracellular segment of PTPζ has been identified as phosphacan, a chondroitin sulfate proteoglycan that binds to the extracellular matrix protein tenascin (Barnea *et al.*, 1994; Maurel *et al.*, 1994). However, it is not yet clear whether the full-length form of PTPζ displays similar binding properties and what the effects of such an interaction would be on the function of PTPζ.

A. PTPμ and Cell–Cell Adhesion

Our interest was piqued by the observation that a common structural theme in the extracellular segment of receptor-like PTPs is the presence of motifs found in proteins involved in cell adhesion, including immunoglobulin (Ig)-like and/or fibronectin type III (FNIII) domains. We have been characterizing one such enzyme, PTPμ (reviewed in Brady-Kalnay and Tonks, 1994b). The extracellular segment of PTPμ comprises one Ig-like and four FNIII domains. In addition, at the extreme N terminus there is a MAM domain; this is a 170-residue motif found in meprins, which are cell surface metalloendoproteases, the A5 transmembrane glycoprotein, which is expressed on neurons of the visual centers of the brain of *Xenopus laevis*, as well as PTPμ and PTPκ, which is a closely related homolog of PTPμ sharing 77% identity throughout its sequence. There is a single transmembrane domain and a large intracellular segment containing two PTP domains preceded by a juxtamembrane segment that is ∼70 residues longer than the equivalent segment in other receptor-like PTPs and that displays homology to the intracellular domain of members of the cadherin family of cell adhesion molecules. This is the segment of cadherins that is essential for their adhesive function and that indirectly, through the binding of proteins termed catenins, directs their association with the actin cytoskeleton. Such homology is unique among both receptor-like PTPs and members of the Ig superfamily (Tonks *et al.*, 1992).

In light of the presence of Ig-like and FNIII motifs in the extracellular segment of PTPμ, it displays features of the Ig superfamily of cell adhesion molecules that includes the neural cell adhesion molecule (N-CAM) (Cunningham, 1995). N-CAM is a sialic acid-rich cell surface glycoprotein that is involved in tissue development and morphogenesis and mediates homophilic cell–cell adhesion, i.e., a molecule of N-CAM on one cell binds to another N-CAM molecule on an apposing cell surface. Based on this structural similarity between the extracellular segment of PTPμ and N-CAM, one would predict that this receptor PTP may participate in homophilic binding interactions. We have tested this prediction using an overexpression system, reconstitution of the binding reaction *in vitro,* and cells in which PTPμ is normally expressed (Brady-Kalnay *et al.*, 1993). Using the baculovirus system, we observed that expression of full-length PTPμ, or mutant forms with an intact extracellular segment, induced Sf9 cells to aggregate (Fig. 2). In addition, we have demonstrated that homophilic binding between molecules of PTPμ can be reconstituted *in vitro*. Fluorescent beads (covaspheres) linked to baculovirus-expressed PTPμ were able to bind to the surface of petri dishes coated with GST/PTPμ (a GST fusion protein containing the extracellular segment of PTPμ purified following expression in *Escherichia coli*) but did not bind to GST alone. In addition, beads coated with the GST/PTPμ fusion protein self-aggregate, whereas those coated with GST alone do not. Thus, the PTPμ-induced aggregation was found to be independent of calcium and glycosylation. Furthermore, these observations indicate that phosphatase activity and the adhesive function are not mutually dependent. We also demonstrated that PTPμ, as it is expressed endogenously on the surface of MvLu cells, retains the capacity to interact with baculovirus-expressed PTPμ linked to beads. The binding of PTPμ-coated beads to the MvLu cell surface is blocked by antibody to the extracellular segment of PTPμ but not by a control preimmune antibody. These data establish homophilic binding between molecules of PTPμ, thus indicating that the ligand for PTPμ is another PTPμ molecule on the surface of an adjacent cell (Brady-Kalnay *et al.*, 1993). Similar data were subsequently obtained independently for PTPμ (Gebbink *et al.*, 1993) and PTPκ (Sap *et al.*, 1994). We have now generated, expressed, and purified various fragments of the extracellular segment of PTPμ and have used covaspheres coated with these fragments in a number of binding assays to establish that the Ig domain is both necessary and sufficient for homophilic binding (Brady-Kalnay and Tonks, 1994a). These data are particularly exciting in light of an observation from Schlessinger's group, who found that when Sf9 cells expressing PTPμ and PTPκ were mixed, they sorted independently (Eijgenraam, 1993). In other words, PTPμ binds to itself but not to PTPκ, suggesting that the homophilic binding interaction is highly specific. Therefore, the reagents we have developed toward the Ig domain of PTPμ may prove very useful in defining the physiological significance of these homophilic binding interactions.

As far as we can tell, aggregation, i.e., ligand binding to the extracellular segment of PTPμ, had no detectable direct effect on the activity of the intracellular PTP domains. However, it is possible that the activity of PTPμ may be controlled indirectly by restricting its distribution in the membrane and thus restricting the spectrum of substrates with which it interacts. Our most recent data suggest that under physiological conditions of expression in mink lung cells, PTPμ is localized to adherens junctions, which are defined points of cell–cell contact at which the actin cytoskeleton is anchored. Within these junctions we have found that PTPμ associates directly with members of the cadherin superfamily and is recovered as part of a complex with cadherin, α-catenin, and β-catenin. Preliminary data suggest that the cad-

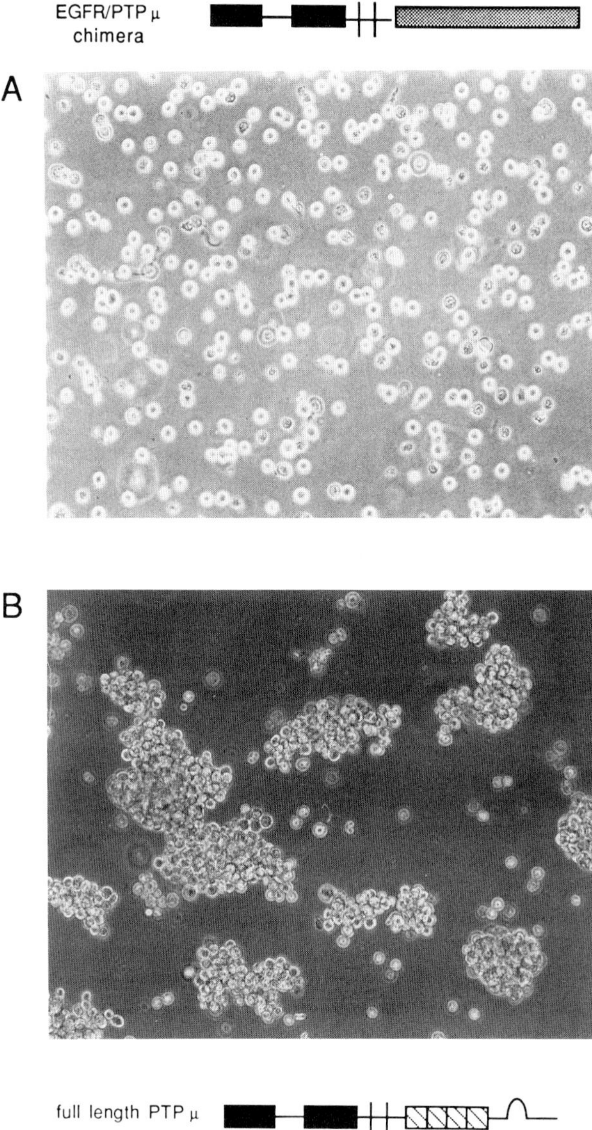

FIGURE 2 Aggregation of Sf9 cells induced by expression of PTPμ from recombinant baculoviruses. The pictures show phase-contrast micrographs of Sf9 cells rotated under low shear conditions for 1 hr. (A) The cells, which are infected with a chimeric molecule comprising the extracellular and transmembrane segments of the EGF receptor fused to the intracellular segment of PTPμ, did not form aggregates. In contrast, in B, cells expressing full-length PTPμ form large aggregates. [Reproduced from Tonks (ed.) (1993), *Sem. Cell Biol.* **4**, 376.]

herins may be endogenous substrates for PTPμ (Brady-Kalnay *et al.*, 1995). These data are provocative because tyrosine phosphorylation of the cadherin/catenin complex has been observed under a variety of physiological conditions and interferes with its normal adhesive function, potentially contributing to transformation and metastasis. It appears that PTPμ may provide the regulatory balance to this phosphorylation event thus maintaining the complex in its functional dephosphorylated state.

B. DEP-1 and Cell–Cell Contact

We isolated cDNA for a novel receptor PTP termed DEP-1 (Östman *et al.*, 1994). The enzyme is broadly expressed and consists of an extracellular segment that is highly glycosylated and comprises eight FNIII domains, a single transmembrane segment, and a single intracellular PTP domain. DEP-1 is therefore a member of the expanding group of receptor PTPs with a single catalytic domain that now also includes PTPβ (Krueger *et al.*, 1990), SAP-1 (Matozaki *et al.*, 1994), GLEPP-1 (Thomas *et al.*, 1994), and the *Drosophila* enzymes, DPTP4E (Oon *et al.*, 1993) and DPTP10D (Tian *et al.*, 1991; Yang *et al.*, 1991). One of the most striking properties of DEP-1 is that in at least two cell lines, W138 and AG1518, the expression of this enzyme is dramatically increased in density relative to sparse cell cultures—hence the name *Density Enhanced PTP*. This modulation of expression is not a general feature of all PTPs because, in contrast to DEP-1, the levels of PTP1B, a ubiquitously expressed enzyme that is the major membrane-associated PTP in many tissues and cell lines, did not change with cell density (Östman *et al.*, 1994). We are currently examining the link between cell contact and expression of DEP-1. In preliminary experiments, we have detected striking morphological changes in cells in which this enzyme is overexpressed that are not seen in cells overexpressing other members of the PTP family (S. Mamajiwalla and N. Tonks, unpublished observation). We are currently trying to identify the important substrates for this effect and preliminary data suggest a number of cytoskeletal targets.

C. PTPs and Contact Inhibition of Cell Growth

As normal cells in culture approach confluence, and adjacent cells touch each other, growth is inhibited. In cancer cells, this process is disrupted. Because tyrosine phosphorylation has been implicated in promoting cell growth and proliferation, PTPs, as the natural antagonists of PTK function, may exert a negative effect on such growth-promoting signals by triggering net dephosphorylation of proteins in the membrane. The advantage of the involvement of receptor PTPs in such a phenomenon is that their extracellular segments may "sense" cell–cell contact directly. Engagement of the extracellular segment of a receptor PTP in this way may either modulate activity directly or target it to a particular junctional complex so as to trigger dephosphorylation of a defined subset of phosphotyrosyl proteins. Furthermore, it is possible that enzymes such as PTPμ may, in addition to mediating homophilic binding interactions, bind distinct ligands. It has been reported that the PTK Axl displays both homophilic interactions and heterophilic binding to GAS6 (Varnum *et al.*, 1995). Following this precedent, it is possible that at high cell density homophilic binding between molecules of PTPμ may predominate, whereas at lower cell densities binding of distinct ligands may occur, thus expanding the repertoire of physiological processes that may be controlled by receptor PTPs. The search for such ligands continues.

III. Nontransmembrane Cytoplasmic PTPs

As it is for the receptor-like enzymes, a striking feature of the cytoplasmic PTPs is their structural diversity. Each contains one copy of the conserved catalytic domain, which typically displays 30–40% identity between individual PTPs. Fused to either the N or C terminus of the catalytic domain are distinct, noncatalytic sequences that distinguish individual PTPs and serve a regulatory function. It now appears that common modes of regulation of PTP function are through control of subcellular distribution and reversible phosphorylation.

A. Targeting of PTPs

PTP1B, the first member of the family to be isolated in homogenous form, comprises 435 amino acids in which the C-terminal, noncatalytic segment of 114 residues has been implicated in controlling the localization of the enzyme. The first ~80 residues in this segment are predominantly hydrophilic and contain three sites of Ser phosphorylation (Flint *et al.*, 1993), whereas the extreme C-terminal 35 amino acids are of a highly hydrophobic nature and are both necessary and sufficient for targeting the enzyme to the cytoplasmic face of membranes of the endoplasmic reticulum (ER) (Frangioni *et al.*, 1992). TCPTP is a close homolog of PTP1B that displays ~75% identity within the catalytic domain. Its noncatalytic segment also shares some features in common with PTP1B, including a hydrophobic extreme C terminus that targets the enzyme to ER membranes, preceded by a predominantly hydrophilic stretch. In the case of TCPTP this hydrophilic segment contains a polybasic nuclear localization motif that is not found in PTP1B. In fact, an alternatively spliced form of TCPTP has been identified in which the hydrophobic sequence is missing; this form is found in the nucleus (Cool and Fischer, 1993; Tillman *et al.*, 1994). Similarly, the *Drosophila* enzyme DPTP61F exists in two alternatively spliced forms that vary at their C termini; in one the extreme C terminus is hydrophobic and directs the enzyme to intracellular membranes, whereas in the other this hydrophobic stretch is missing and the enzyme is targeted to nuclei (McLaughlin and Dixon, 1993). Thus, differential splicing may represent a regulatory mechanism by which a particular PTP may be targeted to different regions of the cell, possibly in response to a signaling event, thereby exerting its effects on discrete pools of substrates. There are now many examples of alternative splicing of PTP transcripts; however, the regulatory consequences largely remain to be defined.

SH2 domains are motifs of ~100 amino acids that recognize phosphotyrosyl residues in defined primary sequence contexts and are found in a variety of proteins that are involved in signaling events downstream of PTKs (Pawson and Schlessinger, 1993). In mammals, two distinct SH2 domain-containing PTPs have been identified. HCP/SH-PTP1 is expressed predominantly in hematopoietic cells, whereas Syp/SH-PTP2 is expressed ubiquitously (Neel, 1993). These enzymes comprise two SH2 domains in the noncatalytic N-terminal segment, a single PTP domain and a C-terminal extension with several consensus phosphorylation sites. The presence of the SH2 domains in these PTPs allows their recruitment into specific signaling complexes in the plasma membrane. For example, HCP/SH-PTP1 binds to Y429 in the cytoplasmic segment of the erythropoietin receptor (Klingmüller *et al.*, 1995) and Syp/SH-PTP2 binds to Y1009 in the C-terminal portion of the PDGF receptor (Kazlauskas *et al.*, 1993; Lechleider *et al.*, 1993). Interestingly, there is an increasing body of evidence to indicate that engagement of the SH2 domains of these PTPs in this way may not only affect their subcellular distribution but may also alter the catalytic activity directly. Thus, addition of phosphotyrosyl peptides corresponding to Y1009 in the PDGF receptor to Syp/SH-PTP2 stimulated activity 5- to 10-fold, whereas other pTyr peptides that were not recognized by the SH2 domains were without effect (Case *et al.*, 1994; Lechleider *et al.*, 1993). The use of diphosphorylated peptides has indicated that simultaneous engagement of both SH2 domains in the PTP produced an even more pronounced activation (Pluskey *et al.*, 1995). Thus, the presence of SH2 domains in these enzymes permits both direct effects on catalytic function and indirect effects through control of subcellular distribution.

Additional examples of targeting of PTPs include PTPH1, which is characterized by the presence of an N-terminal domain of 320 residues with homology to the cytoskeletal-associated proteins of the band 4.1 superfamily (Yang and Tonks, 1991). This domain defines a family of proteins that is targeted to interfaces between the plasma membrane and the cytoskeleton, as illustrated by the association of band 4.1, which promotes the association of actin and spectrin in erythrocytes, with p55 and the intracellular segment of the transmembrane protein glycophorin (Marfatia *et al.*, 1994) and moesin with CD44 (Tsukita *et al.*, 1994). The presence of such a targeting domain implies that PTPH1 may also be directed to cytoskeleton–membrane interfaces. It now appears that there is a subfamily of band 4.1 domain-containing PTPs that includes MEG1 (Gu *et al.*, 1991), PTPD1/PTPRL10/PTP2E (Higashitsuji *et al.*, 1995; L'Abbé *et al.*, 1994; Moller *et al.*, 1994), and PTP-BAS/PTP1E/PTPL1 (Banville *et al.*, 1994; Maekawa *et al.*, 1994; Saras *et al.*, 1994). This latter enzyme has also been described under the name FAP-1 and was shown to associate with Fas, a cell surface receptor linked to signaling pathways that culminate in apoptosis (Sato *et al.*, 1995). We have also characterized two distinct PTPs from *Xenopus laevis*, PTPX1 and PTPX10, that both share sequence identities in their N-terminal segments with two lipid-binding proteins, cellular retinaldehyde-binding protein, and SEC14p, a phospholipid transferase (Del Vecchio and Tonks, 1994). In addition, there is a unique insert in the sequence of PTPX1 that shares identity with PSSA, a protein involved in phosphatidylserine biosynthesis. Sequence comparison suggests that PTPX10 is the *Xenopus* homolog of the human PTP Meg-02 (Gu *et al.*, 1992), whereas PTPX1 is a structurally related yet distinct PTP. This homology between PTPX1 and X10 and lipid-binding proteins, in particular a Golgi-targeting motif in SEC14p, suggests that these enzymes may be associated with intracellular membranes. Interestingly, we have observed association between PTPX1 and membranes in *Xenopus* oocytes and this association is accompanied by an approximately fourfold increase in activity (Del Vecchio and Tonks, 1994).

The take-home message from these structural considerations is that targeting of PTPs to defined subcellular locations will be an important aspect of the control of their function. By restricting the subcellular distribution of these enzymes, the spectrum of substrates with which they may interact is consequently restricted.

B. PTPs as Phosphoproteins

Superimposed on control of PTP function by intracellular targeting is the potential for control by reversible phosphorylation. This is illustrated by our studies of the cytosolic enzyme PTP-PEST, which is expressed ubiquitously in mammalian tissues. PTP-PEST is characterized by an N-terminal catalytic domain followed by a C-terminal segment of ~500 residues that is highly enriched in proline, serine/threonine, and glutamate/aspartate residues, and contains four distinct PEST motifs. Such motifs have been observed in a wide variety of proteins with diverse functions and are believed to mark these proteins for rapid destruction. However, it appears that not all PEST-containing proteins display a short half-life *in vivo* because the half-life of PTP-PEST is >10 hr (A. Garton and N. Tonks, unpublished observations). The sequence of PTP-PEST contains several potential sites of phosphorylation; therefore, we purified the enzyme to homogeneity following expression in Sf9 cells to test its potential for regulation by phosphorylation. We showed that PTP-PEST is phosphorylated *in vitro* on two sites, Ser39 and Ser435, by both PKA and PKC. Furthermore, we found that PTP–PEST is also phosphorylated on the same sites when HeLa cells are treated with TPA, forskolin, or IBMX (Fig. 3). Phosphorylation of Ser39 is directly inhibitory, reducing the affinity of PTP-PEST for its substrate (Garton and Tonks, 1994). More recently, we have observed that phosphorylation of Ser435 appears to affect the subcellular distribution of PTP–PEST, promoting association with an endosomal membrane fraction. We are currently trying to identify the components of the membranes with which PTP-PEST associates and to define which other signal transduction pathways promote the phosphorylation of Ser435 in the phosphatase.

A number of other examples of the phosphorylation of PTPs have been described. PTP1B is phosphorylated *in vivo* in a cell cycle-dependent manner. In asynchronously growing HeLa cells, PTP1B is phosphorylated predominantly on Ser378 by PKC and at two other minor sites. In mitotically arrested cells, the interphase sites are dephosphorylated and PTP1B becomes phosphorylated instead on Ser386 by p34^{cdc2} and on Ser352 by a "proline-directed" kinase. These mitotic phosphorylation events coincide with a 30% inhibition of PTP1B activity (Flint *et al.*, 1993). In addition, phosphorylation of Ser residues in the SH2 domain-containing PTP, HCP/SH-PTP1, was enhanced following TPA-induced differentiation of HL-60 promyelocytic leukemia cells into macrophages, and this was accompanied by translocation of the phosphatase to the membrane (Zhao *et al.*, 1994). These observations are not restricted to cytoplasmic PTPs. Treatment of T cells with ionomycin, which mobilizes Ca^{2+} from intracellular stores, resulted in a dephosphorylation of Ser residues in CD45 coincident with a decrease in its activity (Ostergaard and Trowbridge, 1991). These data suggest that Ser/Thr kinases are capable of phosphorylating and modifying the activity of PTPs *in vivo*, thus offering a mechanism whereby signal transduction pathways acting through

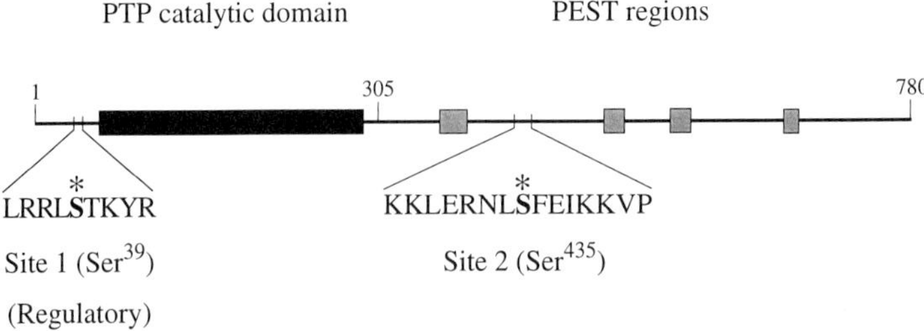

FIGURE 3   Arrangement of PKA/PKC phosphorylation sites on PTP–PEST. The sequences surrounding the major PKA and PKC phosphorylation sites on PTP-PEST and the positions at which they occur in the amino acid sequence of PTP-PEST are shown. The phosphorylated residues are marked with asterisks. Also shown are the positions of the PTP catalytic domain (solid bar) and the four PEST sequences (gray bars). [Reproduced from Garton and Tonks (1993).]

Ser/Thr kinases may influence cellular processes involving reversible tyrosine phosphorylation. This illustrates another point of "cross-talk" between Ser/Thr and Tyr phosphorylation events in signal transduction *in vivo*.

I should point out that there are also examples of the phosphorylation of tyrosyl residues in PTPs *in vivo*. Following recruitment of Syp/SH-PTP2 into the PDGF receptor signaling complex, through interaction between its SH2 domains and the Y1009 autophosphorylation site in the receptor, the phosphatase becomes phosphorylated on a C-terminal tyrosyl residue, Y542 (Bennett *et al.*, 1994). The sequence surrounding this residue matches the consensus for interaction with the SH2 domain of the adaptor protein GRB2, which in turn binds SOS, a guanine nucleotide exchange factor for Ras that converts Ras from an inactive GDP bound form to an active form in which GTP is bound (Pawson and Schlessinger, 1993). Interestingly, binding of GRB2 to Syp/SH-PTP2 that is phosphorylated on Y542 has now been shown both *in vitro* and *in vivo*, with SOS also recovered in the complex *in vivo* (Bennett *et al.*, 1994; Li *et al.*, 1994). Thus, it is possible that Syp/SH-PTP2 may also function as an adaptor molecule to link the PDGF receptor to the activation of Ras, thus facilitating the signaling response to PDGF. In addition, constitutive tyrosine phosphorylation of the receptor-like enzyme PTPα has been observed in NIH-3T3 cells. Again, the site of phosphorylation, Y789, matches the consensus for interaction with GRB2 and association between tyrosine phosphorylated PTPα and GRB2 has been demonstrated (den Hertog *et al.*, 1994; J. Su *et al.*, 1994). However, in this case SOS was not recovered in the complex and thus the physiological significance of the interaction remains unclear.

These observations suggest that tyrosyl phosphorylation of PTPs may play a role in the assembly of multiprotein complexes in the membrane. Perhaps PTPs modulate signaling responses *in vivo* not only through dephosphorylation of key tyrosyl residues in proteins but also through assembling such complexes.

IV. Dual-Specificity Phosphatases

In addition to the classical PTPs described thus far, a rapidly expanding subfamily of dual-specificity phosphatases has been described, of which the prototype is VH1 from the pox virus *Vaccinia* (Guan *et al.*, 1991). These enzymes have now been detected in several viruses, yeast, mammalian cells (Guan and Dixon, 1993), and even in a prokaryote, the cyanobacterium *Nostoc commune* (Potts *et al.*, 1993). They contain the unique signature motif that defines the PTP family but otherwise show little, if any, sequence similarity to the classical PTPs. Furthermore, unlike the bona fide PTPs described thus far, which are absolutely specific for tyrosyl residues in proteins, these enzymes are capable of dephosphorylating Ser/Thr as well as

Tyr residues in proteins. Finally, unlike the classical PTPs that are somewhat promiscuous in their substrate preference, at least *in vitro,* current information indicates that the dual-specificity phosphatases tend to display a restricted substrate specificity. For example, dephosphorylation of Thr14 and Tyr15 in p34^{cdc2} is the key step in activating this protein kinase and driving the transition from the G_2 to M phase of the eukaryotic cell cycle. Both of these dephosphorylation events are catalyzed by a single dual specificity phosphatase, p80^{cdc25}, which appears to be highly selective for members of the cdk family of Ser/Thr kinases (Millar and Russell, 1992). This mechanism of initiation of mitosis is conserved across all eukaryotic species. I illustrate this issue of specificity further with a discussion of another dual-specificity phosphatase, termed MKP-1, which is the product of the immediate early gene 3CH134.

A. The Dual-Specificity Phosphatase MKP-1 Dephosphorylates and Inactivates MAP Kinases *in Vivo*

Serum growth factors stimulate cell proliferation through interactions with their specific cell surface receptors, activating the intrinsic receptor PTK function. This leads to signal transduction events that include the reversible phosphorylation of a number of cellular substrates on tyrosyl residues. Some of these signals are transmitted to the nucleus where the activation of a set of immediate early genes occurs. These immediate early genes encode a diverse array of regulatory proteins including transcription factors and cytokines. Their expression has been hypothesized to mediate the biological effects of the stimulatory growth factor.

3CH134 is a murine immediate early gene activated rapidly and transiently in quiescent fibroblasts treated with serum growth factors (Charles *et al.,* 1992). The human homolog, CL100, was identified in a screen for genes induced by oxidative stress and heat shock (Keyse and Emslie, 1992). *3CH134* is also expressed during liver regeneration soon after partial hepatectomy, further suggesting a link between its expression and the control of cell growth. Its transcription in fibroblasts is detected within minutes of growth factor addition, reaches a peak level by 10–20 min, and is repressed by 1 hr. The *3CH134* mRNA has a very short half-life and accumulates for only 1 or 2 hr during the G_0/G_1 transition. Likewise, the 3CH134 protein is transiently synthesized and has a half-life of 40 min. Lester Lau's group initially noted the presence of the PTP signature motif in the sequence of the protein encoded by the *3CH134* gene. In collaboration with them we expressed the protein in *E. coli,* purified it to homogeneity, and tested it for phosphatase activity against the substrates used routinely in the lab (Charles *et al.,* 1993). Although it did possess intrinsic PTP activity *in vitro,* it expressed very low enzyme activity. Bearing in mind the potential of these enzymes for selective dephosphorylation of substrates, we considered the growth factor receptor signaling pathway for potential targets. Progress has been made by many labs in defining signaling pathways initiated by mitogenic stimuli (Fig. 4) (Blenis, 1993). Growth factor-stimulated autophosphorylation of tyrosyl residues in receptor PTKs induces the formation of multiprotein complexes in the membrane that trigger conversion of Ras from an inactive GDP-bound form to an active GTP-bound state. Activated Ras then initiates a cascade of sequential phosphorylation events in which the Ser/Thr kinase Raf phosphorylates and activates MAP kinase kinase (also known as MEK), which is a dual-specificity kinase that in turn phosphorylates both Thr183 and Tyr185 in MAP kinase. Phosphorylation of both Tyr and Thr regulatory sites is essential for activation. The MAP kinase family of enzymes has been implicated as common and essential components of signaling pathways induced by diverse mitogenic stimuli. Once activated, MAP kinase can phosphorylate a number of substrates, including transcription factors such as TCF/Elk-1, that are essential for triggering the expression of genes required for the mitogenic response, one of these immediate early genes being *3CH134*. When we tested this enzyme against receptor TPKs, we did not detect significant dephosphorylation. However, within the growth factor signaling pathway the MAP kinases are another potential target. In fact, we observed that when the 3CH134 protein was incubated with phosphorylated, activated p42^{mapk} *in vitro*, it readily dephosphorylated both the Tyr and Thr regulatory sites in MAP kinase (Charles *et al.*, 1993; Sun *et al.*, 1993). Furthermore, if a point mutant of the phosphatase was created in which the catalytically essential Cys residue from the signature motif, Cys258, was converted to Ser, the resultant enzyme was inactive and unable to catalyze dephosphorylation of either the Tyr site or the Thr site in MAP kinase. Thus, the dephosphorylation of both regulatory sites is catalyzed by the same active center, indicating that the *3CH134* immediate early gene encodes a dual-specificity phosphatase with the ability to dephosphorylate and inactivate MAP kinase *in vitro*. Hence, we named the protein MAP kinase phosphatase-1 (MKP-1) (Sun *et al.*, 1993).

This observation stimulated us to investigate whether MKP-1 was a physiologically relevant MAP kinase phosphatase. I believe that the following observations clearly establish that this is in fact the case (Sun *et al.*, 1993). In serum-stimulated 3T3 fibroblasts, the kinetics of inactivation of MAP kinases coincides with the appearance of newly synthesized MKP-1 protein. Furthermore, in the presence of the protein synthesis inhibitor cycloheximide, serum stimulation results in persistent activation of MAP kinases. These data indicate that new protein synthesis is required to manifest the activity of the phosphatase responsible for the dephosphorylation and inactivation of MAP kinases—we propose that this requirement for protein synthesis reflects the necessity for production of MKP-1. In light of the apparent selectivity of MKP-1 for MAP kinase *in vitro*, we tested the effect of expressing the phosphatase in COS cells on the global pattern of tyrosine

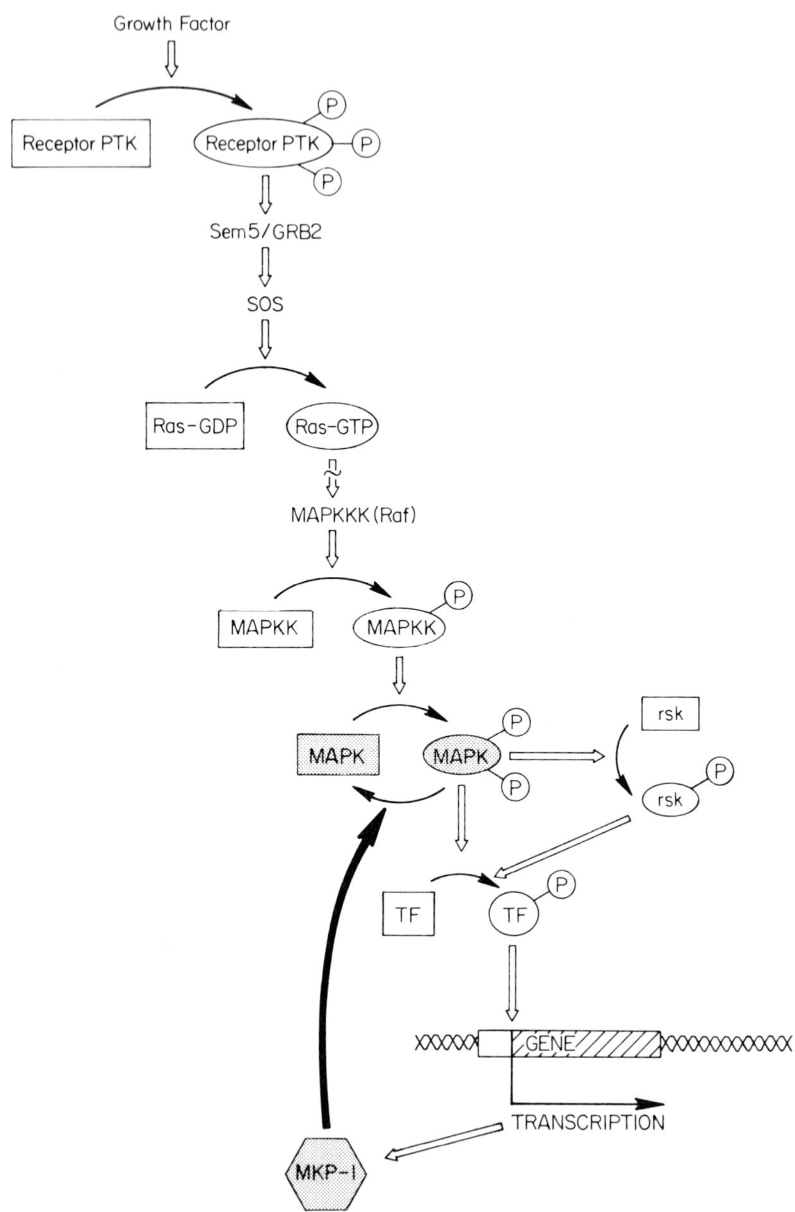

FIGURE 4   A signaling pathway triggered by growth factor receptor PTKs. This figure presents a brief summary of our current appreciation of the cellular responses to activation of a growth factor receptor protein tyrosine kinase (PTK) by binding to its cognate ligand. Autophosphorylation of the receptor creates docking sites that recruit the adaptor protein Grb2, leading to activation of Sos, the GDP–GTP exchanger for Ras. Then a cascade of phosphorylation is triggered, leading to activation of mitogen-activated protein kinase (MAPK). MAPK in turn phosphorylates transcription factors (TF), either directly or through activation

phosphorylation. Whereas serum induced the phosphorylation of tyrosyl residues in several proteins, visualized by immunoblotting with an antiphosphotyrosine antibody, expression of MKP-1 lead to the selective dephosphorylation of MAP kinase. We also observed that expression of MKP-1 in COS cells blocked phosphorylation and activation of $p42^{mapk}$ in response to serum, oncogenic (V^{12}) *Ras*, or activated Raf. Because MKP-1 does not dephosphorylate MAP kinase kinase/MEK, this suggests that MKP-1 exerts its effects on the signaling pathway at the level of MAP kinase. Perhaps most compelling is the observation that the Cys258 → Ser catalytically inactive mutant form of MKP-1 retains the ability to bind to its substrate, MAP kinase, blocking dephosphorylation by native phosphatase. Following expression in COS cells, the mutant phosphatase forms a physical complex specifically with the phosphorylated, activated forms of $p42^{mapk}$ and $p44^{mapk}$ in a serum stimulation-dependent manner. These data clearly suggest that MKP-1 is a physiological MAP kinase phosphatase. Similar data have been presented for CL100 (Alessi *et al.*, 1993).

B. Controversy over the Dephosphorylation of MAP Kinases *in Vivo*

Judging by the tone of recent discussions and publications, it appears that the issue of which phosphatase is responsible for the dephosphorylation of MAP kinases *in vivo* arouses passionate sentiments in some investigators! Although our data attest to the physiological importance of MKP-1 as a MAP kinase phosphatase, they by no means exclude a contribution of other phosphatases *in vivo*. However, one should not form the opinion that MKP-1 only contributes to the dephosphorylation of MAP kinase in fibroblasts. For example, actinomycin D and antisense oligonucleotides directed against MKP-1 both inhibited expression of MKP-1 and triggered prolonged activation of $p42^{mapk}$ and $p44^{mapk}$ in response to angiotensin II stimulation of vascular smooth muscle cells (Duff *et al.*, 1995). In addition, a number of other dual-specificity phosphatases with the potential to dephosphorylate MAP kinases have now been identified, including PAC1 (Ward *et al.*, 1994), hVH2 (Guan and Butch, 1995), and B23/hVH3 (Ishibashi *et al.*, 1994; Kwak and Dixon, 1995). These enzymes show differences in both tissue distribution and time course of expression relative to MKP-1. Further support for a role of dual-specificity phosphatases in the dephosphorylation of

---

of another Ser/Thr kinase, Rsk, thus promoting the transcription of genes required for the growth response. One of the genes induced by growth factor stimulation, presumably through the MAPK pathway, encodes a MAPK phosphatase (MKP-1). As denoted by the solid arrow, MKP-1 may feed back on the pathway by dephosphorylating and inactivating MAPK, thus attenuating the signaling response. [Reproduced from Sun and Tonks (1994), *TIBS* **19**, 480–485.]

MAP kinases has come from yeast genetics. MSG5 has been identified as a suppressor of pheromone-induced cell cycle arrest in *S. cerevisiae* (Doi *et al.*, 1994). Genetic epistasis analysis places the site of action of MSG5 at the level of the FUS3 homolog of MAP kinase in the pheromone-induced signaling pathway. In fact, it has been shown that MSG5, which displays the primary sequence features of a dual-specificity phosphatase, dephosphorylates and inactivates FUS3 *in vitro*. The exquisite selectivity of MSG5 for FUS3 is highlighted by the observation that although native FUS3 is a substrate for the phosphatase, the heat-denatured kinase is not, suggesting that MSG5 recognizes features in its substrate outside the immediate vicinity of the site of phosphorylation. When one considers the diversity of MAP kinase-dependent signaling pathways (Cooper, 1994), the apparent multiplicity of dual-specificity phosphatases raises the possibility that they each may be responsible for the dephosphorylation of distinct MAP kinase isoforms in a cell-type and signaling pathway-specific manner.

Nevertheless, there are clearly situations in which the inactivation of MAP kinase is too rapid to invoke the involvement of an inducible dual-specificity phosphatase such as MKP-1. For example, in many cells the activation of MAP kinase is biphasic comprising a rapid, transient "spike" of activity preceding a longer, sustained period of activation (Meloche *et al.*, 1992). It is well established that activation of MAP kinase requires phosphorylation of both the regulatory Thr and Tyr sites, but dephosphorylation of either site is sufficient for inactivation (Anderson *et al.*, 1990). There is now evidence to indicate that the deactivation phase of this initial burst of MAP kinase activity is catalyzed by protein phosphatase 2A, a Ser/Thr-specific enzyme, and by a tyrosine-specific protein phosphatase (Alessi *et al.*, 1995). A candidate for the latter enzyme was identified in *Xenopus* eggs (Sarcevic *et al.*, 1993). However, it is important to indicate that this initial burst of activation occurs in the cytosol and its physiological function is unclear. It is the sustained period of activation, which is accompanied by translocation of MAP kinase to the nucleus, that is important for the mitogenic response (Gonzalez *et al.*, 1993; Lenormand *et al.*, 1993; Meloche *et al.*, 1992). Interestingly, MKP-1 (Brondello *et al.*, 1995), PAC-1 (Rohan *et al.*, 1993), hVH2 (Guan and Butch, 1995), and B23/hVH3 (Ishibashi *et al.*, 1994; Kwak and Dixon, 1995) have all been shown to be nuclear proteins. There are also conditions in which MAP kinase becomes activated, translocated to the nucleus and, despite the production of MKP-1, the activity is sustained. For example, in PC12 cells stimulation with EGF produces a rapid, transient activation of MAP kinase, expression of MKP-1, and a mitogenic response. When PC12 cells are stimulated with NGF, MAP kinase activation is sustained with translocation to the nucleus even though MKP-1 is produced, and neuronal differentiation ensues (Alessi *et al.*, 1995). Currently, it is unclear what protects MAP kinase from MKP-1 under these conditions. Perhaps there are inhibitor proteins that regulate MKP-1 in a manner analogous to the effects of inhibitors 1 and 2 on protein phosphatase-1 (Shenolikar and Nairn, 1991). In addition, we have noted that MKP-1 is efficiently phosphorylated on multiple sites by MAP kinase (H. Sun and N. K. Tonks, unpublished observations). Similarly, phosphorylation of MSG5 by FUS3 has been reported (Doi *et al.*, 1994). Whether such modifications are of regulatory significance remains to be determined. The identity of the phosphatase that ultimately inactivates MAP kinase in this system—whether it be MKP-1, a distinct dual-specificity enzyme, or single specificity phosphatases—also remains to be ascertained.

C. Expression of MKP-1 Antagonizes MAP Kinase Function

MKP-1 appears to be highly specific for the MAP kinase family of enzymes. To emphasize this point, we have tested the susceptibility of MAP kinase isoforms other than p42/ERK2 and p44/ERK1 to dephosphorylation by MKP-1. JNK-1, which phosphorylates Ser63 and Ser73 at the N terminus of the transcription factor jun, is related to MAP kinase and requires phosphorylation of Tyr and Thr regulatory sites for activation (Dérijard *et al.*, 1994). Its activity is triggered in response to stress stimuli such as UV light. We have shown that although MKP-1 will dephosphorylate and inactivate JNK, it does so with ~30-fold lower efficiency than p42^{mapk} (Sun *et al.*, 1994). This observation further underscores the high degree of substrate selectivity shown by MKP-1 and suggests that it may prove useful as a molecular probe to investigate physiological roles for MAP kinase.

To examine the effect of expression of MKP-1 on signaling responses downstream of MAP kinase, we employed a transcription-based assay in which MAP kinase activation was measured by its ability to transactivate a promoter containing five copies of a serum response element (SRE) linked to a CAT reporter. This effect is presumably mediated by the phosphorylation of the TCF/Elk1 transcription factor by MAP kinases. Using a microinjection strategy in REF52 rat fibroblasts, we have shown that expression of MKP-1 inhibits the ability of growth factors TPA and oncogenic ras, which signal through the MAP kinase pathway, to transactivate the SRE promoter in the CAT reporter. Furthermore, MKP-1 blocked the induction of DNA synthesis in quiescent REF52 cells by the V^{12} oncogenic mutant form of Ras (Fig. 5). Therefore, these data suggest an essential role for the activation of MAP kinases in the transition from quiescence to the DNA synthesis phase of the eukaryotic cell cycle (Sun *et al.*, 1994). Similarly, in collaboration with Pouyssegur's group, we demonstrated that expression of MKP-1 in CCL39 fibroblasts inhibited early and late transcriptional events as well as entry into S phase (Brondello *et al.*, 1995). Finally, in collaboration with Jeremy Minshull and Andrew Murray, we have also used MKP-1 in experiments in *Xenopus* egg extracts to show that the activation of MAP kinase

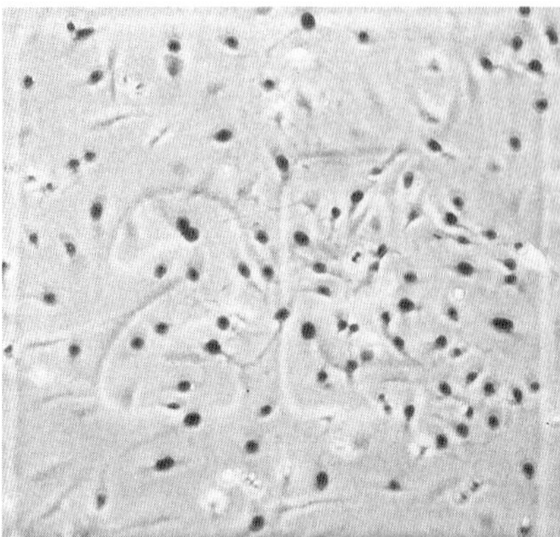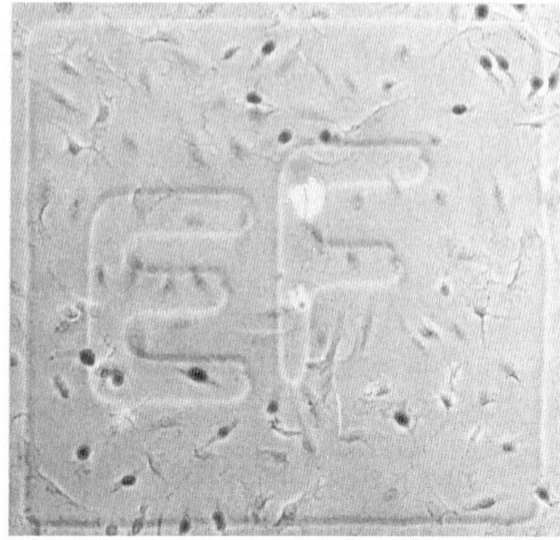

FIGURE 5   Inhibition of V^{12} ras-induced DNA synthesis by MKP-1. Quiescent REF52 cells were microinjected with bacterially expressed V^{12} Ras protein together with control plasmid (*left*) or plasmid expressing MKP-1 (*right*). BrdU incorporation was detected by immunohistochemical staining. The fields were visualized and photographed with phase-contrast microscopy. [Reproduced from Sun *et al.* (1994).]

is required for the establishment and maintenance of the mitotic arrest induced by depolymerization of spindle microtubules (Minshull *et al.*, 1994). I believe these experiments illustrate that MKP-1 will be a powerful tool with which to assess the significance of activation of MAP kinases *in vivo*.

In summary, although the available data clearly point to MKP-1 as a physiologically relevant MAP kinase phosphatase, it by no means represents the only mechanism for inactivation of MAP kinases *in vivo*. It will be important to delineate precisely the relative roles of dual-specificity and single-specificity phosphatases in the dephosphorylation of MAP kinase and to link particular phosphatases to individual MAP kinase isoforms in defined signaling pathways. Large gaps still remain in our understanding of fundamental aspects of the regulation of MKP-1. Although expression of MKP-1 is induced by mitogenic stimulation, and we show MKP-1 as feeding back and inhibiting the growth factor signaling pathway through the dephosphorylation of MAP kinase (Fig. 4), this phosphatase is expressed in response to a wide variety of stimuli. For example, MKP-1 expression is induced particularly effectively in response to cAMP (Noguchi *et al.*, 1993), raising the possibility that it may play a role in cross-talk between different signaling pathways as well as in feedback control. In addition, the characterization of post-translational mechanisms for regulation of MKP-1, including phosphorylation and protein–protein interactions, may reveal important new insights into the physiological function of this enzyme. Nevertheless, despite these gaps in our knowledge the current data serve to emphasize the complexity and importance of "switch off" mechanisms, which in conjunction with the "on switch" act coordinately in regulating cellular signaling responses.

V. Determination of the Crystal Structure of PTP1B

To understand more fully the catalytic mechanism of members of the PTP family and to provide a framework for structure–function analysis we set out, in collaboration with David Barford, to determine the crystal structure of one of the PTPs. We focused on PTP1B, the first member of the family to be isolated in homogeneous form (Barford *et al.*, 1994). PTP1B was initially isolated from human placenta as a monomeric protein of 37 kDa (Tonks *et al.*, 1988b,c), lacking the regulatory C-terminal segment that contains the ER-targeting motif and sites of cell cycle-dependent phosphorylation. It was this truncated protein, comprising predominantly the catalytic domain, that we expressed in *E. coli* and purified to homogeneity for structure determination. The bacterially expressed protein was stable, soluble, and displayed enzymatic characteristics that were indistinguishable from those of the enzyme as purified from placenta. The structure was determined by single isomorphous replacement and anomalous scattering methods using sodium tungstate as the heavy atom derivative. This was important because sodium tungstate is an effective inhibitor of PTP1B (IC_{50} ~10 μM), which binds at the active site and presumably functions as an analog of phosphate. Thus, the structure of the PTP1B–tungstate complex provided us with the first indication of how the enzyme might interact with substrate (Barford *et al.*, 1994).

The structure of PTP1B was determined to 2.8-Å resolution. It is composed of a single domain with the polypeptide chain organized into eight α-helices and 12 β-strands with a 10-stranded mixed β-sheet that adopts a highly twisted conformation spanning the entire length of the molecule (Fig. 6; see color insert). When we aligned the sequences of 27 PTPs for which activity had been convincingly demonstrated, we noted 27 residues that were invariant. These appeared to be spread throughout the catalytic domain. However, when examined in the context of the 3D structure, these residues (shown in yellow in Figure 6; see color insert) were for the most part clustered together on loops connecting elements of secondary structure surrounding the catalytic site. The structure illustrates that the unique signature motif that defines the PTP family of enzymes forms the phosphate recognition site. In PTP1B, the sequence CSAGIGR, from Cys215–Arg221, adopts a rigid cradle structure in which the main chain nitrogen atoms and the side chain of Arg221 encircle the tungstate/phosphate and hold it in place through hydrogen bonding interactions. The structure of the tungstate-bound complex also indicates that the Sγ atom of Cys215 is ideally positioned to act as a nucleophile toward the phosphate moiety of bound substrate. Similar conclusions were drawn from the structure of the *Yersinia* PTP, Yop (Stuckey *et al.*, 1994), and the low M_r acid phosphatase/tyrosine phosphatase (Logan *et al.*, 1994; X.-D. Su *et al.*, 1994; Zhang *et al.*, 1994).

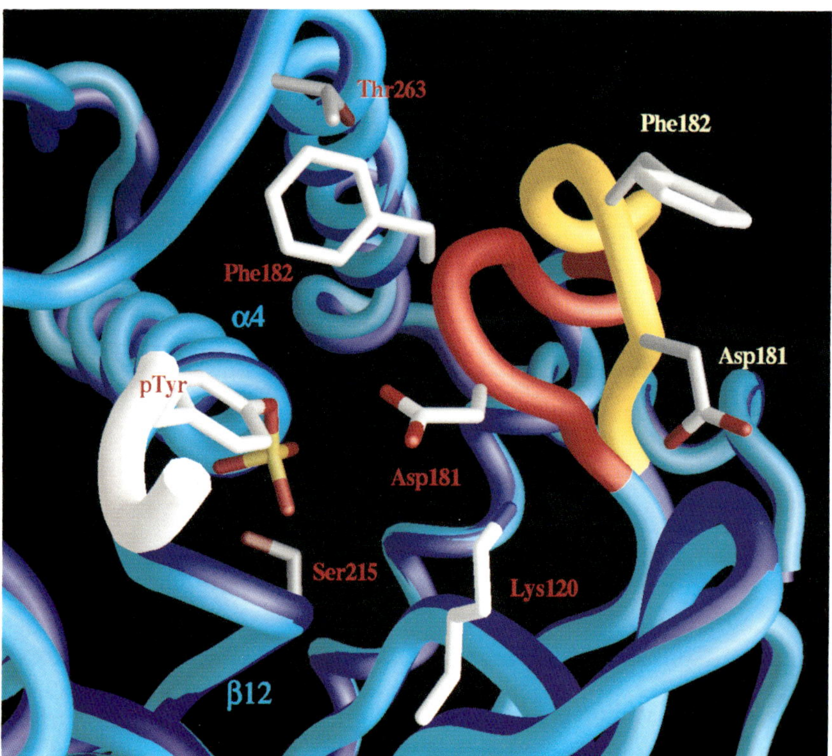

FIGURE 7 View of the conformational change that occurs in the Trp179–Ser187 loop upon binding of substrate to PTP1B. The loop backbone (residues 179–187) is colored yellow in the uncomplexed and red in the peptide substrate-complexed forms of PTP1B. Other regions of the polypeptide chain are colored blue and cyan for the uncomplexed and peptide substrate-complexed forms of PTP1B, respectively. The pTyr peptide backbone is shown in white. The side chains of Asp181 and Phe182 are shown for both structures. The side chains of Lys120, Ser215 (Cys in the wild-type enzyme), Thr263, and pTyr are from the hexapeptide–PTP1B complex. [Reproduced from Jia *et al.* (1995).]

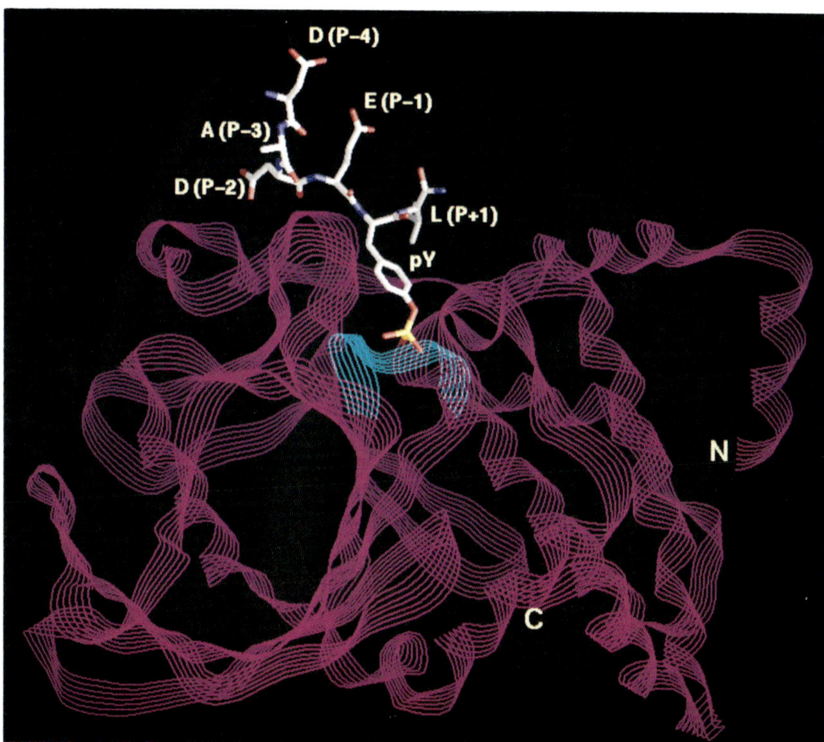

FIGURE 8 Overall view of the structure of PTP1B complexed with the peptide substrate DADEpYL. The main chain backbone of PTP1B is represented by the purple ribbon with the exception of the PTP signature motif, which forms the phosphatate binding loop and which is shown in blue. The side chain of pTyr extends 9 Å into the recognition pocket. This anchors the substrate in the peptide binding site wherein hydrogen bonding interactions between peptide main chain atoms and the protein contribute to binding affinity and specific interactions between acidic side chains in the peptide and basic residues on the surface of the enzyme contribute to sequence specificity.

that are important for substrate recognition. Preliminary studies suggest that some of these mutants retain a high affinity for substrate but display an exceedingly poor ability to catalyze dephosphorylation. We are currently expressing such mutants to see if we can trap and thus identify physiological substrates for PTP1B.

VI. Conclusions

As the natural antagonists of PTK function, PTPs obviously have the potential to exert a negative influence on tyrosine phosphorylation-dependent signaling pathways. For example, there have been several reports of the ability of PTPs to counteract the effect of oncogenic PTKs (Cool and Fischer, 1993). In addition, mutations in an SH2 domain-containing PTP, HCP/SH-PTP1, have been shown to be the cause of the *motheaten* phenotype in mice. The phenotype of these *motheaten* mice suggests that HCP/SH-PTP1 is an important negative regulator of cytokine signaling in general—its loss resulting in sustained tyrosine phosphorylation and enhanced proliferation (Neel, 1993). However, PTPs also act positively to cooperate with PTKs, as illustrated in the control of the Src family of PTKs. These nonreceptor PTKs feature a C-terminal site of tyrosine phosphorylation, which results in inhibition of catalytic activity. Dephosphorylation of this inhibitory phosphotyrosine by a PTP would result in activation of the Src family PTK and generation of a signaling response. An example would be the positive role of CD45 in promoting signaling through the T cell receptor as a result of dephosphorylation and activation of $p56^{lck}$ and/or $p59^{fyn}$ (Woodford-Thomas and Thomas, 1993). In addition, unlike HCP/SH-PTP1, Syp/SH-PTP2 has been shown to be a positive mediator of growth factor signaling. Genetic epistasis analyses have also shown that *corkscrew*, a *Drosophila* homolog of Syp/SH-PTP2, functions positively in conjunction with D-Raf to transduce the signal from the *torso* receptor PTK (Neel, 1993). The definition of PTPs as a large, structurally diverse family comprising both receptors with the ability to transmit a signal directly across the membrane and cytoplasmic enzymes, and with the ability to act both positively and negatively in the control of cell function, further emphasizes the important concept that it is the coordinated actions of both PTKs and PTPs that are responsible for the regulatory effects of reversible tyrosine phosphorylation in cell signaling. The PTPs represent crucial targets for future research to define precisely the physiological roles of tyrosine phosphorylation.

Acknowledgments

I am indebted to all my colleagues, particularly those whose work is described here. Work in my lab is supported by NIH Grants CA53840 and CA64593, the Tobacco Research Council, the Mellam Family Foundation, and the Pew Scholars Program in the Biomedical Sciences. I thank Carol Marcincuk for typing the manuscript.

References ________________________________

Alessi, D. R., Gomez, N., Moorhead, G., Lewis, T., Keyse, S. M., and Cohen, P. (1995). Inactivation of p42 MAP kinase by protein phosphatase 2A and a protein tyrosine phosphatase, but not CL100, in various cell lines. *Curr. Biol.* **5**, 283–295.

Alessi, D. R., Smythe, C., and Keyse, S. M. (1993). The human CL100 gene encodes a Tyr/Thr-protein phosphatase which potently and specifically inactivates MAP kinase and suppresses its activation by oncogenic ras in *Xenopus* oocyte extracts. *Oncogene* **8**, 2015–2020.

Anderson, N. G., Maller, J. L., Tonks, N. K., and Sturgill, T. W. (1990). Requirement for integration of signals from two distinct phosphorylation pathways for activation of MAP kinase. *Nature* **343**, 651–653.

Banville, D., Ahmad, S., Stocco, R., and Shen, S.-H. (1994). A novel protein–tyrosine phosphatase with homology to both the cytoskeletal proteins of the band 4.1 family and junction-associated guanylate kinases. *J. Biol. Chem.* **269**, 22320–22327.

Barford, D., Flint, A. J., and Tonks, N. K. (1994). Crystal structure of human protein tyrosine phosphatase 1B. *Science* **263**, 1397–1404.

Barnea, G., Grumet, M., Milev, P., Silvennoinen, O., Levy, J. B., Sap, J., and Schlessinger, J. (1994). Receptor tyrosine phosphatase β is expressed in the form of proteoglycan and binds to the extracellular matrix protein tenascin. *J. Biol. Chem.* **269**, 14349–14352.

Bennett, A. M., Tang, T. L., Sugimoto, S., Walsh, C. T., and Neel, B. G. (1994). Protein-tyrosine-phosphatase SHPTP2 couples platelet-derived growth factor receptor β to Ras. *Proc. Natl. Acad. Sci. U.S.A.* **91**, 7335–7339.

Blenis, J. (1993). Signal transduction via the MAP kinases: Proceed at your own RSK. *Proc. Natl. Acad. Sci. U.S.A.* **90**, 5889–5892.

Brady-Kalnay, S. M., Flint, A. J., and Tonks, N. K. (1993). The receptor-type protein tyrosine phosphatase PTPμ mediates cell–cell aggregation. *J. Cell Biol.* **122**, 961–972.

Brady-Kalnay, S. M., Rimm, D. L., and Tonks, N. K. (1995). The receptor protein tyrosine phosphatase PTPμ associates with cadherins and catenins *in vivo*. *J. Cell Biol.* **130**, 977–986.

Brady-Kalnay, S. M., and Tonks, N. K. (1994a). Identification of the homophilic binding site of the receptor protein–tyrosine phosphatase PTPμ. *J. Biol. Chem.* **269**, 28472–28477.

Brady-Kalnay, S. M., and Tonks, N. K. (1994b). Receptor protein tyrosine phosphatases, cell adhesion and signal transduction. *Adv. Prot. Phosphatases* **8**, 241–271.

Brondello, J. M., McKenzie, F. R., Sun, H., Tonks, N. K., and Pouysségur, J. (1995). Constitutive MAP kinase phosphatase (MKP-1) expression blocks G1 specific gene transcription and S-phase entry in fibroblasts. *Oncogene* **10**, 1895–1904.

Case, R. D., Piccione, E., Wolf, G., Benett, A. M., Lechleider, R. J., Neel, B. G., and Shoelson, S. E. (1994). SH-PTP2/Syp SH2 domain binding specificity is defined by direct interactions with platelet-derived growth factor β-receptor, epidermal growth factor receptor, and insulin receptor substrate-1-derived phosphopeptides. *J. Biol. Chem.* **269**, 10467–10474.

Charbonneau, H., Tonks, N. K., Kumar, S., Diltz, C. D., Harrylock, M., Cool, D. E., Krebs, E. G., Fischer, E. H., and Walsh, K. A. (1989). Human placenta protein-tyrosine-phosphatase: Amino acid sequence and relationship to a family of receptor-like proteins. *Proc. Natl. Acad. Sci. U.S.A.* **86**, 5252–5256.

Charbonneau, H., Tonks, N. K., Walsh, K. A., and Fischer, E. H. (1988). The leukocyte common antigen (CD45): A putative receptor-linked protein tyrosine phosphatase. *Proc. Natl. Acad. Sci. U.S.A.* **85**, 7182–7186.

Charles, C. H., Abler, A. S., and Lau, L. F. (1992). cDNA sequence of a growth factor-inducible immediate early gene and characterization of its encoded protein. *Oncogene* **7**, 187–190.

Charles, C. H., Sun, H., Lau, L. F., and Tonks, N. K. (1993). The growth factor inducible immediate early gene 3CH134 encodes a protein tyrosine phosphatase. *Proc. Natl. Acad. Sci. U.S.A.* **90**, 5292–5296.

Cool, D. E., and Fischer, E. H. (1993). Protein tyrosine phosphatases in cell transformation. *Sem. Cell Biol.* **4**, 443–453.

Cooper, J. A. (1994). Straight and narrow or tortuous and intersecting? *Curr. Biol.* **4**, 1118–1121.

Cunningham, B. A. (1995). Cell adhesion molecules as morphoregulators. *Curr. Opin. Cell Biol.* **7**, 628–633.

Del Vecchio, R. L., and Tonks, N. K. (1994). Characterization of two structurally related *Xenopus laevis* protein tyrosine phosphatases with homology to lipid-binding proteins. *J. Biol. Chem.* **269**, 19639–19645.

den Hertog, J., Tracy, S., and Hunter, T. (1994). Phosphorylation of receptor protein-tyrosine phosphatase α on Tyr789, a binding site for the SH3-SH2-SH3 adaptor protein GRB-2 *in vivo*. *EMBO J.* **13**, 3020–3032.

Dérijard, B., Hibi, M., Wu, I.-H., Barrett, T., Su, B., Deng, T., Karin, M., and Davis, R. J. (1994). JNK1: A protein kinase stimulated by UV light and Ha-Ras that binds and phosphorylates the c-Jun activation domain. *Cell* **76**, 1025–1037.

Doi, K., Gartner, A., Ammerer, G., Errede, B., Shinkawa, H., Sugimoto, K., and Matsumoto, K. (1994). MSG5, a novel protein phosphatase promotes adaptation to pheromone response in *S. cerevisiae*. *EMBO J.* **13**, 61–70.

Duff, J. L., Monia, B. P., and Berk, B. C. (1995). Mitogen-activated protein (MAP) kinase is regulated by the MAP kinase phosphatase (MKP-1) in vascular smooth muscle cell. *J. Biol. Chem.* **270**, 7161–7166.

Eijgenraam, F. (1993). Things start getting sticky for a cell surface enzyme. *Science* **261**, 833.

Flint, A. J., Gebbink, M. F. B. G., Franza, B. R., Jr., Hill, D. E., and Tonks, N. K. (1993). Multi-site phosphorylation of the protein tyrosine phosphatase, PTP1B: Identification of cell cycle regulated and phorbol ester stimulated sites of phosphorylation. *EMBO J.* **12**, 1937–1946.

Frangioni, J. V., Beahm, P. H., Shifrin, V., Jost, C. A., and Neel, B. G. (1992). The nontransmembrane tyrosine phosphatase PTP-1B localizes to the endoplasmic reticulum via its 35 amino acid C-terminal sequence. *Cell* **68**, 545–560.

Garton, A. J., and Tonks, N. K. (1994). PTP-PEST: A protein tyrosine phosphatase regulated by serine phosphorylation. *EMBO J.* **13**, 3763–3771.

Gebbink, M. F. B. G., Zondag, G. C. M., Wubbolts, R. W., Beijersbergen, R. L., van Etten, I., and Moolenaar, W. H. (1993). Cell–cell adhesion mediated by a receptor-like protein tyrosine phosphatase. *J. Biol. Chem.* **268**, 16101–16104.

Gonzalez, F. A., Seth, A., Raden, D. L., Bowman, D. S., Fay, F. S., and Davis, R. J. (1993). Serum-induced translocation of mitogen-activated protein kinase to the cell surface ruffling membrane and the nucleus. *J. Cell Biol.* **122**, 1089–1101.

Gu, M., Warshawsky, I., and Majerus, P. W. (1992). Cloning and expression of a cytosolic megakaryocyte protein tyrosine phosphatase with sequence homology to retinaldehyde-binding protein and yeast SEC14p. *Proc. Natl. Acad. Sci. U.S.A.* **89**, 2980–2984.

Gu, M., York, J. D., Warshawsky, L., and Majerus, P. W. (1991). Identification, cloning, and expression of a cytosolic megakaryocyte protein–tyrosine–phosphatase with sequence homology to cytoskeletal protein 4.1. *Proc. Natl. Acad. Sci. U.S.A.* **88**, 5867–5871.

Guan, K.-L., Broyles, S. S., and Dixon, J. E. (1991). A Tyr/Ser protein phosphatase encoded by vaccinia virus. *Nature* **350**, 359–362.

Guan, K.-L., and Butch, E. (1995). Isolation and characterization of a novel dual specific phosphatase, HVH2, which selectively dephosphorylates the mitogen-activated protein kinase. *J. Biol. Chem.* **270**, 7197–7203.

Guan, K.-L., and Dixon, J. E. (1991). Evidence for protein–tyrosine–phosphatase catalysis proceeding via a cysteine–phosphate intermediate. *J. Biol. Chem.* **266**, 17026–17030.

Guan, K.-L., and Dixon, J. E. (1993). Bacterial and viral protein tyrosine phosphatases. *Sem. Cell Biol.* **4**, 389–396.

Higashitsuji, H., Arii, S., Furutani, M., Imamura, M., Kaneko, Y., Takenawa, J., Nakayama, H., and Fujita, J. (1995). Enhanced expression of multiple protein tyrosine phosphatases in the regenerating mouse liver: Isolation of PTP-RL10, a novel cytoplasmic-type phosphatase with sequence homology to cytoskeletal protein 4.1. *Oncogene* **10**, 407–414.

Ishibashi, T., Bottaro, D. P., Michieli, P., Kelley, C. A., and Aaronson, S. A. (1994). A novel dual specificity phosphatase induced by serum stimulation and heat shock. *J. Biol. Chem.* **269**, 29897–29902.

Jia, Z., Barford, D., Flint, A. J., and Tonks, N. K. (1995). Structural basis for phosphotyrosine peptide recognition by protein tyrosine phosphatase 1B. *Science* **268**, 1754–1758.

Kazlauskas, A., Feng, G.-S., Pawson, T., and Valius, M. (1993). The 64-kDa protein that associates with the platelet-derived growth factor receptor β subunit via Tyr-1009 is the SH2-containing phosphotyrosine phosphatase Syp. *Proc. Natl. Acad. Sci. U.S.A.* **90**, 6939–6942.

Keyse, S. M., and Emslie, E. A. (1992). Oxidative stress and heat shock induce a human gene encoding a protein–tyrosine phosphatase. *Nature* **359**, 644–647.

Klingmüller, U., Lorenz, U., Cantley, L. C., Neel, B. G., and Lodish, H. F. (1995). Specific recruitment of SH-PTP1 to the erythropoietin receptor causes inactivation of JAK2 and termination of proliferative signals. *Cell* **80**, 729–738.

Krueger, N. X., Streuli, M., and Saito, H. (1990). Structural diversity and evolution of human receptor-like protein tyrosine phosphatases. *EMBO J.* **9**, 3241–3252.

Kwak, S. P., and Dixon, J. E. (1995). Multiple dual specificity protein tyrosine phosphatases are expressed and regulated differentially in liver cell lines. *J. Biol. Chem.* **270**, 1156–1160.

L'Abbé, D., Banville, D., Tong, Y., Stocco, R., Masson, S., Ma, S., Fantus, G., and Shen, S.-H. (1994). Identification of a novel protein tyrosine phosphatase with sequence homology to the cytoskeletal proteins of the band 4.1 family. *FEBS Lett.* **356**, 351–356.

Lechleider, R. J., Sugimoto, S., Bennett, A. M., Kashishian, A. S., Cooper, J. A., Shoelson, S. E., Walsh, C. T., and Neel, B. G. (1993). Activation of the SH2-containing phosphotyrosine phosphatase SH-PTP2 by its binding site, phosphotyrosine 1009, on the human platelet-derived growth factor receptorβ. *J. Biol. Chem.* **268**, 21478–21481.

Lenormand, P., Sardet, C., Pagès, G., L'Allemain, G., Brunet, A., and Pouysségur, J. (1993). Growth factors induce nuclear translocation of MAP kinases (p42^{mapk} and p44^{mapk}) but not of their activator MAP kinase kinase (p45^{mapkk}) in fibroblasts. *J. Cell Biol.* **122**, 1079–1088.

Li, W., Nishimura, R., Kashishian, A., Batzer, A. G., Kim, W. J. H., Cooper, J. A., and Schlessinger, J. (1994). A new function for a phosphotyrosine phosphatase: Linking GRB2-Sos to a receptor tyrosine kinase. *Mol. Cell. Biol.* **14**, 509–517.

Logan, T. M., Zhou, M.-M., Nettesheim, D. G., Meadows, R. P., Van Etten, R. L., and Fesik, S. W. (1994). Solution structure of a low molecular weight protein tyrosine phosphatase. *Biochemistry* **33**, 11087–11096.

Maekawa, K., Imagawa, N., Nagamatsu, M., and Harada, S. (1994). Molecular cloning of a novel protein-tyrosine phosphatase containing a membrane-binding domain and GLGF repeats. *FEBS Lett.* **337**, 200–206.

Marfatia, S. M., Lue, R. A., Branton, D., and Chishti, A. H. (1994). *In vitro* binding studies suggest a membrane-associated complex between erythroid p55, protein 4.1, and glycophorin C. *J. Biol. Chem.* **269**, 8631–8634.

Matozaki, T., Suzuki, T., Uchida, T., Inazawa, J., Ariyama, T., Matsuda, K., Horita, K., Noguchi, H., Mizuno, H., Sakamoto, C., and Kasuga, M. (1994). Molecular cloning of a human transmembrane-type protein tyrosine phosphatase and its expression in gastrointestinal cancers. *J. Biol. Chem.* **269**, 2075–2081.

Maurel, P., Rauch, U., Flad, M., Margolis, R. K., and Margolis, R. U. (1994). Phosphacan, a chondroitin sulfate proteoglycan of brain that interacts with neurons and neural cell-adhesion molecules, is an extracellular variant of a receptor-type protein tyrosine phosphatase. *Proc. Natl. Acad. Sci. U.S.A.* **91**, 2512–2516.

McLaughlin, S., and Dixon, J. E. (1993). Alternative splicing gives rise to a nuclear protein tyrosine phosphatase in *Drosophila. J. Biol. Chem.* **268**, 6839–6842.

Meloche, S., Seuwen, K., Pagès, G., and Pouysségur, J. (1992). Biphasic and synergistic activation of p44^{mapk} (ERK1) by growth factors: Correlation between late phase activation and mitogenicity. *Mol. Endocrinol.* **6**, 845–854.

Millar, J. B. A., and Russell, P. (1992). The cdc25 M-phase inducer: An unconventional protein phosphatase. *Cell* **68**, 407–410.

Minshull, J., Sun, H., Tonks, N. K., and Murray, A. W. (1994). A MAP kinase-dependent spindle assembly checkpoint in *Xenopus* egg extracts. *Cell* **79**, 475–486.

Moller, N. P. H., Moller, K. B., Lammers, R., Kharitonenkov, A., Sures, I., and Ullrich, A. (1994). Src kinase associates with a member of a distinct subfamily of protein-tyrosine phosphatases containing an ezrin-like domain. *Proc. Natl. Acad. Sci. U.S.A.* **91**, 7477–7481.

Neel, B. G. (1993). Structure and function of SH2-domain containing tyrosine phosphatases. *Sem. Cell Biol.* **4**, 419–432.

Noguchi, T., Metz, R., Chen, L., Mattéi, M.-G., Carrasco, D., and Bravo, R. (1993). Structure, mapping, and expression of *erp,* a growth factor-inducible gene encoding a nontransmembrane protein tyrosine phosphatase, and effect of *erp* on cell growth. *Mol. Cell. Biol.* **13**, 5195–5205.

Oon, S. H., Hong, A., Yang, X., and Chia, W. (1993). Alternative splicing in a novel tyrosine phosphatase gene (DPTP4E) of *Drosophila melanogaster* generates two large receptor-like proteins which differ in their carboxyl termini. *J. Biol. Chem.* **268**, 23964–23971.

Ostergaard, H. L., and Trowbridge, I. S. (1991). Negative regulation of CD45 protein tyrosine phosphatase activity by ionomycin in T cells. *Science* **253**, 1423–1425.

Östman, A., Yang, Q., and Tonks, N. K. (1994). Expression of DEP-1, a receptor-like protein-tyrosine-phosphatase, is enhanced with increasing cell density. *Proc. Natl. Acad. Sci. U.S.A.* **91**, 9680–9684.

Pawson, T., and Gish, G. D. (1992). SH2 and SH3 domains: From structure to function. *Cell* **71**, 359–362.

Pawson, T., and Schlessinger, J. (1993). SH2 and SH3 domains. *Curr. Biol.* **3**, 434–442.

Pluskey, S., Wandless, T. J., Walsh, C. T., and Shoelson, S. E. (1995). Potent stimulation of SH-PTP2 phosphatase activity by simultaneous occupancy of both SH2 domains. *J. Biol. Chem.* **270**, 2897–2900.

Potts, M., Sun, H., Mockaitis, K., Kennelly, P. J., Reed, D., and Tonks, N. K. (1993). A protein-tyrosine/serine phosphatase encoded by the genome of the cyanobacterium *Nostoc commune* UTEX 584. *J. Biol. Chem.* **268**, 7632–7635.

Rohan, P. J., Davis, P., Moskaluk, C. A., Kearns, M., Krutzsch, H., Siebenlist, U., and Kelly, K. (1993). PAC-1: A mitogen-induced nuclear protein tyrosine phosphatase. *Science* **259**, 1763–1766.

Sap, J., Jiang, Y.-P., Friedlander, D., Grumet, M., and Schlessinger, J. (1994). Receptor tyrosine phosphatase R-PTP-κ mediates homophilic binding. *Mol. Cell. Biol.* **14**, 1–9.

Saras, J., Claesson-Welsh, L., Heldin, C.-H., and Gonez, L. J. (1994). Cloning and characterization of PTPL1, a protein tyrosine phosphatase with similarities to cytoskeletal-associated proteins. *J. Biol. Chem.* **269**, 24082–24089.

Sarcevic, B., Erikson, E., and Maller, J. L. (1993). Purification and characterization of a mitogen-activated protein kinase tyrosine phosphatase from *Xenopus* eggs. *J. Biol. Chem.* **268**, 25075–25083.

Sato, T., Irie, S., Kitada, S., and Reed, J. C. (1995). FAP-1: A protein tyrosine phosphatase that associates with Fas. *Science* **268**, 411–415.

Sgroi, D., Varki, A., Braesch-Andersen, S., and Stamenkovic, I. (1993). CD22, a B cell-specific immunoglobulin superfamily member, is a sialic acid-binding lectin. *J. Biol. Chem.* **268**, 7011–7018.

Shenolikar, S., and Nairn, A. C. (1991). Protein phosphatases: Recent progress. *Adv. Second Messenger Phosphoprotein Res.* **23**, 1–121.

Stamenkovic, I., Sgroi, D., Aruffo, A., Sy, M. S., and Anderson, T. (1991). The B lymphocyte adhesion molecule CD22 interacts with leukocyte common antigen CD45 RO on T cells and α2-6 sialyltransferase, CD75 on B cells. *Cell* **66**, 1133–1144.

Stone, R. L., and Dixon, J. E. (1994). Protein–tyrosine phosphatases. *J. Biol. Chem.* **269**, 31323–31326.

Stover, D. R., and Walsh, K. A. (1994). Protein–tyrosine phosphatase activity of CD45 is activated by sequential phosphorylation by two kinases. *Mol. Cell. Biol.* **14**, 5523–5532.

Streuli, M., Krueger, N. X., Thai, T., Tang, M., and Saito, H. (1990). Distinct functional roles of the two intracellular phosphatase like domains of the receptor-linked protein tyrosine phosphatases LCA and LAR. *EMBO J.* **9**, 2399–2407.

Stuckey, J. A., Schubert, H. L., Fauman, E. B., Zhang, Z.-Y., Dixon, J. E., and Saper, M. A. (1994). Crystal structure of *Yersinia* protein tyrosine phosphatase at 2.5Å and the complex with tungstate. *Nature* **370**, 571–575.

Su, J., Batzer, A., and Sap, J. (1994). Receptor tyrosine phosphatase R-PTP-α is tyrosine-phosphorylated and associated with the adaptor protein Grb2. *J. Biol. Chem.* **269**, 18731–18734.

Su, X.-D., Taddei, N., Stefani, M., Ramponi, G., and Nordlund, P. (1994). The crystal structure of a low-molecular-weight phosphotyrosine protein phosphatase. *Nature* **370**, 575–578.

Sun, H., Charles, C. H., Lau, L. F., and Tonks, N. K. (1993). MKP-1 (3CH134), an immediate early gene product, is a dual specificity phosphatase that dephosphorylates MAP kinase in vivo. *Cell* **75**, 487–493.

Sun, H., and Tonks, N. K. (1994). The coordinated action of protein tyrosine phosphatases and kinases in cell signalling. *TIBS* **19**, 480–485.

Sun, H., Tonks, N. K., and Bar-Sagi, D. (1994). Inhibition of Ras-induced DNA synthesis by expression of the phosphatase MKP-1. *Science* **265**, 285–288.

Thomas, P. E., Wharram, B. L., Goyal, M., Wiggins, J. E., Holzman, L. B., and Wiggins, R. C. (1994). GLEPP1, a renal glomerular epithelial cell (podocyte) membrane protein–tyrosine phosphatase. *J. Biol. Chem.* **269**, 19953–19962.

Tian, S.-S., Tsoulfas, P., and Zinn, K. (1991). Three receptor-linked protein–tyrosine phosphatases are selectively expressed on central nervous system axons in the *Drosophila* embryo. *Cell* **67**, 675–685.

Tillman, U., Wagner, J., Boerboom, D., Westphal, H., and Tremblay, M. L. (1994). Nuclear localization and cell cycle regulation of a murine protein tyrosine phosphatase. *Mol. Cell. Biol.* **14**, 3030–3040.

Tonks, N. K. (ed.) (1993). Protein tyrosine phosphatases. *Sem. Cell Biol.* **4**, 373–453.

Tonks, N. K., Charbonneau, H., Diltz, C. D., Fischer, E. H., and Walsh, K. A. (1988). Demonstration that the leukocyte common antigen CD45 is a protein tyrosine phosphatase. *Biochemistry* **27**, 8695–8701.

Tonks, N. K., Diltz, C. D., and Fischer, E. H. (1988a). Purification of the major protein-tyrosine phosphatases of human placenta. *J. Biol. Chem.* **263**, 6722–6730.

Tonks, N. K., Diltz, C. D., and Fischer, E. H. (1988b). Characterization of the major protein tyrosine phosphatases of human placenta. *J. Biol. Chem.* **263**, 6731–6737.

Tonks, N. K., Yang, Q., Flint, A. J., Gebbink, M. F. B. G., Franza, B. R. J., Hill, D. E., Sun, H., and Brady-Kalnay, S. (1992). Protein tyrosine phosphatases: The problems of a growing family. *Cold Spring Harbor Symp. Quantitative Biol.* **57**, 87–94.

Tsukita, S., Oishi, K., Sato, N., Sagara, J., Kawai, A., and Tsukita, S. (1994). ERM family members as molecular linkers between the cell surface glycoprotein CD44 and actin-based cytoskeletons. *J. Cell Biol.* **126**, 391–401.

Varnum, B. C., Young, C., Elliott, G., Garcia, A., Bartley, T. D., Fridell, Y.-W., Hunt, R. W., Trail, G., Clogston, C., Toso, R. J., Yanagihara, D., Bennett, L., Sylber, M., Merewether, L. A., Tseng, A., Escobar, E., Liu, E. T., and Yamane, H. K. (1995). Axl receptor tyrosine kinase stimulated by the vitamin K-dependent protein encoded by growth-arrest-specific gene 6. *Nature* **373**, 623–626.

Walton, K. M., and Dixon, J. E. (1993). Protein tyrosine phosphatases. *Annu. Rev. Biochem.* **62**, 101–120.

Wang, Y., and Pallen, C. J. (1991). The receptor-like protein tyrosine phosphatase HPTPa has two active catalytic domains with distinct substrate specificities. *EMBO J.* **10**, 3231–3237.

Ward, Y., Gupta, S., Jensen, P., Wartmann, M., Davis, R. J., and Kelly, K. (1994). Control of MAP kinase activation by the mitogen-induced threonine/tyrosine phosphatase PAC1. *Nature* **367**, 651–654.

Woodford-Thomas, T., and Thomas, M. L. (1993). The leukocyte common antigen, CD45 and other protein tyrosine phosphatases in hematopoietic cells. *Sem. Cell Biol.* **4**, 409–418.

Yang, Q., and Tonks, N. K. (1991). Isolation of a cDNA clone encoding a human protein-tyrosine phosphatase with homology to the cytoskeletal-associated proteins band 4.1, ezrin, and talin. *Proc. Natl. Acad. Sci. U.S.A.* **88**, 5949–5953.

Yang, X., Seow, K. T., Bahri, S. M., Oon, S. H., and Chia, W. (1991). Two *Drosophila* receptor-like tyrosine phosphatase genes are expressed in a subset of developing axons and pioneer neurons in the embryonic CNS. *Cell* **67**, 661–673.

Zhang, M., Van Etten, R. L., and Stauffacher, C. V. (1994). Crystal structure of bovine heart phosphotyrosyl phosphatase at 2.2-Å resolution. *Biochemistry* **33**, 11097–11105.

Zhao, Z., Shen, S.-H., and Fischer, E. H. (1994). Phorbol ester-induced expression, phosphorylation, and translocation of protein-tyrosine-phosphatase 1C in HL-60 cells. *Proc. Natl. Acad. Sci. U.S.A.* **91**, 5007–5011.

Tetsuo Moriguchi
Yukiko Gotoh
Eisuke Nishida
Department of Genetics and Molecular Biology
Institute for Virus Research
Kyoto University
Kyoto 606-01, Japan

Roles of the MAP Kinase Cascade in Vertebrates

I. Introduction

Mitogen-activated protein (MAP) kinases are serine/threonine kinases that mediate intracellular phosphorylation events triggered by a variety of extracellular stimuli including growth factors and differentiation factors (reviewed in Cobb *et al.*, 1991; Sturgill and Wu, 1991; Pelech and Sanghera, 1992; Davis, 1993; Nishida and Gotoh, 1993; Blumer and Johnson, 1994). MAP kinases are also activated by cytokines, hormones, heat shock, hyperosmolarity and integrin-mediated cell attachment, and during *Xenopus* oocyte maturation. One of the most outstanding characteristics of MAP kinases is their activation mechanism. Activation of MAP kinases requires the phosphorylation of both threonine and tyrosine residues in the sequence "TEY" located between kinase subdomains VII and VIII (Payne *et al.*, 1991). Phosphorylation of these residues is brought about by the dual-specificity kinase,

Advances in Pharmacology, Volume 36

MAP kinase kinases (MAPKKs, also called MEKs) (Ahn *et al.*, 1991; Gomez and Cohen, 1991; Matsuda *et al.*, 1992; Nakielny *et al.*, 1992; Seger *et al.*, 1992a; Shirakabe *et al.*, 1992). cDNA cloning of MAPKKs (Ashworth *et al.*, 1992; Crews *et al.*, 1992; Seger *et al.*, 1992b; Brott *et al.*, 1993; Kosako *et al.*, 1993; Wu *et al.*, 1993a,b; Zheng and Guan, 1993) and MAP kinases (Boulton *et al.*, 1990, 1991; Gotoh *et al.*, 1991b) shows that several MAPKK homologs and MAP kinase homologs exist in yeast (Nishida and Gotoh, 1993). This may suggest that the MAPKK/MAP kinase cascade (the MAP kinase cascade) functions universally in eukaryotic systems. In fact, the MAP kinase cascade functions in pathways of the pheromone response, the osmolarity glycerol response and the cell wall integrity in budding yeast (Herskowitz, 1995; Levin and Errede, 1995), in the auxin-like signal transduction pathway in plant cells (Mizoguchi *et al.*, 1994), during R7 photoreceptor induction and gap gene expression of *Drosophila* (reviewed in Perrimon and Desplan, 1994), and during vulval induction of *C. elegans* (reviewed in Eisenmann and Kim, 1994).

MAPKK is activated by phosphorylation of one or two serine residue(s) between kinase subdomains VII and VIII (Alessi *et al.*, 1994; Gardner *et al.*, 1994; Gotoh *et al.*, 1994; Zheng and Guan, 1994), catalyzed by a MAPKK-activating kinase (MAPKK-K) (Matsuda *et al.*, 1993). A number of candidates for MAPKK-Ks have been reported. A protooncogene product c-Raf-1 was the first molecule identified as MAPKK-K in mammalian cultured cells (Dent *et al.*, 1992; Howe *et al.*, 1992; Kyriakis *et al.*, 1992). Subsequently, c-Mos (Nebreda and Hunt, 1993; Posada *et al.*, 1993), B-raf (Traverse and Cohen, 1994), and MEKK (Lange-Carter *et al.*, 1993) have been shown to function as MAPKK-K in *Xenopus* oocytes, neuronal tissues, and various mammalian cultured cells. Other molecules were also reported as MAPKK-K in mammalian fat cells and fibroblasts (Haystead *et al.*, 1994; Reuter *et al.*, 1995). Although it is clear that these MAPKK-Ks can activate the same MAP kinase cascade, it is also possible that each MAPKK-K might activate signaling pathways other than the MAP kinase cascade.

Deactivation of MAP kinase *in vitro* can be achieved by treatment with either protein phosphatase 2A or tyrosine phosphatase CD45 (Anderson *et al.*, 1990). Dual-specificity protein phosphatases such as human CL-100 or mouse 3CH134 (MPK-1), human PAC-1, and yeast MSG5 were identified that can dephosphorylate MAP kinase family molecules (reviewed in Keyse, 1995). 3CH134 was discovered as an immediate early gene, and CL-100 as a stress-inducible gene. Inhibition of MKP-1 induction by cycloheximide blocked the deactivation of MAP kinase in serum-stimulated NIH3T3 cells (Sun *et al.*, 1993). However, rapid deactivation of MAP kinase in many cases cannot be blocked by cycloheximide treatment in many cell lines. (Alessi *et al.*, 1995). The relative importance of these dual-specificity phosphatases and other phosphatases in various signaling pathways remains to be determined.

MAP kinase is commonly activated by a large number of extracellular stimuli and is supposed to play a key role in various intracellular signal transduction pathways because MAP kinase can phosphorylate many target substrates *in vitro* (Davis, 1993). However, few have been shown to be substrates for MAP kinase *in vivo*. In yeast, *Drosophila,* and *C. elegans,* genetic studies revealed a crucial role of the MAP kinase cascade in several signal transduction pathways. In higher eukaryotes, the question of whether the activation of the MAP kinase cascade is required or sufficient to initiate biological effects remains to be answered. By using several techniques, such as transfection or microinjection of cells with constitutively active or dominantly negative forms of components of the MAP kinase cascade or with MAP kinase phosphatase (Fig. 1), some of the roles of the MAP kinase cascade have been demonstrated. In this chapter, we focus on the roles of the MAP kinase cascade in mammalian cultured cells and *Xenopus* systems.

Recently, novel subgroups of the MAP kinase superfamily, stress-activated protein kinase (SAPK)/c-Jun N-terminal kinase (JNK) (Derijard *et al.,* 1994; Kyriakis *et al.,* 1994) and p38/MPK2 (Han *et al.,* 1994; Rouse *et al.,* 1994) were identified. SAPKs/JNKs contain the sequence TPY and p38/MPK2 contains the sequence TGY between kinase subdomain VII and VIII, in contrast to the TEY sequence on MAP kinases. Both the SAPK/JNK family and the p38/MPK2 family need phosphorylation of both threonine and tyrosine residues for their kinase activity, similar to the MAP kinase. We also describe some features of protein kinase cascades of the MAP kinase superfamily.

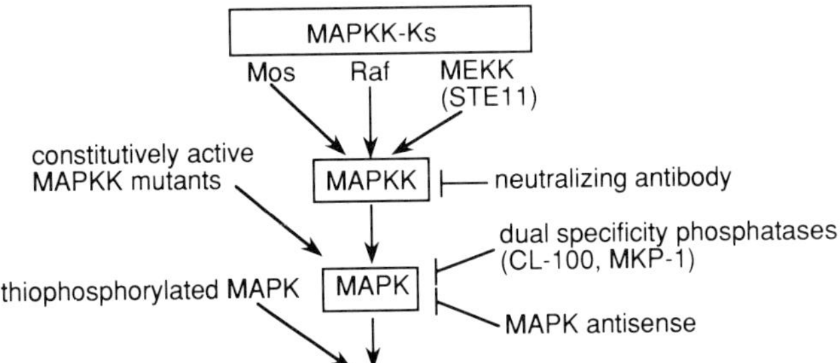

FIGURE I   Several approaches for studying the role of the MAP kinase cascade. Microinjection of mRNAs or proteins of constitutively active mutants of MAPKK elevates the activity of MAP kinase. Thiophosphorylated MAP kinase (MAPK) is resistant to phosphatases in cells. The signaling through the MAP kinase cascade is inhibited by MAP kinase (MAPK) antisense expression, dual-specificity phosphatases, and/or the anti-MAPKK neutralizing antibody.

II. Roles of the MAP Kinase Cascade in Mammalian Cultured Cells

A. Differentiation

The adrenal pheochromocytoma cell line PC12 undergoes growth arrest and differentiates to sympathetic neuron-like cells in response to extracellular factors such as basic fibroblast growth factor (bFGF) and nerve growth factor (NGF). In PC12 cells, both NGF and epidermal growth factor (EGF) activate 43-kDa MAP kinase (ERK1) and 41-kDa MAP kinase (ERK2) (Gotoh *et al.*, 1990). The EGF-induced activation of MAP kinases is transient, whereas the NGF-induced activation is sustained (Gotoh *et al.*, 1990) and is accompanied by nuclear translocation of MAP kinases (Traverse *et al.*, 1992). Activation of MAPKK1 and MAPKK2 roughly parallels the activation of MAP kinases in NGF- or EGF-stimulated PC12 cells (Traverse *et al.*, 1994; Moriguchi *et al.*, 1995b). Because NGF, but not EGF, induces neuronal differentiation of PC12 cells, the difference in the mode of activation of MAP kinases between NGF and EGF stimulations might be involved in their difference in the cellular response. Then, to examine the role of the MAP kinase cascade in neuronal differentiation of PC12 cells, several groups utilized techniques of microinjection of a variety of MAP kinase cascade components that can result in changes in the activity of MAP kinases. Cowley *et al.* (1994) replaced two serine residues of the activating phosphorylation sites of MAPKK1 by alanine or glutamic acid to generate dominant-negative or constitutive-active mutants of MAPKK1 and microinjected PC12 cells with plasmids containing these mutant MAPKKs. When injected with constitutive-active mutant constructs, PC12 cells developed neurites within 48 hr. However, PC12 cells microinjected with dominant-negative MAPKK1 mutant constructs failed to differentiate in response to stimuli (Cowley *et al.*, 1994). To examine the effect of constitutive activation of MAPKK-K and MAP kinase, we used an N-terminal truncated STE11 (one of yeast MAPKK-Ks) and thiophosphorylated MAP kinase (Fukuda *et al.*, 1995). The N-terminal truncated STE11, which was bacterially produced, could activate MAPKK and MAP kinase potently in cell-free extracts obtained from *Xenopus* oocytes (K. Takenaka *et al.*, unpublished results). Thiophosphorylated proteins are generally resistant to dephosphorylation by protein phosphatases. In fact, thiophosphorylated MAP kinase, which was produced by using recombinant *Xenopus* MAP kinase, purified MAPKK and ATP-γS, was active and the thiophosphorylated MAP kinase, purified from the above mixture was not inactivated. Microinjection of either the N-terminal truncated STE11 or the purified thiophosphorylated MAP kinase into PC12 cells induced neurite outgrowth. Moreover, the neurite-inducing activity of the N-terminal truncated STE11 was canceled by coinjection of the dual-specificity phosphatase CL-100. Together, these results suggest that activation of the MAP kinase cascade is necessary and sufficient for neural differentiation of PC12 cells. Because Ras p21, a protooncogene product has also been shown to be necessary and sufficient for the neurite outgrowth (Bar-Sagi and Feramisco, 1985; Noda *et al.*, 1985; Hagag *et al.*, 1986; Szeberenyi *et al.*, 1990), the Ras/MAP kinase pathway may constitute a major route for PC12 cell differentiation.

B. Proliferation

MAP kinase was first described as a microtubule-associated protein 2 (MAP2)-phosphorylation activity activated by insulin (Ray and Sturgill, 1987, 1988) and by various growth factors and tumor promoters (Hoshi *et al.*, 1988). Moreover, the MAP kinase cascade is activated downstream of multiple oncogene products such as Ras, Raf, and Src (reviewed in Avruch *et al.*, 1994; Daum *et al.*, 1994). Therefore, the MAP kinase cascade is thought to play a pivotal role in growth control. Pages *et al.* (1993) revealed that transient expression of either phosphorylation site mutants of MAP kinase or antisense RNA inhibited MAP kinase activation and cell growth. Overexpression of MKP-1 was also shown to inhibit fibroblast proliferation (Sun *et al.*, 1994; Brondello *et al.*, 1995). Furthermore, constitutively active MAPKK mutants were capable of promoting cell transformation (Cowley *et al.*, 1994; Mansour *et al.*, 1994).

III. Roles of the MAP Kinase Cascade in *Xenopus* ___________

A. Oocyte Maturation

Fully grown *Xenopus* oocytes are naturally arrested at prophase of meiosis I. Progesterone releases this arrest and induces progression through meiosis, leading to the production of the unfertilized eggs that are arrested at metaphase of meiosis II. During this transition, called oocyte maturation, maturation-promoting factor (MPF) is activated, followed by germinal vesicle breakdown (GVBD). MPF is a complex of p34^{cdc2} kinase and cyclin B and is stored in *Xenopus* immature oocytes as an inactive complex called pre-MPF (reviewed by Lohka, 1989; Nurse, 1990; Maller, 1991). The conversion of pre-MPF into active MPF requires phosphorylation on Thr161 and dephosphorylation on Tyr15 and Thr14 in p34^{cdc2} kinase. Phosphorylation and dephosphorylation are catalyzed by CAK (Frequet *et al.*, 1993; Poon *et al.*, 1993; Solomon *et al.*, 1993), which is composed of MO15 and cyclin H (Fisher and Morgan, 1994; Makela *et al.*, 1994), and protein phosphatase cdc25 (Gautier *et al.*, 1991; Kumagai and Dunphy, 1991), respectively. We previously showed that injection of purified MPF into immature oocytes or addition of purified MPF into interphase extracts activated both MAPKK and MAP kinase (Gotoh *et al.*, 1991b; Matsuda *et al.*, 1992). The activities of MAPKK and MAP kinase are elevated at about the same time as MPF during progesterone-induced oocyte maturation (Gotoh *et al.*, 1991a; Matsuda *et al.*, 1992; Nebreda and Hunt, 1993), suggesting the possibility of a positive feedback loop between the MAP kinase cascade and MPF.

The product of the c-*mos* protooncogene is a 39-kDa serine/threonine protein kinase that is synthesized from maternal mRNA in response to progesterone. The synthesis of Mos is necessary for the progesterone-induced oocyte maturation (Sagata *et al.*, 1988) and microinjection of mos mRNA or bacterially expressed Mos protein is able to induce maturation in the absence of progesterone (Sagata *et al.*, 1988; Freeman *et al.*, 1989; Yew *et al.*, 1992). In our experiments, Mos is a major MAPKK-K in *Xenopus* mature oocyte extracts (S. Matsuda *et al.*, unpublished results). Thus, Mos can work as MAPKK-K during oocyte maturation.

We produced several anti-*Xenopus* MAPKK antibodies, one of which is capable of inhibiting *Xenopus* MAPKK activity *in vitro* (Kosako *et al.*, 1994a). Microinjection of this neutralizing antibody into immature oocytes prevented progesterone- or Mos-induced activation of MAP kinase (Kosako *et al.*, 1994a). Furthermore, the inhibition of the activation of the MAP kinase by this antibody blocked activation of H1 kinase and GVBD (Kosako *et al.*, 1994a). Injection of CL-100 mRNA also inhibited progesterone-induced GVBD as well as MAP kinase activation (Gotoh *et al.*, 1995a). These results suggested that activation of the MAP kinase cascade is required for oocyte maturation. Moreover, the N-terminal truncated STE11 (a MAPKK-K), the constitutively active mutant of MAPKK, or thiophosphorylated MAP kinase can induce GVBD, although these effects were partially blocked by inhibition of protein synthesis (Gotoh *et al.*, 1995a; Haccard *et al.*, 1995; Huang *et al.*, 1995). Other signaling pathways, such as the cAMP-dependent pathway, may be involved in progesterone-induced MPF activation (Daar *et al.*, 1993), but activation of the MAP kinase cascade alone is sufficient for at least the onset of oocyte maturation.

B. Metaphase Arrest

Unfertilized *Xenopus* eggs are arrested at metaphase II and the activity of MPF remains high. Fertilization releases this meiotic arrest and brings about the degradation of cyclin B and the inactivation of MPF. A cytostatic factor (CSF) present in the unfertilized egg is believed to stabilize MPF and cause the metaphase II arrest. Mos, which functions at oocyte maturation, also functions as a component of CSF. The microinjection of *mos* mRNA or bacterially expressed mos protein into one blastomere of a two-cell embryo arrests cleavage of the injected half of the embryo at metaphase (Sagata *et al.*, 1989; Freeman *et al.*, 1990; Yew *et al.*, 1992). MAP kinase, which is activated downstream of mos during oocyte maturation, is activated by microinjection of mos protein into two-cell embryos (Posada *et al.*, 1993). Injection of thiophosphorylated MAP kinase into two-cell embryos induced metaphase arrest (Haccard *et al.*, 1993). Moreover, mos-induced metaphase arrest was prevented by coinjection of the anti-MAPKK neutralizing antibody (Kosako *et al.*, 1994b). Thus, the MAP kinase cascade may play a pivotal role in the cell cycle arrest at metaphase in mature oocytes.

C. Early Development

During early embryogenesis, the animal region gives rise to ectoderm and the vegetal region becomes endoderm. The mesoderm was induced in the marginal zone of blastula by an inductive signal(s) from a vegetal hemisphere. Nieuwkoop and Faber (1967) showed that animal pole cells explanted in the blastula stage can be induced to mesoderm only when in contact with vegetal pole cells. By using animal cap assays, which are based on Nieuwkoop and Faber's animal–vegetal recombinants, several factors have been identified as mesoderm-inducing factors; the members of the transforming growth factor-β family such as Vg1, activin and bone morphogenic proteins, and the members of fibroblast growth factor (FGF) family such as bFGF and eFGF (reviewed in Kessler and Melton, 1994). Dominant-negative mutants of Ras or Raf-1 have been shown to block the induction of mesoderm markers by activin or by bFGF (Whitman and Melton, 1992; Labonne and Whitman, 1994). Then, we examined the role of the MAP kinase cascade in mesoderm induction. Microinjection of either mRNA encoding a constitutively active mutant of MAPKK or a constitutively active form of STE11 leads to the induction of mesoderm in animal cap explants as judged by morphology, histology, and expression of tissue-specific markers (Gotoh *et al.*, 1995b). Moreover, injection of CL-100 mRNA inhibited FGF-induced and activin-induced mesoderm induction in animal cap explants (Gotoh *et al.*, 1995b). Furthermore, injection of CL-100 mRNA into two-cell stage embryos causes severe defects in posterior development similar to those obtained with a dominant-negative FGF receptor or a dominant-negative Raf-1 (Amaya *et al.*, 1991; MacNicol *et al.*, 1993; Gotoh *et al.*, 1995b). These results suggest that the MAP kinase cascade is functioning mainly in the FGF receptor-mediating signaling pathway downstream of Ras during mesoderm induction. However, it is still unknown whether the MAP kinase cascade is directly involved in the activin-mediated signaling pathway.

IV. MAP Kinase Superfamily

A. Subfamilies of MAP Kinase—SAPK/JNK and p38/MPK2

New members of the MAP kinase superfamily, SAPK/JNK and p38/MPK2, have been identified in vertebrates (Fig. 2). SAPK was identified first as a 54-kDa MAP2-phosphorylating enzyme, which was isolated from cycloheximide-treated rat liver (Kyriakis and Avruch, 1990). This kinase, like the classical MAP kinase mentioned previously, requires both threonine and tyrosine phosphorylation for its activity but it is distinct from MAP kinase in its substrate specificity (Kyriakis and Avruch, 1990; Kyriakis *et al.*, 1991). JNKs were identified in human cells as 46- and 55-kDa kinases (named JNK1 and JNK2, respectively), which were activated by UV irradiation. They have been shown to be capable of binding to the c-Jun amino-terminal sequence and phosphorylating serine 63 and serine 73 of c-Jun (Hibi *et al.*, 1993). Molecular cloning of SAPK and JNK revealed a unique subfamily of MAP kinase superfamily that is about 40–45% identical to the classical MAP kinase in amino acid sequence (Derijard *et al.*, 1994; Kyriakis *et al.*, 1994; Sluss *et al.*, 1994). Rat SAPKγ and SAPKα correspond to human JNK1 and JNK2, respectively. p49[3F12] kinase, which was discovered as a novel MAP kinase expressed in the human nervous system, may probably be a human homolog of rat SAPKβ (Mohit *et al.*, 1995). The SAPK/JNK subfamily is defined as containing the TPY sequence instead of the TEY sequence of the classical MAP kinase.

p38 was described as a tyrosine-phosphorylated protein in the lipopolysaccharide (LPS)-stimulated mouse pre-B cells, which were transfected with LPS receptor CD14 (Han *et al.*, 1994). Other studies have identified the

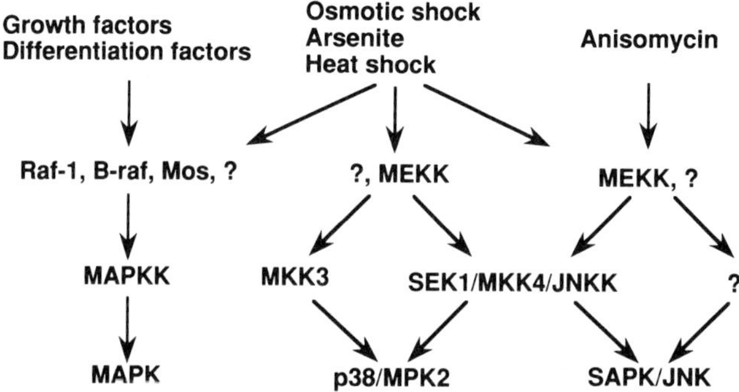

FIGURE 2 Comparison of the MAP kinase superfamily signal transduction pathways. There are at least three types of MAP kinase superfamily molecules, and the hypothetical pathways are illustrated schematically. Growth factors and differentiation factors are the major stimuli that activate MAP kinase (MAPK). Anisomysin activates SAPK/JNK. All three MAP kinase family molecules can be activated by osmotic shock, arsenite, and heat shock.

kinase of this family as a direct activator for MAPKAP kinase 2 in response to heat shock or sodium arsenite (Rouse *et al.*, 1994) or as a kinase activated in the interleukin-1-induced signaling pathway (Freshney *et al.*, 1994). Rouse *et al.* (1994) revealed that a MAPKAP kinase 2 activator is immunologically closely related to MPK2, which is a *Xenopus* homolog of mammalian p38. p38/MPK2 was also identified as a target of the inhibitor for production of interleukin-1 and tumor necrosis factor (Lee *et al.*, 1994).

B. Upstream of the Cascades

Two different approaches have been used to identify direct upstream components of these new members of the MAP kinase superfamily. One is searching for the kinases encoding protein kinases related to the classical MAPKK by polymerase chain reaction and the other is dissecting upstream pathways by fractionating cell extracts.

The first approach revealed two MAPKK-related kinases, mouse SEK1 (Sanchez *et al.*, 1994) and human MKK4 (Derijard *et al.*, 1995) and human MKK3 (Derijard *et al.*, 1995). SEK1 was identified as a murine homolog of XMEK2, which was previously isolated from the *Xenopus* cDNA library (Yashar *et al.*, 1993). MKK4 (human homolog of SEK1/XMEK2) and MKK3 were isolated by using degenerate PCR primers against PBS2, which is one of the yeast MAPKK family molecules acting upstream of HOG1 (a yeast homolog of p38). SEK1/MKK4 can act as a direct activator for SAPK/ JNK when expressed in cells and MKK3 can act as an activator for p38/ MPK2.

Cohen and co-workers (Rouse *et al.*, 1994) used the second approach. They fractionated extracts from sodium arsenite-stimulated PC12 cells by a Mono-Q column and found the existence of an upstream activator for p38/MPK2 that is distinct from classical MAPKK. It is unknown whether this kinase is the same as MKK3. We fractionated extracts obtained from fibroblastic cells exposed to hyperosmolar media on Q-Sepharose column chromatography and found that the SAPK-activating activity was resolved into unadsorbed and adsorbed fractions (Moriguchi *et al.*, 1995a). By using several anti-XMEK2 antibodies, we identified the SAPK-activating activity in the adsorbed fractions as XMEK2/SEK1/MKK4. The unadsorbed fractions in the Q-Sepharose chromatography were further resolved into several peaks upon sequential chromatography on heparin–Sepharose and hydroxylapatite (Moriguchi *et al.*, 1995a). These results suggest that there exist several activators for SAPK/JNK, including XMEK2/SEK1/MKK4 (Fig. 2).

A direct activator for SEK1 was identified by a transfection method. When cultured cells were transfected with MEKK, which was originally identified as a candidate for MAPKK-K, SAPK/JNK was preferentially activated (Minden *et al.*, 1994; Yan *et al.*, 1994) through activation of SEK1

(Yan *et al.*, 1994). Consistent with these results, we have found that endogenous MEKK, which was activated by osmotic shock, was capable of phosphorylating and activating recombinant XMEK2 (Matsuda *et al.*, 1995). Moreover, a mutant XMEK2 that has activating phosphorylation sites that have changed to nonphosphorylatable amino acids was not phosphorylated and activated by MEKK (T. Moriguchi *et al.*, unpublished results). A number of papers have appeared that suggest that Rac1 and Cdc42 lie upstream of the SAPK/JNK and p38/MPK2 pathways (Coso *et al.*, 1995; Minden *et al.*, 1995). Further studies are needed to fully elucidate the activating pathways for the SAPK/JNK subfamily and the p38/MPK2 subfamily.

V. Conclusion and Perspectives

The MAP kinase cascade is activated by various extracellular stimuli. Evidence is accumulating suggesting the involvement of the MAP kinase cascade in a wide variety of cellular responses including differentiation of PC12 cells, proliferation of fibroblasts, *Xenopus* oocyte maturation, metaphase arrest, and mesoderm induction as summarized here. In addition, the requirement of the MAP kinase cascade in various biological processes has been suggested by a large number of experiments. For example, the role of the MAP kinase cascade in thymocyte differentiation was demonstrated by transgenic mice that expressed a dominant-negative MAPKK in thymocytes (Alberola-Ila *et al.*, 1995). The requirement of MAP kinase activation in PC12 cell differentiation was also shown by utilizing a chemical inhibitor of MAPKK (Pang *et al.*, 1995), and a crucial role of MAP kinase during differentiation of 3T3 L1 fibroblasts into adipocytes was suggested by the antisense technique (Sale *et al.*, 1995). Why does activation of the same MAP kinase cascade bring about different biological effects? Efforts for pursuing answers to this question will probably lead to understanding of molecular mechanisms for cell proliferation, cell differentiation, development, and other biological phenomena.

Recent studies revealed the existence of novel members of the MAP kinase superfamily in vertebrate (Davis, 1994), but the roles of these members have yet to be clarified. An interesting result was reported by Su *et al.* (1994), who suggested that signals from T cell receptor and CD28 appear to be integrated at the level of SAPK/JNK. Moreover, there may be additional members of the MAP kinase superfamily. Recently, MEK5, a MAPKK superfamily molecule, and ERK5, a MAP kinase family molecule, were identified (Zhou *et al.*, 1995). MEK5 and ERK5 were suggested to constitute a new kinase cascade because ERK5 was shown to selectively bind to MEK5. The N terminus of MEK5 has a sequence similar to those of CDC24 and scd1 and thus may be involved in some GTPase signaling pathway. In addition, there exist several dual-specificity phosphatases (Guan and Butch, 1995;

Kwak and Dixon, 1995; Misra-Press *et al.*, 1995). It is likely that multiple dual-specificity phosphatases selectively deactivate certain members of the MAP kinase superfamily molecules.

In this chapter, we emphasize the roles of the MAP kinase cascade that have been revealed by various approaches. Of course, care must be taken to interpret the results obtained by using just one of the approaches. For example, dual-specificity phosphatases may deactivate not only classical MAP kinase but also SAPK/JNK and p38/MPK2, and constitutively active mutants might activate other incorrect downstream molecules because mutations might cause their loss of substrate specificity. However, selecting suitable methods or a combination of multiple approaches should reduce false interpretations. Developing new approaches may also be required for a complete understanding of the signal transduction pathways and the functions of the MAP kinase superfamily molecules.

References

Ahn, N., Seger, R., Bratlien, R. L., Diltz, C. D., Tonks, N. K., and Krebs, E. G. (1991). Multiple components in an epidermal growth factor-stimulated protein kinase cascade. *J. Biol. Chem.* **266**, 4220–4227.

Alberola-Ila, J., Forbush, K. A., Seger, R., Krebs, E. G., and Perlmutter, R. M. (1995). Selective requirement for MAP kinase activation in thymocyte differentiation. *Nature* **373**, 620–623.

Alessi, D. R., Gomez, N., Moorhead, G., Lewis, T., Keyse, S. M., and Cohen, P. (1995). Inactivation of p42 MAP kinase by protein phosphatase 2A and a protein tyrosine phosphatase, but not CL100, in various cell lines. *Curr. Biol.* **5**, 283–295.

Alessi, D. R., Saito, Y., Campbell, D. G., Cohen, P., Sithanandam, G., Rapp, U., Ashworth, A., Marshall, C. J., and Cowley, S. (1994). Identification of the sites in MAP kinase kinase-1 phosphorylated by p74^{raf-1}. *EMBO J.* **13**, 1610–1619.

Amaya, E., Musci, T. J., and Kirschner, M. W. (1991). Expression of a dominant negative mutant of the FGF receptor disrupts mesoderm formation in Xenopus embryos. *Cell* **66**, 257–270.

Anderson, N. G., Maller, J. L., Tonks, N. K., and Sturgill, T. W. (1990). Requirement for integration of signals from two distinct phosphorylation pathways for activation of MAP kinase. *Nature* **343**, 651–653.

Ashworth, A., Nakielny, S., Cohen, P., and Marshall, C. (1992). The amino acid sequence of a mammalian MAP kinase kinase. *Oncogene* **7**, 2555–2556.

Avruch, J., Zhang, X.-F., and Kyriakis, J. M. (1994). Raf meets Ras: Completing the framework of a signal transduction pathway. *Trends Biochem. Sci.* **19**, 279–283.

Bar-Sagi, S., and Feramisco, J. R. (1985). Microinjection of the ras oncogene protein into PC12 cells induces morphological differentiation. *Cell* **42**, 841–848.

Blumer, K. J., and Johnson, G. L. (1994). Diversity in function and regulation of MAP kinase pathways. *Trends Biochem. Sci.* **19**, 236–240.

Boulton, T. G., Nye, S. H., Robbins, D., Ip, N. Y., Radziejewska, E., Morgenbesser, S. D., DePinho, R. A., Panayotatos, N., Cobb, M. H., and Yancopoulos, G. D. (1991). ERKs: A family of protein-serine/threonine kinases that activated and tyrosine phosphorylated in response to insulin and NGF. *Cell* **65**, 663–675.

Boulton, T. G., Yancopoulos, G. D., Gregory, J. S., Slaughter, C., Moomow, C., Hsu, J., and Cobb, M. H. (1990). An insulin-stimulated protein kinase similar to yeast kinases involved in cell cycle control. *Science* **249**, 64–67.

Brondello, J. M., McKenzie, F. R., Sun, H., Tonks, N. K., and Pouyssegur, J. (1995). Constitutive MAP kinase phosphatase (MPK-1) expression blocks G1 specific gene transcription and S-phase entry in fibroblasts. *Oncogene* **10**, 1895–1904.

Brott, B. K., Alessandrini, A., Largaespada, D. A., Copeland, N. G., Jenkins, N. A., Crews, C. M., and Erikson, R. L. (1993). MEK2 is a kinase related to MEK1 and is differentially expressed in murine tissues. *Cell Growth Differ.* **4**, 921–929.

Cobb, M. H., Boulton, T. G., and Robbins, D. J. (1991). Extracellular signal-regulated kinases: ERKs in progress. *Cell Regul.* **2**, 965–978.

Coso, O. A., Chiariello, M., Yu, J.-C., Teramoto, H., Crespo, P., Xu, N., Miki, T., and Gutkind, J. S. (1995). The small GTP-binding proteins Rac1 and Cdc42 regulate the activity of the JNK/SAPK signaling pathway. *Cell* **81**, 1137–1146.

Cowley, S., Paterson, H., Kemp, P., and Marshall, C. J. (1994). Activation of MAP kinase kinase is necessary and sufficient for PC12 differentiation and for transformation of NIH3T3 cells. *Cell* **77**, 841–852.

Crews, C. M., Alessandrini, A., and Erikson, R. L. (1992). The primary structure of MEK, a protein kinase that phosphorylates the *ERK* gene product. *Science* **258**, 478–480.

Darr, I., Yew, N., and Vande Woude, G. F. (1993). Inhibition of mos-induced oocyte maturation by protein kinase A. *J. Cell Biol.* **120**, 1197–1202.

Daum, G., Eisenmann-Tappe, I., Fries, H.-W., Troppmair, J., and Rapp, U. R. (1994). The ins and outs of Raf kinases. *Trends Biochem. Sci.* **19**, 474–480.

Davis, R. J. (1993). The mitogen-activated protein kinase signal transduction pathway. *J. Biol. Chem.* **268**, 14553–14556.

Davis, R. J. (1994). MAPKs: New JNK expands the group. *Trends Biochem. Sci.* **19**, 470–473.

Dent, P., Haser, W., Haystead, T. A. J., Vincent, L. A., Roberts, T. M., and Sturgill, T. W. (1992). Activation of mitogen-activated protein kinase kinase by v-raf in NIH 3T3 cells and in vitro. *Science* **257**, 1404–1406.

Derijard, B., Hibi, M., Wu, I.-H., Barrett, T., Su, B., Deng, T., Karin, M., and Davis, R. J. (1994). JNK1: A protein kinase stimulated by UV light and Ha-ras that binds and phosphorylates the c-Jun activation domain. *Cell* **76**, 1025–1037.

Derijard, B., Raingeaud, J., Barrett, T., Wu, I.-H., Han, J., Ulevitch, R. J., and Davis, R. J. (1995). Independent human MAP kinase signal transduction pathways defined by MEK and MKK isoforms. *Science* **267**, 682–685.

Eisenmann, D. M., and Kim, S. K. (1994). Signal transduction and cell fate specification during Caenorhabditis elegans vulval development. *Curr. Opin. Genet. Dev.* **4**, 508–516.

Fisher, R. P., and Morgan, D. O. (1994). A novel cyclin associates with MO15/CDK7 to form the CDK-activating kinase. *Cell* **78**, 713–724.

Freeman, R. S., Kanki, J. P., Ballantyne, S. M., Pickham, K. M., and Donoghue, D. J. (1990). Effects of the v-mos oncogene on Xenopus development: Meiotic induction in oocytes and mitotic arrest in cleaving embryos. *J. Cell Biol.* **111**, 533–541.

Frequet, D., Labbe, J. C., Derancourt, J., Capony, J. P., Galas, S., Girard, F., Lorca, T., Shuttleworth, J., Doree, M., and Cavadore, J. C. (1993). The MO15 gene encodes the catalytic subunit of a protein kinase that activates cdc2 and other cyclin-dependent kinases (CDKs) through phosphorylation of Thr161 and its homologues. *EMBO J.* **12**, 3111–3121.

Freshney, N. W., Rawlinson, L., Guesdon, F., Jones, E., Cowley, S., Hsuan, J., and Saklatvala, J. (1994). Interleukin-1 activates a novel protein kinase cascade that results in the phosphorylation of hsp27. *Cell* **78**, 1039–1049.

Fukuda, M., Gotoh, Y., Tachibana, T., Dell, K., Hattori, S., Yoneda, Y., and Nishida, E. (1995). Induction of neurite outgrowth by MAP kinase in PC12 cells. *Oncogene* **11**, 239–244.

Gardner, A. M., Vaillancourt, R. R., Lange-Carter, C. A., and Johnson, G. L. (1994). MEK-1 phosphorylation by MEK kinase, Raf, and mitogen-activated protein kinase: Analysis of phosphopeptides and regulation of activity. *Mol. Biol. Cell* **5**, 193–201.

Gautier, J., Solomon, M. J., Booher, R. N., Bazan, J. F., and Kirschner, M. W. (1991). cdc25 is a specific tyrosine phosphatase that directly activates p34cdc2. *Cell* **67**, 197–211.

Gomez, N., and Cohen, P. (1991). Dissection of the protein kinase cascade by which nerve growth factor activates MAP kinases. *Nature* **353**, 170–173.

Gotoh, Y., Masuyama, N., Dell, K., Shirakabe, K., and Nishida, E. (1995a). Initiation of Xenopus oocyte maturation by activation of the mitogen-activated protein kinase cascade. *J. Biol. Chem.* **270**, 25898–25904.

Gotoh, Y., Masuyama, N., Suzuki, A., Ueno, N., and Nishida, E. (1995b). Involvement of the MAP kinase cascade in Xenopus mesoderm induction. *EMBO J.* **14**, 2491–2498.

Gotoh, Y., Matsuda, S., Takenaka, K., Hattori, S., Iwamatsu, A., Ishikawa, M., Kosako, H., and Nishida, E. (1994). Characterization of recombinant Xenopus MAP kinase kinases mutated at potential phosphorylation sites. *Oncogene* **9**, 1891–1898.

Gotoh, Y., Moriyama, K., Matsuda, S., Okumura, E., Kishimoto, T., Kawasaki, H., Suzuki, K., Yahara, I., Sakai, H., and Nishida, E. (1991b). Xenopus M phase MAP kinase: Isolation of its cDNA and activation by MPF. *EMBO J.* **10**, 2661–2668.

Gotoh, Y., Nishida, E., Matsuda, S., Shiina, N., Kosako, H., Shiokawa, K., Akiyama, T., Ohta, K., and Sakai, H. (1991a). In vitro effects on microtubule dynamics of purified Xenopus M phase-activated MAP kinase. *Nature* **349**, 251–254.

Gotoh, Y., Nishida, E., Yamashita, T., Hoshi, M., Kawakami, M., and Sakai, H. (1990). Microtubule-associated-protein (MAP) kinase activated by nerve growth factor and epidermal growth factor in PC12 cells. *Eur. J. Biochem.* **193**, 661–669.

Guan, K.-L., and Butch, E. (1995). Isolation and characterization of a novel dual specific phosphatase, HVH2, which selectively dephosphorylates the mitogen-activated protein kinase. *J. Biol. Chem.* **270**, 7197–7203.

Haccard, O., Lewellyn, A., Hartley, R. S., Erikson, E., and Maller, J. L. (1995). Induction of Xenopus oocyte meiotic maturation by MAP kinase. *Dev. Biol.* **168**, 677–682.

Haccard, O., Sarcevic, B., Lewellyn, A., Hartley, R., Roy, L., Izumi, T., Erikson, E., and Maller, J. L. (1993). Induction of metaphase arrest in cleaving Xenopus embryos by MAP kinase. *Science* **262**, 1262–1265.

Hagag, N., Halegoua, S., and Viola, M. (1986). Inhibition of growth factor-induced differentiation of PC12 cells by microinjection of antibody to ras p21. *Nature* **319**, 680–682.

Han, J., Lee, J.-D., Bibbs, L., and Ulevitch, R. J. (1994). A MAP kinase targeted by endotoxin and hyperosmolarity in mammalian cells. *Science* **265**, 808–811.

Haystead, C. M. M., Gregory, P., Shirazi, A., Fadden, P., Mosse, C., Dent, P., and Haystead, T. A. J. (1994). Insulin activates a novel adipocyte mitogen-activated protein kinase kinase kinase that shows rapid phasic kinetics and is distinct from c-Raf. *J. Biol. Chem.* **269**, 12804–12808.

Herskowitz, I. (1995). MAP kinase pathways in yeast: For mating and more. *Cell* **80**, 187–197.

Hibi, M., Lin, A., Smeal, T., Minden, A., and Karin, M. (1993). Identification of an oncoprotein- and UV-responsive protein kinase that binds and potentiates the c-Jun activation domain. *Genes Dev.* **7**, 2135–2148.

Hoshi, M., Nishida, E., and Sakai, H. (1988). Activation of a Ca^{2+}-inhibitable protein kinase that phosphorylates microtubule-associated protein 2 in vitro by growth factors, phorbol esters, and serum in quiescent cultured human fibroblasts. *J. Biol. Chem.* **263**, 5396–5401.

Howe, L. R., Leevers, S. J., Gomez, N., Nakielny, S., Cohen, P., and Marshall, C. J. (1992). Activation of the MAP kinase pathway by the protein kinase raf. *Cell* **71**, 335–342.

Huang, W., Kessler, D. S., and Erikson, R. L. (1995). Biochemical and biological analysis of Mek1 phosphorylation site mutants. *Mol. Biol. Cell* **6**, 237–245.

Kessler, D. S., and Melton, D. A. (1994). Vertebrate embryonic induction: Mesodermal and neural patterning. *Science* **266**, 596–604.

Keyse, S. M. (1995). An emerging family of dual specificity MAP kinase phosphatases. *Biochim. Biophys. Acta* **1265**, 152–160.

Kosako, H., Gotoh, Y., and Nishida, E. (1994a). Requirement for the MAP kinase kinase/ MAP kinase cascade in Xenopus oocyte maturation. *EMBO J.* **13**, 2131–2138.

Kosako, H., Gotoh, Y., and Nishida, E. (1994b). Mitogen-activated protein kinase kinase is required for the Mos-induced metaphase arrest. *J. Biol. Chem.* **269**, 28354–28358.

Kosako, H., Nishida, E., and Gotoh, Y. (1993). cDNA cloning of MAP kinase kinase reveals kinase cascade pathways in yeast to vertebrates. *EMBO J.* **12**, 787–794.

Kumagai, A., and Dunphy, W. G. (1991). The cdc25 protein controls tyrosine dephosphorylation of the cdc2 protein in a cell-free system. *Cell* **64**, 903–914.

Kwak, S. P., and Dixon, J. E. (1995). Multiple dual specificity protein tyrosine phosphatases are expressed and regulated differentially in liver cell lines. *J. Biol Chem.* **270**, 1156–1160.

Kyriakis, J. M., and Avruch, J. (1990). pp54 microtubule-associated protein 2 kinase. A novel serine/threonine protein kinase regulated by phosphorylation and stimulated by poly-L-lysine. *J. Biol. Chem.* **265**, 17355–17363.

Kyriakis, J. M., Brautigan, D. L., Ingebritsen, T. S., and Avruch, J. (1991). pp54 microtubule-associated protein-2 kinase requires both tyrosine and serine/threonine phosphorylation for activity. *J. Biol. Chem.* **266**, 10043–10046.

Kyriakis, J. M., App, H., Zhang, Z.-F., Banerjee, P., Brautigan, D. L., Rapp, U., and Avruch, J. (1992). Raf-1 activates MAP kinase-kinase. *Nature* **358**, 417–421.

Kyriakis, J. M., Banerjee, P., Nikolakaki, E., Dai, T., Rubie, E. A., Ahmad, M. F., Avruch, J., and Woodgett, J. R. (1994). The stress-activated protein kinase subfamily of c-Jun kinases. *Nature* **369**, 156–160.

Labonne, C., and Whitman, M. (1994). Mesoderm induction by activin requires FGF mediated intracellular signals. *Development* **120**, 463–472.

Lange-Carter, C. A., Pleiman, C. M., Gardner, A. M., Blumer, K. J., and Johnson, G. L. (1993). A divergence in the MAP kinase regulatory network defined by MEK kinase and Raf. *Science* **260**, 315–319.

Lee, J. C., Laydon, J. T., McDonnell, P. C., Gallagher, T. F., Kumar, S., Green, D., McNulty, D., Blumenthal, M. J., Heys, J. R., Landvatter, S. W., Strickler, J. E., McLaughlin, M. M., Siemens, I. R., Fisher, S. M., Livi, G. P., White, J. R., Adams, J. L., and Young, P. R. (1994). A protein kinase involved in the regulation of inflammatory cytokine biosynthesis. *Nature* **372**, 739–746.

Levin, D. E., and Errede, B. (1995). The proliferation of MAP kinase signaling pathways in yeast. *Curr. Opin. Cell Biol.* **7**, 197–202.

Lohka, M. J. (1989). Mitotic control by metaphase-promoting factor and cdc proteins. *J. Cell Sci.* **92**, 131–135.

MacNicol, A. M., Muslin, A. J., and Williams, L. T. (1993). Raf-1 kinase is essential for early Xenopus development and mediates the induction of mesoderm by FGF. *Cell* **73**, 571–584.

Makela, T. P., Tassan, J. P., Nigg, E. A., Frutiger, S., Hughes, G. J., and Weinberg, R. A. (1994). A cyclin associated with the CDK-activating kinase MO15. *Nature* **371**, 254–257.

Maller, J. L. (1991). Mitotic control. *Curr. Opin. Cell Biol.* **3**, 269–275.

Mansour, S. J., Matten, W. T., Hermann, A. S., Candia, J. M., Rong, S., Fukasawa, K., Vande Woude, G. F., and Ahn, N. G. (1994). Transformation of mammalian cells by constitutively active MAP kinase kinase. *Science* **265**, 966–970.

Matsuda, S., Gotoh, Y., and Nishida, E. (1993). Phosphorylation of Xenopus mitogen-activated protein (MAP) kinase kinase by MAP kinase kinase kinase and MAP kinase. *J. Biol. Chem.* **268**, 3277–3281.

Matsuda, S., Kawasaki, H., Moriguchi, T., Gotoh, Y., and Nishida, E. (1995). Activation of protein kinase cascades by osmotic shock. *J. Biol. Chem.* **270**, 12781–12786.

Matsuda, S., Kosako, H., Takenaka, K., Moriyama, K., Sakai, H., Akiyama, T., Gotoh, Y., and Nishida, E. (1992). Xenopus MAP kinase activator: Identification and function as a key intermediate in the phosphorylation cascade. *EMBO J.* **11**, 973–982.

Minden, A., Lin, A., Claret, F.-X., Abo, A., and Karin, M. (1995). Selective activation of the JNK signaling cascade and c-Jun transcriptional activity by the small GTPases Rac and Cdc42Hs. *Cell* **81**, 1147–1157.

Minden, A., Lin, A., McMahon, M., Lange-Carter, C., Derijard, B., Davis, R. J., Johnson, G. L., and Karin, M. (1994). Differential activation of ERK and JNK mitogen-activated protein kinases by Raf-1 and MEKK. *Science* **266**, 1719–1723.

Misra-Press, A., Rim, C. S., Yao, H., Roberson, M. S., and Stork, P. J. S. (1995). A novel mitogen-activated protein kinase phosphatase. *J. Biol. Chem.* **270**, 14587–14596.

Mizoguchi, T., Gotoh, Y., Nishida, E., Yamaguchi-Shinozaki, K., Hayashida, N., Iwasaki, T., Kawada, H., and Shinozaki, K. (1994). Characterization of two cDNAs that encode MAP kinases in Arabidopsis thaliana and the role of auxin in activating such kinases in plants. *Plant J.* **5**, 111–122.

Mohit, A. A., Martin, J. H., and Miller, C. A. (1995). p49[3F12] kinase: A novel MAP kinase expressed in a subset of neurons in the human nervous system. *Neuron* **14**, 67–78.

Moriguchi, T., Gotoh, Y., and Nishida, E. (1995b). Activation of two isoforms of mitogen-activated protein kinase kinase in response to epidermal growth factor and nerve growth factor. *Eur. J. Biochem.* **234**, 32–38.

Moriguchi, T., Kawasaki, H., Matsuda, S., Gotoh, Y., and Nishida, E. (1995a). Evidence for multiple activators for stress-activated protein kinase/c-Jun amino-terminal kinases. Existence of novel activators. *J. Biol. Chem.* **270**, 12969–12972.

Nakielny, S., Campbell, D. G., and Cohen, P. (1992). MAP kinase kinase from rabbit skeletal muscle. *FEBS Lett.* **308**, 183–189.

Nebreda, A. R., and Hunt, T. (1993). The c-mos proto-oncogene protein kinase turns on and maintains the activity of MAP kinase, but not MPF, in cell-free extracts of *Xenopus* oocytes and eggs. *EMBO J.* **12**, 1979–1986.

Nieuwkoop, P. D., and Faber, J. (1967). "Normal Table of Xenopus laevis (Daudin)." North-Holland, Amsterdam.

Nishida, E., and Gotoh, Y. (1993). The MAP kinase cascade is essential for diverse signal transduction pathways. *Trends Biochem. Sci.* **18**, 128–131.

Noda, M., Ko, M., Ogura, A., Liu, D., Amano, T., Takano, T., and Ikawa, Y. (1985). Sarcoma viruses carrying ras oncogenes induce differentiation-associated properties in a neuronal cell line. *Nature* **318**, 73–75.

Nurse, P. (1990). Universal control mechanism regulating onset of M-phase. *Nature* **344**, 503–508.

Pages, G., Lenormand, P., L'Allemain, G., Chambard, J.-C., Meloche, S., and Pouyssegur, J. (1993). Mitogen-activated protein kinases p42[mapk] and p44[mapk] are required for fibroblast proliferation. *Proc. Natl. Acad. Sci. U.S.A.* **90**, 8319–8323.

Pang, L., Sawada, T., Decker, S. J., and Saltiel, A. R. (1995). Inhibition of MAP kinase kinase blocks the differentiation of PC-12 cells induced by nerve growth factor. *J. Biol. Chem.* **270**, 13585–13588.

Payne, D. M., Rossomando, A. J., Martino, P., Erickson, A. K., Her, J.-H., Shabanowitz, J., Hunt, D. F., Weber, M. J., and Sturgill, T. W. (1991). Identification of the regulatory phosphorylation sites in pp42/mitogen-activated protein kinase (MAP kinase). *EMBO J.* **10**, 885–892.

Pelech, S. L., and Sanghera, J. S. (1992). MAP kinases: Charting the regulatory pathways. *Science* **257**, 1355–1356.

Perrimon, N., and Desplan, C. (1994). Signal transduction in the early Drosophila embryo: When genetics meets biochemistry. *Trends Biochem. Sci.* **19**, 439–518.

Poon, R. Y. C., Yamashita, K., Adamczewski, J. P., Hunt, T., and Shuttleworth, J. (1993). The cdc2-related protein p40MO15 is the catalytic subunit of a protein kinase that can activate p33cdk2 and p34cdc2. *EMBO J.* **12**, 3123–3132.

Posada, J., Yew, N., Ahn, N. G., Vande Woude, G. F., and Cooper, J. A. (1993). Mos stimulates MAP kinase in *Xenopus* oocytes and activates a MAP kinase kinase in vitro. *Mol. Cell. Biol.* **13**, 2546–2553.

Ray, L. B., and Sturgill, T. W. (1987). Rapid stimulation by insulin of a serine/threonine kinase in3T3-L1 adipocytes that phosphorylates microtubule-associated protein 2 in vitro. *Proc. Natl. Acad. Sci. U.S.A.* **84**, 1502–1506.

Ray, L. B., and Sturgill, T. W. (1988). Characterization of insulin-stimulated microtubule-associated protein kinase. Rapid isolation and stabilization of a novel serine/threonine kinase from 3T3-L1 cells. *J. Biol. Chem.* **263**, 12721–12727.

Reuter, C. W. M., Catling, A. D., Jelinek, T., and Weber, M. J. (1995). Biochemical analysis of MEK activation in NIH3T3 fibroblasts. Identification of B-Raf and other activators. *J. Biol. Chem.* **270**, 7644–7655.

Rouse, J., Cohen, P., Trigon, S., Morange, M., Alonso-Llamazares, A., Zamanillo, D., Hunt, T., and Nebreda, A. R. (1994). A novel kinase cascade triggered by stress and heat shock that stimulates MAPKAP kinase-2 and phosphorylation of the small heat shock proteins. *Cell* **78**, 1027–1037.

Sagata, N., Oskarsson, M., Copeland, T., Brumbaugh, J., and Vande Woude, G. F. (1988). Function of c-mos proto-oncogene product in meiotic maturation in Xenopus oocytes. *Nature* **335**, 519–525.

Sagata, N., Watanabe, N., Vande Woude, G. F., and Ikawa, Y. (1989). The c-mos proto-oncogene product is a cytostatic factor responsible for meiotic arrest in vertebrate eggs. *Nature* **342**, 512–518.

Sale, E. M., Atkinson, P. G. P., and Sale, G. J. (1995). Requirement of MAP kinase for differentiation of fibroblasts to adipocytes, for insulin activation of p90 S6 kinase and for insulin or serum stimulation of DNA synthesis. *EMBO J.* **14**, 674–684.

Sanchez, I., Hughes, R. T., Mayer, B. J., Yee, K., Woodgett, J. R., Avruch, J., Kyriakis, J. M., and Zon, L. I. (1994). Role of SAPK/ERK kinase-1 in the stress-activated pathway regulating transcription factor c-Jun. *Nature* **372**, 794–798.

Seger, R., Ahn, N. G., Posada, J., Munar, E. S., Jensen, A. M., Cooper, J. A., Cobb, M. H., and Krebs, E. G. (1992a). Purification and characterization of mitogen-activated protein kinase activator(s) from epidermal growth factor-stimulated A431 cells. *J. Biol. Chem.* **267**, 14373–14381.

Seger, R., Seger, D., Lozeman, F. J., Ahn, N. G., Graves, L. M., Campbell, J. S., Ericsson, L., Harrylock, M., Jensen, A. M., and Krebs, E. G. (1992b). Human T-cell mitogen-activated protein kinase kinases are related to yeast signal transduction kinases. *J. Biol. Chem.* **267**, 25628–25631.

Shirakabe, K., Gotoh, Y., and Nishida, E. (1992). A mitogen-activated protein (MAP) kinase activating factor in mammalian mitogen-stimulated cells is homologous to *Xenopus* M phase MAP kinase activator. *J. Biol. Chem.* **267**, 16685–16690.

Sluss, H. K., Barrett, T., Derijard, B., and Davis, R. J. (1994). Signal transduction by tumor necrosis factor mediated by JNK protein kinases. *Mol. Cell. Biol.* **14**, 8376–8384.

Solomon, M. J., Harper, J. W., and Shuttleworth, J. (1993). CAK, the p34cdc2-activating kinase, contains a protein identical or closely related to p40MO15. *EMBO J.* **12**, 3133–3142.

Sturgill, T. W., and Wu, J. (1991). Recent progress in characterization of protein kinase cascades for phosphorylation of ribosomal protein S6. *Biochim. Biophys. Acta* **1092**, 350–357.

Su, B., Jacinto, E., Hibi, M., Kallunki, T., Karin, M., and Ben-Neriah, Y. (1994). JNK is involved in signal integration during costimulation of T lymphocytes. *Cell* **77**, 727–736.

Sun, H., Charles, C. H., Lau, L. F., and Tonks, N. K. (1993). MKP-1 (3CH134), an immediate early gene product, is a dual specificity phosphatase that dephosphorylates MAP kinase in vivo. *Cell* **75**, 487–493.

Sun, H., Tonks, N. K., and Bar-Sagi, D. (1994). Inhibition of Ras-induced DNA synthesis by expression of the phosphatase MKP-1. *Science* **266**, 285–288.

Szeberenyi, J., Cai, H., and Cooper, G. M. (1990). Effect of a dominant inhibitory Ha-ras mutation on neuronal differentiation of PC12 cells. *Mol. Cell. Biol.* **10**, 5324–5332.

Traverse, S., and Cohen, P. (1994). Identification of latent MAP kinase kinase kinase in PC12 cells as B-raf. *FEBS Lett.* **350**, 13–18.

Traverse, S., Gomez, N., Paterson, H., Marshall, C., and Cohen, P. (1992). Sustained activation of the mitogen-activated protein (MAP) kinase cascade may be required for differentiation of PC12 cells. Comparison of the effects of nerve growth factor and epidermal growth factor. *Biochem. J.* **288**, 351–355.

Traverse, S., Seedorf, K., Paterson, H., Marshall, C. J., Cohen, P., and Ullrich, A. (1994). EGF triggers neuronal differentiation of PC12 cells that overexpress the EGF receptor. *Curr. Biol.* **4**, 694–701.

Whitman, M., and Melton, D. A. (1992). Involvement of p21 ras in Xenopous mesoderm induction. *Nature* **357**, 252–255.

Wu, J., Harrison, J. K., Dent, P., Lynch, K. R., Weber, M. J., and Sturgill, T. W. (1993b). Identification and characterization of a new mammalian mitogen-activated protein kinase kinase, MKK2. *Mol. Cell. Biol.* **13**, 4539–4548.

Wu, J., Harrison, J. K., Vincent, L. A., Haystead, C., Haystead, T., Michel, H., Hunt, D. F., Lynch, K. R., and Sturgill, T. W. (1993a). Molecular structure of a protein-tyrosine/threonine kinase activating p42 mitogen-activated protein (MAP) kinase: MAP kinase kinase. *Proc. Natl. Acad. Sci. U.S.A.* **90**, 173–177.

Yan, M., Sai, T., Deak, J. C., Kyriakis, J. M., Zon, L. I., Woodgett, J. R., and Templeton, D. J. (1994). Activation of stress-activated protein kinase by MEKK1 phosphorylation of its activator SEK1. *Nature* **372**, 798–800.

Yashar, B. M., Kelley, C., Yee, K., Errede, B., and Zon, L. I. (1993). Novel members of the mitogen-activated protein kinase activator family in *Xenopus laevis. Mol. Cell. Biol.* **13**, 5738–5748.

Yew, N., Mellini, M. L., and Vande Woude, G. F. (1992). Meiotic initiation by the mos protein in Xenopus. *Nature* **355**, 649–652.

Zheng, C.-F., and Guan, K.-L. (1993). Cloning and characterization of two distinct human extracellular signal-regulated kinase activator kinases, MEK1 and MEK2. *J. Biol. Chem.* **268**, 11435–11439.

Zheng, C.-F., and Guan, K.-L. (1994). Activation of MEK family kinases requires phosphorylation of two conserved Ser/Thr residues. *EMBO J.* **13**, 1123–1131.

Zhou, G., Bao, Z. Q., and Dixon, J. E. (1995). Components of a new human protein kinase signal transduction pathway. *J. Biol. Chem.* **270**, 12665–12669.

Tadaomi Takenawa*
Kenji Miura†
Hiroaki Miki*
Kazutada Watanabe‡

*Department of Biochemistry
Institute of Medical Science
University of Tokyo
Shirokanedai, Minato-ku, Tokyo 108, Japan

†Department of Biochemistry
Defense Medical College
Namiki, Tokorozawa, Saitama 359, Japan

‡Department of Experimental Biology
Tokyo Metropolitan Institute of Gerontology
Sakae-cho, Itabashi-ku, Tokyo 173, Japan

Signal Transductions of SH2/SH3: Ash/Grb-2 Downstream Signaling

I. Introduction

Extracellular stimulation of receptor tyrosine kinases elicits a wide range of intracellular changes leading to cell proliferation or differentiation. Binding of growth factors to receptors triggers homodimerization of the receptors, followed by the autophosphorylation at specific tyrosine residues in the receptors. Then the phosphotyrosine residues are directly recognized by the SH2 domains of cytoplasmic signaling proteins. This step is essential for tyrosine kinase signaling. In addition to SH2 domains, many proteins regulated by receptor tyrosine kinases have SH3 domains that associate with the molecules downstream of tyrosine kinases in the signaling pathways. A variety of proteins have been shown to have SH2 and SH3 domains (Fig. 1). These proteins are divided into two groups. One is SH2/SH3 proteins that have enzymatic activities, such as phospholipase Cγ, Src family tyrosine

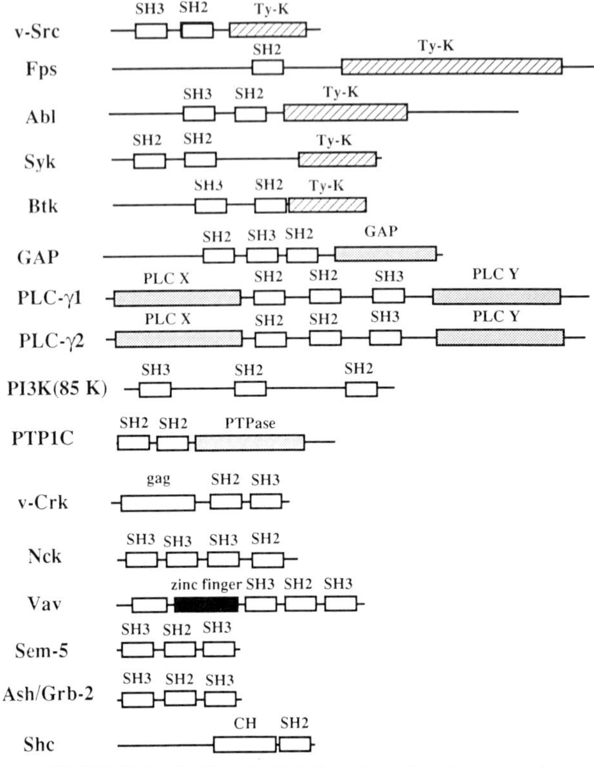

FIGURE I   SH2 and SH3 domains of various proteins.

kinase, and Ras GAP. The other group is proteins that only consist of SH2 and SH3 domains such as Nck, c-Crk, and Ash/Grb-2.

A molecule mediating the signal from a receptor tyrosine kinase to the Ras guanine nucleotide-binding protein was first identified genetically in the nematode worm *Caenorhabditis elegans*. Vulval formation of *C. elegans* is initiated by a signal from a gonadal cell to vulval precursor cells. Many genes, including *let-23* and *let-60,* are required for this differentiation event (Horvitz and Sternberg, 1991). *let-23* and *let-60* genes encode an EGF receptor-like tyrosine kinase (Aroian *et al.*, 1990) and a Ras protein (Han and Sternberg, 1990), respectively. Thus, signal transduction in vulval formation has been implicated in the Ras signaling. A novel gene isolated by Clark *et al.* (1992) was a key molecule in the signaling. The gene, *sem-5*, was essential for vulval formation and encoded Sem-5 protein, which is almost exclusively composed of a single SH2 domain and two SH3 domains. Because SH2 domains recognized phosphotyrosine residues of tyrosine kinase receptors, Sem-5 became the most likely candidate for coupling the tyrosine kinase receptor encoded by *let-23* with Ras signaling. A mammalian counterpart of Sem-5, Ash/Grb-2, was identified independently by two groups using different strategies. Matuoka *et al.* (1992) cloned the cDNAs from rat and human by hybridization with mixed oligonucleotides encoding a consensus sequence to SH2 domain. However, Lowenstein *et al.* (1992) utilized the CORT method for the human cDNA in which the phosphory-lated tyrosine residues in the EGF receptors were used as a probe and λgt11 phages were screened for the proteins binding to the probe. The evidence that Ash/Grb-2 links tyrosine kinase receptors with Ras signaling has been obtained from the direct association of Ash/Grb-2 with Sos (a guanine nucleotide exchange factor for Ras) (Chardin, 1993; Egan *et al.*, 1993; Li *et al.*, 1993; Olivier *et al.*, 1993; Rozakis-Adcock *et al.*, 1993; Simon *et al.*, 1993). Thus, it has been demonstrated that Ash/Grb-2 is the adaptor mole-cule that physically associates with Sos, allowing ligand-activated tyrosine kinase receptors to modulate Ras activity.

Studies have shown that injection of the antibody against Ash/Grb-2 into cells abolishes the reorganization of actin stress fibers (Matsuoka *et al.*, 1993) that is controlled by Rho and Rac proteins (Ridley and Hall, 1992; Ridley *et al.*, 1992). In addition, dynamin is a GTPase that is required for synthetic vesicle endocytosis as indicated by its identity to the product of the *Drosophila* shibire gene. It has been found to bind to the Ash SH3 domain through its carboxy-terminal proline-rich region (Gout *et al.*, 1993; Herskovits *et al.*, 1993; Miki *et al.*, 1994). These findings raise the possibility that Ash/Grb2 functions not only in Ras signaling but also in other pathways. When Ash-binding proteins in rat brain were investigated, two other brain-specific proteins, synapsin I and p145, were found to bind the Ash SH3 domain (McPherson *et al.*, 1994a). Synapsin I is a synaptic vesicle-associated nerve-terminal protein that interacts with actin and is thought to mediate the interactions of synaptic vesicles with the presynaptic cytomatrix. How-ever, p145 is present exclusively in neurons and colocalizes with dynamin, suggesting the important roles in endocytosis (McPherson *et al.*, 1994b).

II. Ash Downstream Signaling in Brain

A. Isolation of Ash-Binding Proteins from Bovine Brain

To clarify the downstream signalings of Ash, we tried to discover pro-teins specifically bound to Ash using bovine brain. Bovine brain cytosol fractions were applied to a GST–Ash–glutathione agarose column to purify Ash-binding proteins. Proteins that bound to the column were eluted with glutathione and separated by SDS–PAGE. As shown in Figure 2, several proteins were bound to the GST–Ash fusion protein affinity column. 180-, 150-, 130-, 110-, 100-, and 65-kDa proteins were found to be the major

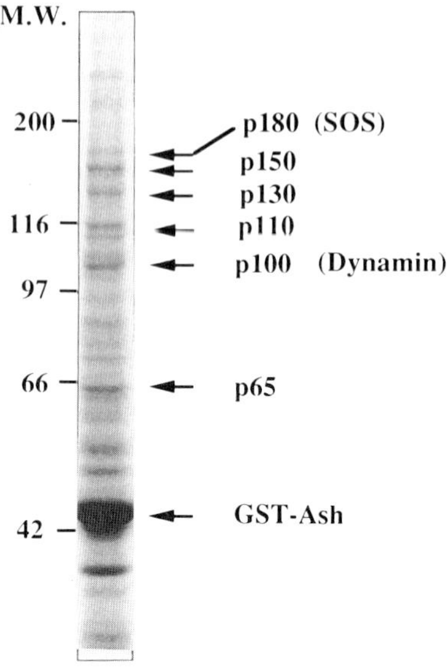

FIGURE 2   Proteins bound to the GST–Ash affinity column. Bovine brain cytosol fractions were applied to a GST–Ash affinity column and the bound proteins were eluted with glutathione. The proteins were separated by SDS–PAGE and stained with Coomassie brilliant blue.

Ash-binding proteins in bovine brain cytosol. Because the 180- and 100-kDa proteins have been identified as Sos and dynamin, respectively, we tried to determine the amino acid sequences of the 150-kDa (p150) and 65-kDa (p65) proteins. After Ash-binding proteins were subjected to SDS–PAGE and transferred to PVDF membranes, the membranes were stained with Ponceau S. The protein bands of p150 and p65 were cut out and digested with lysyl-endopeptidase. The resultant peptides were separated by C18 reverse-phase column chromatography. Two peptides (A and B) from p150 and two peptides (C and D) from p65 were sequenced, and we obtained amino acid sequences from p150 (A,DGARSVSRTIQNNFFD; B,QEAIDVL-LLGNTLNSDLA) and from p65 (C,NPEITTNRFYGPQINNISHT; D,RRD-PPNGPNLPMATV). None of these amino acid sequences showed any homology to other proteins in the protein sequence database.

Therefore, for further characterization of these proteins we prepared antibodies against p150 and p65. Antibodies against p150 and p65 were made by immunizing rabbits with PVDF membranes containing p150 and p65. The sera were affinity purified using p150 or p65 blotted on PVDF membranes. The anti-p150 antibody reacted with a 150-kDa band on immunoblotting of crude bovine brain homogenates and GST–Ash-binding proteins. Similarly, the anti-p65 antibody reacted only with the 65-kDa band on immunoblotting of crude homogenate and GST–Ash-binding proteins. These antibodies also reacted with rat brain p150 and p65.

B. Tissue Distribution of Ash, p65, and p150

To examine the tissue distribution of p150 and p65, various rat tissues were homogenized and centrifuged at 100,000g for 1 hr. The supernatant fractions were subjected to SDS–PAGE and Western blot analyses with anti-Ash, anti-p150, and anti-p65 antibodies. It was found that p150 was expressed exclusively in the brain and was not detected in other tissues such as kidney, stomach, lung, skeletal muscle, or spleen. Similarly, p65 was only detected in brain and not in other tissues. However, Ash was expressed ubiquitously in all tissues examined. These results suggest that p150 and p65 function specifically in the brain, whereas Ash is expressed widely. To further examine the brain distribution of p150 and p65, immunohistochemical studies on rat brain were performed using these antibodies. Cerebral cortex, cerebellum, and hippocampus were stained strongly by anti-Ash, anti-p150, and anti-p65 antibodies. However, there are differences in the subcellular distribution of the three proteins. Ash was present predominantly in cerebellar Purkinje cells with their dendric shaft and satellite cells and hippocampal pyramidal neurons with their apical dendrites. Intense immunoreactivity of p150 was observed in the cell bodies of Purkinje neurons, the molecular layer, satellite cells, basket cells, and mossy fiber terminals in the cerebellum. In the hippocampus, the cell bodies of pyramineurons and interneurons were stained heavily for p150. The immunoreactivity of p65 was strong in the molecular layer with the terminals of the parallel fibers. In the hippocampus, p65 was present predominantly in the pyramidal neurons and their dendrites.

C. p150 and p65 Bind to the SH3 Domain of Ash

To examine the domain of Ash to which p150 and p65 bind, we prepared GST fusion proteins such as GST–Ash (N-SH3), GST–Ash (C-SH3), and GST–Ash (SH2). Affinity columns were prepared with these GST fusion proteins and the proteins binding to each SH domain were checked. p150 and p65 bound to both SH3 domains (N terminal and C terminal), but not to the SH2 domain. Next, to determine whether the binding of these proteins is specific to Ash SH3, we examined binding to the PLCγ1 SH3 and PI 3-kinase SH3 domains. As shown in Figure 3, p150 and p65 bound only to the Ash SH3 domain, suggesting the specific binding of p150 and p65 to the Ash SH3 domain.

D. Ash SH3 Domain-Binding Proteins *in Vivo*

To determine whether p150 and p65 bind to Ash *in vivo*, we prepared a synthetic peptide containing the EGF receptor Ash binding site (PVPEYIN-QSVPK, Y is phosphorylated). The EGF-receptor Ash-binding motif peptide

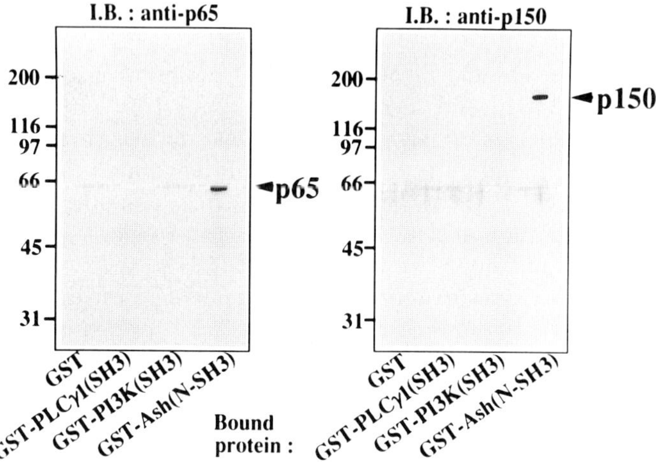

FIGURE 3   Specific binding of p150 and p65 to Ash–SH3. Bovine brain cytosol fractions were applied to GST–Ash(N-SH3), GST–PLC γ1(SH3), and GST–PI 3-kinase (SH3) beads and the bound proteins were analyzed by Western blotting with anti-p150 and anti-p65 antibodies.

was coupled to agarose beads, and bovine brain cytosol fractions were applied to the peptide column. The peptide can be expected to recognize Ash and bind through the Ash SH2 domain. Therefore, proteins other than Ash that absorb to this column are considered to bind through a SH3 domain of Ash. Peptide-bound proteins were separated by SDS–PAGE, transferred to PVDF membranes, and stained with Coomassie brilliant blue. Various protein bands were detected with the major bands at 180, 150, 110, 100, 65, 55, and 28 kDa. These proteins are comparable to those that bound to the GST–Ash affinity column except for Ash. Moreover, we found that the 180- and 100-kDa proteins were Sos and dynamin, respectively, by antibodies against Sos and dynamin. The 150- and 65-kDa proteins were identified as p150 and p65 by their specific antibodies. These data show that all these proteins form complexes with Ash *in vivo*.

E. Possible Function of Ash in Synaptosomes

Considering the fact that dynamin, p150, and p65 are localized in nerve terminals, these proteins may play important roles in the exo/endocytotic cycle of synaptic vesicles. Synapsin I has multiphosphorylation sites and its phosphorylation is increased by the stimuli of Ca^{2+}-dependent release of neurotransmitter such as the depolarization of synaptosomes (Greengard *et al.*, 1993). The phosphorylation of synapsin I causes inhibition of its binding to actin and synaptic vesicles. In contrast to its effect on synapsin I, depolarization of synaptosomes causes the dephosphorylation of dynamin (Robinson *et al.*, 1993). This opposite action may reflect participation in opposite

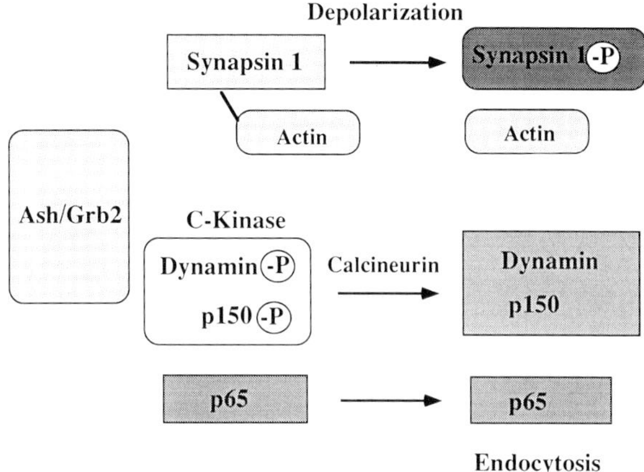

FIGURE 4   Possible functions of Ash-bound proteins in synaptosomes.

stages of the exo/endocytotic cycle of synaptic vesicles. McPherson *et al.* (1994b) reported that p145 is dephosphorylated in parallel with dynamin after synaptosomal depolarization, suggesting that p145/p150, like dynamin, participates in synaptic vesicle endocytosis (Fig. 4). We do not know whether p65 is phosphorylated by depolarization, but it is possible that p65, together with dynamin and p145/p150, is involved in synaptic vesicle transport. If dynamin, p145/p150, and p65 play important roles in synaptic vesicle endocytosis, Ash regulates the functions of these proteins in brain. In the nervous system, Ash forms complexes with Sos, p145/p150, dynamin, synapsin I, and p65 enabling them to function in harmony. In this case, it remains to be determined whether there are any tyrosine kinases upstream of Ash and whether these tyrosine kinases generate signals through Ash leading to collaborative functions of the exo/endocytotic cycle of synaptic vesicles.

III. Splicing Isoforms of Rat Ash/Grb-2

The findings discussed previously raise the possibility that Ash functions not only in Ras signaling but also in other pathways. Because the SH3 domains function to mediate signals downstream of Ash, differences in the binding specificities of the two SH3 domains to various signaling molecules will regulate the signal direction. Therefore, we thought that it would be worthwhile to search for Ash isoforms, especially those having different SH3 domains from the original. Here, we describe the isolation of splicing isoforms of Ash, the generation mechanism of the isoforms as revealed by genomic cloning of the Ash-encoding gene, and the roles of these isoforms in mitogenic signaling.

A. Isolation and Characterization of Ash/Grb-2 Splicing Isoforms

Initial screening of 2×10^6 recombinant phages with ^{32}P-labeled Ash/Grb-2(Ash-l) cDNA as a probe yielded 57 positive clones. Ten recombinant phage DNAs that hybridized to the probe were purified. Among these cDNA clones, 3 did not hybridize to the oligonucleotide derived from the 3'-coding region of the Ash-l nucleotide sequence. Nucleotide sequence analysis showed that 2 clones were identical to each other except for a difference in size and shared 93% identity in nucleotide sequence with Ash-l cDNA. However, the clones do not have long open reading frames because of the frequent appearance of stop codons. We concluded, therefore, that these cDNAs, designed Ash-ψ, were derived from unusual transcribed genes. The third clone, Ash-s cDNA, had an identical nucleotide sequence to that of Ash-l cDNA in the first SH3 domain. However, the coding sequence was terminated at the end of the SH3 domain by a stop codon, after which the sequence differed from that of Ash-l cDNA. Accordingly, the Ash-s isoform is solely composed of a single SH3 domain identical to that of Ash-l. To search for other isoform types, we purified 47 additional phage DNAs that had given positive signals in the first library screening. Then we performed polymerase chain reaction experiments on the DNAs, and these gave 506 bp fragments for Ash-l cDNA. We detected 3 clones generating about 450 bp fragments. Because these clones were identical to one another, we designated them Ash-m cDNA. The coding sequence of Ash-m cDNA is 42 nucleotides shorter than that of Ash-l cDNA. The defective region of Ash-m corresponds to 14 amino acid residues between 157 and 170 of Ash-l, which comprise the amino-terminal region of the second SH3 domain. According to the length of the cDNAs, Ash/Grb-2 and two isoforms are named Ash-l, Ash-m, and Ash-s (Fig. 5).

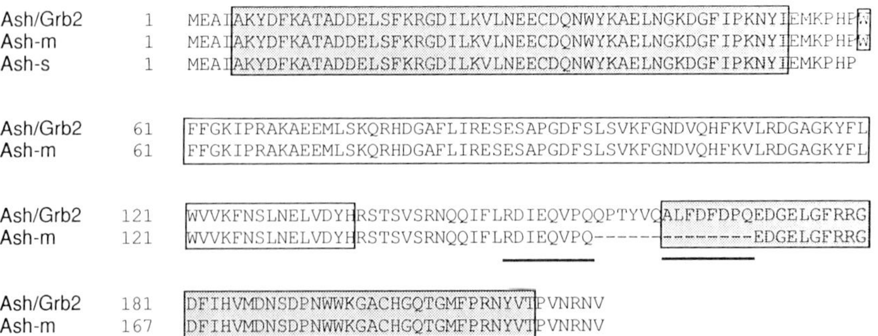

FIGURE 5 Comparison of the amino acid sequences among Ash-l, Ash-m, and Ash-s. The amino acid sequences in open and shaded boxes are the SH2 and SH3 domains of Ash-l, respectively. Ash-m lacks the amino acid sequence of QPTYVQALFDFDPQ.

B. Cloning and Organization of the Ash-l-Encoding Gene and Generation of Ash Isoforms

To confirm that the three isoforms, including Ash-l, arise from a single gene by alternative splicing, we isolated genomic clones of the rat Ash-l encoding gene and analyzed gene organization by restriction mapping and nucleotide sequencing together with genomic Southern blot analysis. Screening of a rat genomic DNA library yielded 37 genomic clones hybridizing to the Ash cDNA probe. These clones were classified into three types by digesting the recombinant phage DNAs with restriction enzymes. One of these, λG–Ash-ψ, corresponded to genomic DNA encoding Ash-ψ. The other two, λG–Ash1 and λG–Ash2, encoded Ash/Grb-2 (Ash-l). The amino- and carboxy-terminal regions were encoded in λG–Ash1 and λG–Ash2, respectively. Although λG–Ash1 and λG–Ash2 do not overlap, the nucleotide sequences of the exons in these two clones account for the entire sequence of Ash-l cDNA. Ash-l gene has a length of more than 16 kb and consists of six exons and five introns (Fig. 6). The nucleotide sequences at the exon–intron boundaries show that all introns start with GT at the 5′ end and terminate with AG at the 3′ end. This conforms to the "GT–AG rule" (Breathnach and Chambon, 1981). The genomic cloning of the gene revealed

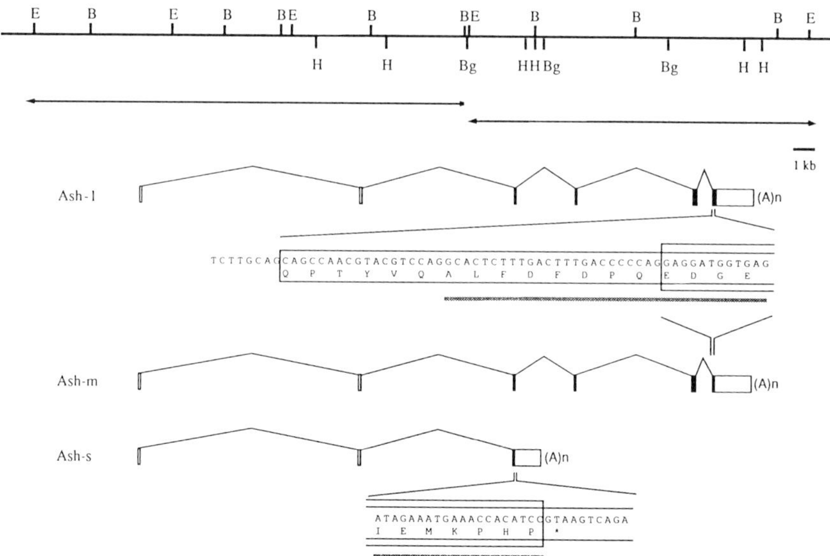

FIGURE 6   Restriction map and organization of the Ash isoform-encoding gene. Exons and introns are shown. Solid and open boxes indicate coding and noncoding exons, respectively. The exon–intron boundaries generating the Ash isoforms are enlarged to indicate the partial sequences around the splicing sites. The boxed nucleotide sequences are parts of the exons. Long black arrows represent the genomic regions cloned in λ-phages. Shaded bars indicate regions in the first and the second SH3 domains. (A)n, poly(A) in the transcripts.

that the two isoforms, Ash-s and -m, are generated from a single gene by unusual alternative splicings. Ash-s mRNAs are composed of the first two exons and a part of the second intron of the gene. A splicing donor site for the third exon was not recognized in Ash-s transcripts. Thus, a poly(A) sequence is attached in the transcribed region of the intron. Because a stop codon appears in frame at the exon–intron boundary of the gene, a part of the third intron is recognized as a 3′-flanking sequence in Ash-s mRNAs. Thus, Ash-s consists solely of the first SH3 domain of Ash-l. However, splicing for Ash-m mRNAs occurred at the sequence 5′-GACTTTGACCCC-CAGG-3′, which is located within the sixth exon, instead of at 5′-CTCTTC-TCTCTTGCAGC-3′ as for Ash-l mRNAs. The first 42 nucleotide residues of the sixth exon of Ash-l are recognized as a part of the fifth intron in Ash-m. Thus, the sixth exon of Ash-m starts inside the Ash-l exon and 14 amino acid residues encoding the amino terminal of the second SH3 domain are defective in Ash-m.

C. Tissue Distribution of Ash-l and Ash-m

Ash-m and Ash-s mRNAs were detected in various tissues by the reverse-transcribed polymerase chain reaction. The ratio between Ash-m and Ash-l mRNAs varied among tissues. Higher contents of Ash-m mRNA than that of Ash-l were observed in lung, stomach, and skeletal muscle. By contrast, Ash-m mRNAs were barely detectable in cerebrum, spleen, kidney, and testis. However, Ash-s mRNAs were detected abundantly in spleen and kidney but rarely expressed in stomach and skeletal muscle.

D. Binding Proteins to Ash-l and Ash-m

To compare the binding ability of Ash isoforms, the concentrations of GST-fusion proteins were adjusted and binding assays were carried out. *In vitro* binding experiments using the GST-fusion proteins of Ash isoforms demonstrated that Ash-m does not associate effectively with proteins such as dynamin and Sos in PC12 cell lysates compared to Ash-l (Fig. 7a). The bands of dynamin and Sos were identified by anti-dynamin and anti-Sos antibodies. The binding ability of Ash-m to Sos was very low. However, Ash-s formed a significant complex with dynamin and Sos in PC12 cell lysates, although binding to other proteins, including 55, 130, and 180-kDa proteins, appeared not as marked as that of Ash-l. These results were unaffected by treating the cells with NGF. Ash-s is composed exclusively of a single SH3 domain. Hence, the binding ability of Ash-s to Sos and dynamin was compared with that of SH3 domains from phospholipase Cγ 1 (PLC) and phosphatidylinositol 3-kinase (PI 3-kinase). As shown in Figure 7b, the SH3 domains from PLC and PI 3-kinase bound less effectively to SOS or dynamin in cell lysates than to Ash-s, whereas the Ash-s SH3 domain associ-

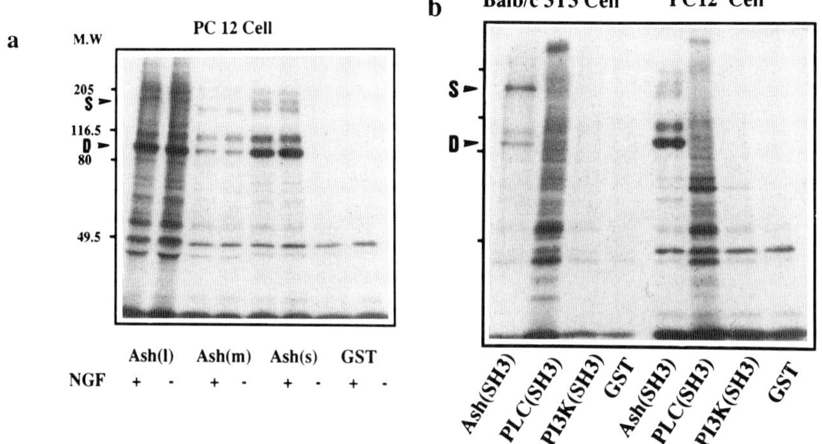

FIGURE 7   Binding specificity of Ash isoforms to the proteins in cell lysates. (a) Binding of Ash isoforms to Sos and dynamin in PC12 cell lysates. PC12 cells were metabolically labeled with [^{35}S]methionine. Cell lysates harvested in the presence and absence of NGF were applied to GST-fusion protein columns and bound proteins were subjected to SDS–PAGE. (b) Binding of various SH3 domains to Sos and dynamin. The SH3 domains from Ash-s, PLC, and PI 3-kinase were prepared as GST-fusion proteins in *E. coli*. Lysates from [^{35}S]methionine-labeled Balb/c cells and PC12 cells are used.

ated with both molecules. The binding of Ash-s to dynamin in PC12 cell lysates was more marked than that to Sos. In contrast, Ash-s bound to Sos more than to dynamin in Balb/c 3T3 cell lysates (Fig. 7b). These results suggest that signaling molecules bound to Ash-s differ among cells, although it only reflects the existing amounts of proteins.

E. Binding Proteins to the Carboxy-Terminal SH3 Domain of Ash-m

As the result of splicing, carboxy-terminal SH3 domain of Ash-m forms a new type of SH3. Therefore, the new SH3 may bind different proteins from those bound to carboxy-terminal SH3 of Ash-l. To examine the possibility, we compared the proteins bound to the carboxy-terminal SH3 of Ash-m to those bound to amino-terminal and carboxy-terminal SH3 domains of Ash-l using bovine brain. As shown in Figure 8a, the carboxy-terminal SH3 of Ash-m did not bind Sos or dynamin. However, it was found to be able to bind several new proteins (60- and 170-kDa proteins shown by arrows). Because Ash-m was found to be abundantly expressed in skeletal muscle, we also examined proteins bound to the carboxy-terminal Ash-m SH3 domain using chicken skeletal muscle. The protein-binding pattern to the carboxy-terminal Ash-m SH3 domain was different from that of Ash-l (Fig. 8b). Proteins (90- and 170-kDa proteins shown by arrows) were specifically bound to carboxy-

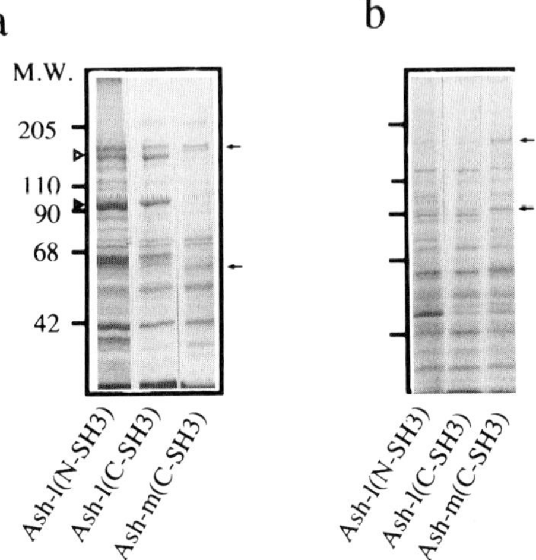

FIGURE 8   Different proteins bind to carboxy-terminal SH3 domains of Ash-m and Ash-l. (a) Bovine brains (100,000g supernatant fractions) or (b) chicken skeletal muscle supernatant fractions were applied to GST-fusion protein columns. Bound proteins were analyzed by SDS–PAGE and stained with Coomassie blue. Lane 1, GST amino-terminal SH3 domain of Ash-l; lane 2, GST carboxy-terminal SH3 domain of Ash-l; lane 3, GST carboxy-terminal SH3 domain of Ash-m. Open triangle indicates Sos and closed triangle indicates dynamin.

terminal Ash-m SH3 domain. These data suggest that Ash-m generates new signals different from those of Ash-l.

F. Effects of Ash-l, Ash-m, and Ash-s on PDGF-Induced DNA Synthesis

To examine the effect of Ash-l, Ash-m, and Ash-s on DNA synthesis, the proteins were microinjected into Balb/c 3T3 cells. After microinjection with various Ash proteins or control GST protein, the cells were stimulated with PDGF. The cells were then labeled with bromodeoxyuridine (BrdUrd), fixed, and immunostained with anti-BrdUrd antibody. Stimulation with PDGF led to a marked increase in the proportion of cells staining positive for BrdUrd incorporation, and this was unaffected by the control GST proteins (Fig. 9). However, injection of Ash-l protein caused a further increase in BrdUrd incorporation in PDGF-stimulated cells from 56 to 70%, although the basal incorporation did not increase without PDGF. Injection of Ash-s or Ash-m, on the contrary, inhibited the enhanced incorporation of BrdUrd with PDGF. However, the inhibition by Ash-m was less effective than that by Ash-s. Similar to Ash-l, neither Ash-m nor Ash-s affected the basal incorporation of BrdUrd without PDGF. These results suggest that Ash-m and Ash-s act as negative regulators in cells and inhibit the signaling of Ash-l to DNA synthesis.

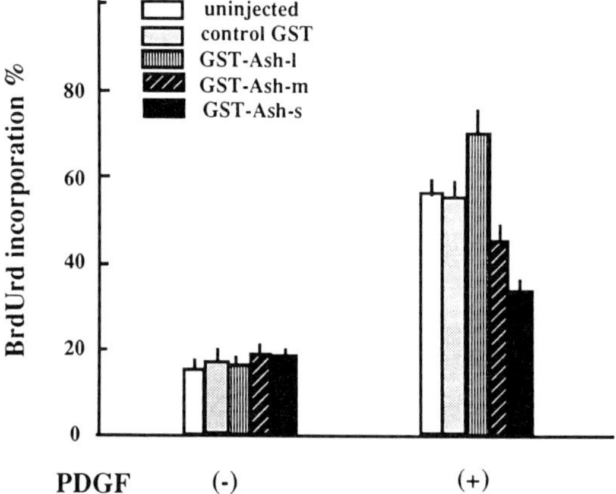

FIGURE 9   Effect of Ash-l, Ash-m, and Ash-s on PDGF-induced DNA synthesis. Ash-l, Ash-m, Ash-s, and GST protein were microinjected into serum-starved Balb/c 3T3 cells. Each protein was injected into more than 100 cells. One hour later, the cells were stimulated with PDGF (5 U/ml), labeled with bromodeoxyuridine, and incubated for 20 hr. The cells were stained with antibromodeoxyuridine antibody and positive cells were counted. Results are shown as mean ± SD in four separate experiments.

G. Downstream Signaling of Ash/Grb-2 and Its Isoforms

Ash-m is defective in the second SH3 domain. Although Ash-l associates with Sos and dynamin by virtue of its SH3 domains, the significance of the two SH3 domains is not necessarily clear. Our results show that the binding ability of Ash-m to proteins, including Sos and dynamin, is weaker than that of Ash-l. Ash-m is not able to bind with Sos effectively. In contrast, Ash-s binds to both Sos and dynamin despite the fact that Ash-s is composed almost exclusively of the first SH3 domain of Ash-l. Accordingly, the second SH3 of Ash-l is not necessarily essential for binding to dynamin and Sos, whereas the first SH3 domain seems to play an important role in binding. Steric hindrance caused by the altered conformation of the defective SH3 domain in Ash-m may account for the loss of association with Sos. It has been proposed that the two SH3 domains in Ash-l are not redundant and might have different functions because mutations in either SH3 domain alter the *sem-5* gene activity in *C. elegans* (Clark *et al.*, 1992). We are now considering the following two possibilities for the role of Ash-m. Because Ash-m does not associate with Sos effectively because of its defective SH3 domain, the signals through Ash-m may be transduced mainly to dynamin and other proteins without activating Ras. Then the signals possibly change cell functions involving dynamin without inducing mitogenesis. If this is the case, the direction of the signals may be regulated in the cytoplasm by the ratio between the amounts of Ash-m and Ash-l. Another possibility is that the second SH3 domain in Ash-m generates signals differing from those of Ash-l. The amino acid residues of the deleted region are similar to those ahead of the region. The newly created amino acid sequence produced by the deletion is somewhat similar to the amino termini of SH3 domains, although, as in Nck (Lehmann, 1990) it deviates from typical SH3 domains (Koyama *et al.*, 1993). If the second SH3 domain of Ash-m is functional, the signals may be transduced to other pathways by unknown molecules. Indeed, the newly formed SH3 domain by splicing binds different proteins from those bound to the carboxy-terminal SH3 domain of Ash-l.

As for Ash-s, it associates with both Sos and dynamin. The characteristics of its associations with proteins in PC12 cell lysates differ from those of its associations with Balb/c 3T3 cell lysate proteins. Because a single SH3 domain constitues Ash-s, it neither binds to tyrosine kinase receptors nor has any other functional domain to respond to extracellular signals. Accordingly, Ash-s may exert inhibitory effects on Ash-l by competition.

References

Aroian, R. V., Koga, M., Mendel, J. E., Ohshima, Y., and Sternberg, P. W. (1990). *Nature* **348**, 693–699.

Breathnach, R., and Chambon, P. (1981). *Annu. Rev. Biochem.* **50**, 349–383.

Chardin, P., Camonis, J. H., Gale, N. W., van Aelst, L., Schlessinger, J., Wigler, M. H., and Bar-Sagi, D. (1993). *Science* **260**, 1338–1343.

Clark, S. G., Stern, M. J., and Horvitz, H. R. (1992). *Nature* **356**, 340–344.

Egan, S. E., Giddings, B. W., Brooks, M. W., Buday, L., Siezland, A. M., and Weinberg, R. A. (1993). *Nature* **363**, 45–51.

Gout, I., Dhand, R., Hiles, I. D., Fry, M. J., Panayotou, G., Das, P., Truong, O., Totty, N. F., Hsuan, J., Booker, G. W., Campbell, I. D., and Waterfield, M. D. (1993). *Cell* **75**, 25–36.

Greengard, P., Valtorta, F., Czernik, A. J., and Benfenati, F. (1993). *Science* **259**, 780–785.

Han, M., and Sternberg (1990). *Cell* **63**, 921–931.

Herskovits, J. S., Shpetner, H. S., Burgess, C. C., and Vallee, R. B. (1993). *Proc. Natl. Acad. Sci. U.S.A.* **90**, 11468–11472.

Horvitz, H. R., and Sternberg, P. W. (1991). *Nature* **351**, 535–541.

Lehmann, J. M., Riethmuller, G., and Johnson, J. P. (1990). *Nucleic Acids Res.* **18**, 1048.

Li, N., Batzer, A., Daly, R., Yajnik, V., Skolnik, E., Chardin, P., Bar-Sagi, D., Margolis, B., and Schlessinger, J. (1993). *Nature* **363**, 85–88.

Lowenstein, E. J., Daly, R. J., Datzer, A. Z., Li, W., Margolis, B., Lammers, R., Ullrich, A. L., Skolnik, F. Y., Bar-Sagi, D., and Shlessinger, J. (1992). *Cell* **70**, 431–442.

Matuoka, K., Shibasaki, F., Shibata, M., and Takenawa, T. (1993). *EMBO J.* **12**, 3467–3473.

Matuoka, K., Shibata, M., Yamakawa, A., and Takenawa, T. (1992). *Proc. Natl. Acad. Sci. U.S.A.* **89**, 9015–9019.

McPherson, P. S., Czernik, A. J., Chilcote, T. J., Onofri, F., Benfenati, F., Greengard, P., Schlessinger, J., and Camilli, P. D. (1994a). *Proc. Natl. Acad. Sci. U.S.A.* **91**, 6486–6490.

McPherson, P. S., Takei, K., Schmid, S. L., and Camilli, P. D. (1994b). *J. Biol. Chem.* **269**, 30132–30139.

Miki, H., Miura, K., Matuoka, K., Nakata, T., Hirokawa, N., Orita, S., Kaibuchi, K., Takai, Y., and Takenawa, T. (1994). *J. Biol. Chem.* **269**, 5489–5492.

Olivier, J. P., Raabe, T., Henkemeyer, M., Dickson, B., Mbamalu, G., Margolis, B., Schlessinger, J., Hafen, E., and Pawson, T. (1993). *Cell* **73**, 179–191.

Ridley, A. J., and Hall, A. (1992). *Cell* **70**, 389–399.

Ridley, A. J., Paterson, H. F., Johnston, C. L., Diekmann, D., and Hall, A. (1992). *Cell* **70**, 401–410.

Robinson, P. J., Sontag, J.-M., Liu, J. P., Fyske, M., Slaughter, C., McMahon, H., and Sudhof, T. C. (1993). *Nature* **365**, 163–166.

Rozakis-Adcock, M., Fernley, R., Wade, J., Pawson, T., and Bowtell, D. (1993). *Nature* **363**, 83–85.

Simon, M. A., Dodson, G. S., and Rubin, G. M. (1993). *Cell* **73**, 169–177.

Sen-itiroh Hakomori

Division of Biomembrane Research
Pacific Northwest Research Foundation
and Departments of Pathobiology and Microbiology
University of Washington
Seattle, Washington 98122

Sphingolipid-Dependent Protein Kinases

I. Introduction

Glycosphingolipids (GSLs) in cellular membranes play two major functional roles: as mediators of cell–cell or cell–substratum interaction and as modulators of transmembrane signaling (Fig. 1). In this chapter, studies focused on the second role are summarized, i.e., how do GSLs and sphingolipids (SLs) modulate transmembrane signaling?

GSLs are known to show dramatic changes during differentiation, development, and oncogenic transformation. Many GSLs have been identified as tumor-associated antigens through extensive studies by the author and his colleagues (hereafter referred to as "we") and others (Hakomori, 1984). The major question is why do GSL synthesis and degradation change so dramatically in association with the previously mentioned processes? GSLs presumably have some essential function in maintenance of not only cell social activity (i.e., cell–cell interaction) but also cell growth, cell cycle, cell

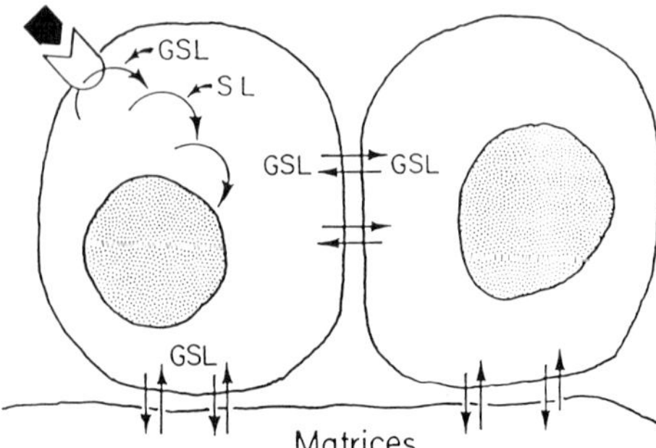

FIGURE I   Bifunctional role of GSLs. Cellular GSLs function to (i) control transmembrane signaling cascade, and (ii) mediate cell–cell or cell–substratum (–matrix) interaction. The first function is performed by GSL-susceptible growth factor receptor and SL-sensitive molecules that transmit information from receptor to nucleus. The second function is maintained directly by GSL–GSL interaction or by integrin receptors that are highly sensitive to GSL.

proliferation, and cell death, as indicated by numerous cell biological studies during the past 10 years. We evaluated the possible function of GSLs as modulators of transmembrane signal transduction through (i) tyrosine ki-nases associated with growth factor receptors or insulin receptors, (ii) protein kinase C (PKC), and (iii) integrin receptors that mediate cell adhesion as well as signal transduction. Many of our studies were based on hematoside (now called GM3) and its derivatives. This ganglioside was originally discovered by Yamakawa and Suzuki (1951) as a ubiquitous ganglioside present in essentially all types of extraneural cells. GM3 is now recognized as a melanoma-associated antigen (Hirabayashi *et al.*, 1985; Nores *et al.*, 1987) and an important cell adhesion and recognition site through interaction with complementary GSLs such as Gg3Cer, LacCer, and Gb4Cer (Kojima and Hakomori, 1989, 1991). In 1982, we found that this ganglioside inhibits fibroblast growth factor (FGF)-dependent cell growth and stops internaliza-tion of FGF when exogenously added, even though GM3 does not bind to FGF (Bremer and Hakomori, 1982). We subsequently found that GM3 inhibits epidermal growth factor (EGF) receptor tyrosine phosphorylation, and exogenous GM3 blocks EGF-dependent cell proliferation (Bremer *et al.*, 1986). In 1993, we found that GM3 interacts with $\alpha5\beta1$ integrin receptor and enhances fibronectin-dependent cell adhesion (Zheng *et al.*, 1993). The primary degradation products of GM3 (de-N-acetyl-GM3 and lyso-GM3) are present (in small quantities) as physiological components in A431 cells and various other types of cells. De-N-acetyl-GM3 enhances cell prolifera-tion (Hanai *et al.*, 1988a), whereas lyso-GM3 strongly inhibits cell growth through inhibition of PKC (Hanai *et al.*, 1988b). These findings are summa-rized in Figure 2. Various examples are presented in the following sections.

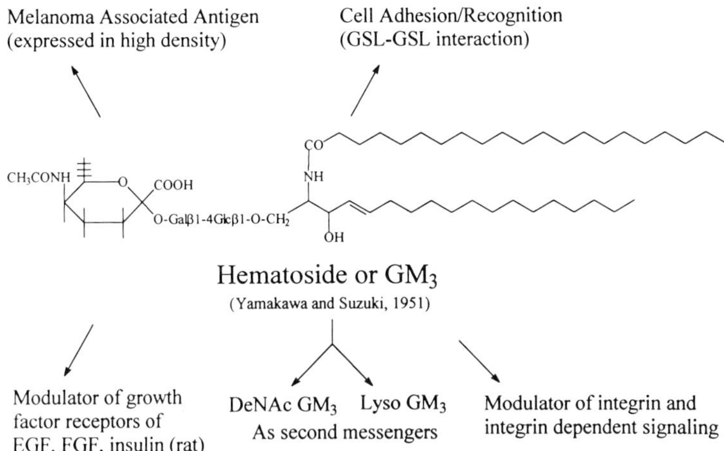

FIGURE 2   Multiple functions of membrane gangliosides. Hematoside (now called GM3) is used as an example. Five distinct functional roles of GM3 are shown schematically: (i) GM3 expressed at high density at the cell surface is recognized as an antigen (typically melanoma-associated antigen); (ii) GM3 is an adhesion molecule recognized by complementary structures such as Gg3Cer, LacCer, and globoside via GSL–GSL interaction; (iii) GM3 modulates growth factor receptors to inhibit tyrosine kinase; (iv) GM3 enhances integrin function within a defined, narrow concentration range, and thereby modulates cell adhesion and adhesion-dependent signaling; and (v) the immediate degradation product of GM3, de-N-acetyl-GM3, promotes activity of various kinases (see Figs. 5 and 6) and enhances cell proliferation. Another direct catabolite, lyso-GM3, inhibits PKC and inhibits cell growth. These two catabolites act as second messengers.

II. Ganglioside-Dependent Modulation of Growth Factor- or Insulin-Dependent Tyrosine Kinase

GM3-dependent cell growth and associated inhibition of tyrosine phosphorylation in A431 cells and KB cells were the starting point for our own series of studies (Bremer *et al.*, 1986). The original data are shown in Figure 3. Further evidence that GM3 modulates growth factor-dependent cell growth was provided by two different experiments, one by Weis and Davis (1990) using an epimeraseless mutant of Chinese hamster ovary cells (LA1D), and another by my colleague Dr. Tsuruoka using a GM3 mutant FUA169 derived from F28-7 cells (Tsuruoka *et al.*, 1993). LA1D was incapable of converting UDP-Glc to UDP-Gal. Therefore, in the absence of Gal, the cells could not synthesize LacCer or GM3. On addition of Gal in culture medium, the cells became capable of synthesizing LacCer or GM3. This mutant did not have EGF receptor and did not show EGF-dependent cell growth. After transfection of the EGF receptor gene, LA1D⁻EGFR⁺ cells were isolated. These cells grew well in the presence of EGF, and showed EGF receptor tyrosine phosphorylation. When the LA1D⁻EGFR⁺ cells were grown in the presence of Gal, whereby GM3 was synthesized, EGF-

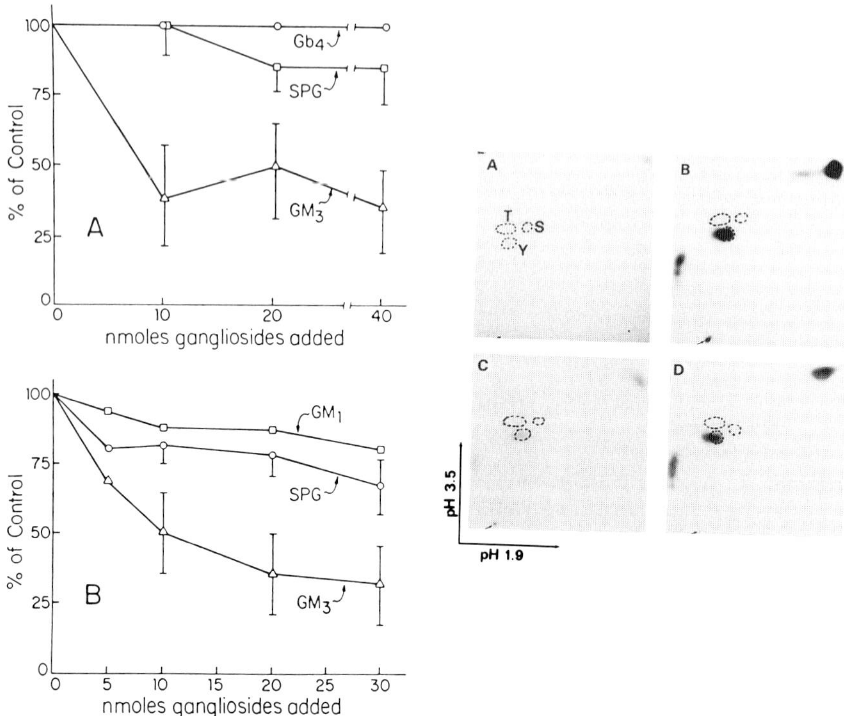

FIGURE 3   GM3 specifically inhibits EGF receptor tyrosine kinase activity. (Left) Inhibition of EGF receptor tyrosine kinase activity by GM3 but not by sialosylparagloboside, GM1, or Gb3. (A) A431 cell EGF receptor; (B) KB cell EGF receptor. (Right) Phosphoamino acid phosphorylation pattern of isolated EGF receptor from A431 cells. (A) Nonstimulated cells; (B) stimulated with EGF; (C) stimulated with EGF plus 20 nmol GM3; (D) stimulated with EGF plus 20 nmol GM1. This is one of the earliest demonstrations that gangliosides modulate transmembrane signaling. [Reproduced with permission from Bremer *et al.* (1986). *J. Biol. Chem.* **261**, 2434–2440.]

dependent cell growth and EGF receptor tyrosine phosphorylation were both inhibited (Fig. 4, top).

GM3 mutant FUA169 was isolated from F28-7 cells. FUA169 did not show insulin-dependent cell growth. Mouse insulin receptor is highly sensitive to GM3, and insulin-dependent growth of F28-7 cells was strongly inhibited in the presence of GM3 (Tsuruoka *et al.*, 1993) (Fig. 4, bottom). However, in human cells insulin receptor is most strongly inhibited by sialosylparagloboside (Nojiri *et al.*, 1991).

A crucial mechanism for explaining how growth factor binding to receptors creates a signal in the form of tyrosine phosphorylation has been suggested by Ullrich and Schlessinger (1990). They postulated that receptor–receptor interaction (i.e., dimerization) occurs when growth factor binds to the cell surface binding site of receptor. GM3-dependent inhibition of tyro-

Epimerase-less mutant of CHO cells (lalD)

$$(\text{UDP Glc} \longrightarrow\!\!\!| \cdots\!\!\rightarrow \text{UDP Gal})$$
do not synthesize LacCer and GM₃
$$+ \text{Gal} \downarrow \qquad \uparrow - \text{Gal}$$
synthesize GM₃

- lalD transfectant with cDNA clone of EGF receptor
  → EGFR expression and shows EGF-dependent cell growth (lalD-EGFR⁺)

- lalD-EGFR⁺ cells

  (1) + Gal → GM3 ↑ → EGFR Y-P ↓

      → EGF-dependent cell growth ↓

  (2) − Gal → GM₃ ↓ → EGFR Y-P ↑

      → EGF dependent cell growth ↑

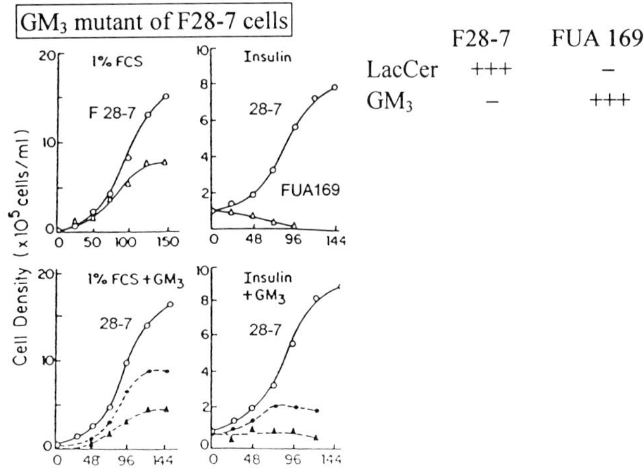

FIGURE 4 Further evidence for modulation by GM3 of growth factor (or hormone)-dependent cell growth. *In vitro* observations that growth factor receptor tyrosine kinase activity is strongly modulated by GM3, and that exogenous GM3 addition inhibits cell growth, were further substantiated by various types of *in vivo* experiments. Two examples using mutant cells with regard to GM3 expression are summarized. (Top) Experiment using an epimerase-less mutant of Chinese hamster ovary (CHO) cells (La1D) that is only capable of synthesizing GM3 in the presence of galactose in culture medium. In the absence of galactose, the mutant is not capable of synthesizing LacCer or GM3 because UDP-Glc in the cells is incapable of being converted into UDP-Gal. Because the mutant lacks EGF receptor, EGF receptor gene was transfected into the cells (La1D⁻EGFR⁺). The transfected cells showed inhibition of EGF-dependent cell growth in the presence (but not in the absence) of Gal. (Bottom) Experiment with a GM3 mutant, FUA169, derived from F28-7 cells. FUA169 was characterized by a high level of GM3, in contrast to parent F28-7 cells that have no GM3. Growth of F28-7 in medium containing 1% FCS was strongly inhibited by exogenous addition of GM3 (lower left). FUA169 did not show insulin-dependent growth because of strong inhibition of insulin receptor function by GM3 (upper right). Insulin-dependent growth of F28-7 was strongly inhibited by exogenous GM3. GM3 is now known to interact with and inhibit insulin receptor kinase of rat.

sine kinase may inhibit receptor–receptor interaction. However, this possibility was ruled out by a recent study (Zhou *et al.*, 1994). EGF-dependent tyrosine phosphorylation decreases in both monomeric and dimeric receptors upon addition of GM3. The quantity of monomeric and dimeric forms of the EGF receptor is unchanged upon addition of GM3. Therefore, GM3 may not interfere with dimerization, but rather have a direct action on kinase activity, possibly through membrane perturbation. However, the exact mechanism remains unknown.

III. Functional Roles of De-N-Acetyl-Gangliosides and Lyso-Gangliosides

The immediate catabolites of GM3 (de-*N*-acetyl-GM3 and lyso-GM3) have been found as physiological components in various types of cells. We detected de-*N*-acetyl-GM3 in A431, B16 melanoma, and 3T3 cells using a specific monoclonal antibody DH5 directed to this novel ganglioside. Exogenous addition of de-*N*-acetyl-GM3 promotes growth of these cells (Hanai *et al.*, 1988a). De-*N*-acetyl-GM3 significantly enhances not only tyrosine phosphorylation of EGF receptor, but also serine (Ser) phosphorylation (Zhou *et al.*, 1994). The overall mechanism of cell growth promotion is still unidentified. Lyso-GM3 strongly inhibits PKC (Igarashi *et al.*, 1989b) and EGF receptor kinase (RK) (Hanai *et al.*, 1988b). These two immediate GM3 metabolites may act as second messengers in control of transmembrane signaling (Fig. 5). The association of de-*N*-acetyl-GM3 with cancer cell growth has been demonstrated. Interestingly, the enzyme *N*-acetylase, which converts GM3 to de-*N*-acetyl-GM3 (or GD3 to de-*N*-acetyl-GD3) by removal of the *N*-acetyl group, is greatly enhanced in various types of tumor cells and is susceptible to tyrosine kinase. Genistein, which inhibits tyrosine phosphorylation, promotes these conversions (Sjoberg *et al.*, 1995). The effects of de-*N*-acetyl-GM3 and lyso-GM3 on transmembrane signaling are summarized in Figure 6.

IV. Transmembrane Signal Control by Sphingosine and Its Derivatives

Modulation of transmembrane signaling was found to be regulated by simpler, basic SLs such as sphingosine (Sph) and its derivatives. Hannun and Bell originally observed that Sph inhibits PKC (Hannun *et al.*, 1986; Hannun and Bell, 1987). The inhibitory effect, however, was not stereospecific (Merrill *et al.*, 1989). We found that Sph can be *N*-methylated, and *N*,*N*-dimethyl-Sph (DMS) is formed by a specific methyltransferase (Igarashi and Hakamori, 1989). The inhibition of PKC by DMS is highly stereospecific, i.e., observed for only *N*,*N*-dimethyl-D-erythrosphingenine, but not

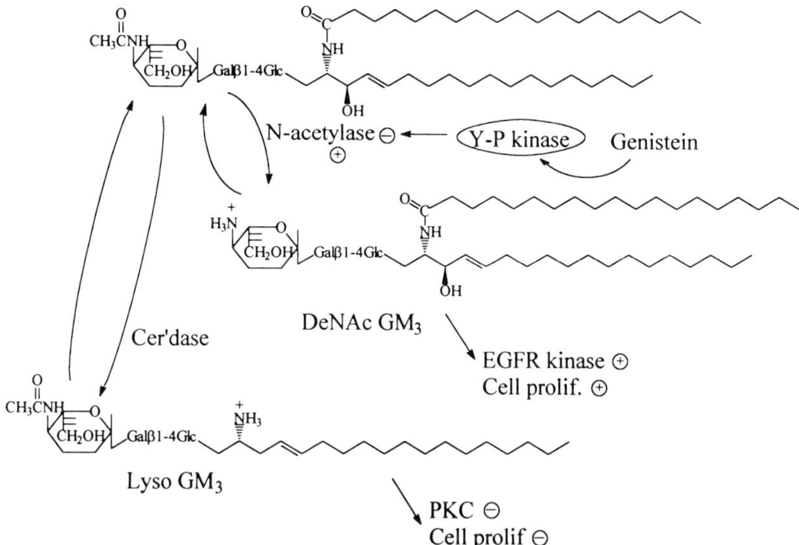

FIGURE 5   Two catabolites of GM3 act as second messengers. De-*N*-acetyl-GM3 (yielded by de-*N*-acetylation of sialic acid moiety of GM3) enhances EGF receptor tyrosine and serine phosphorylation and induces cell proliferaton. The de-*N*-acetylation process is apparently controlled by tyrosine kinase, because the process is inhibited by genistein. Lyso-GM3 (yielded by ceramidase action on GM3) strongly inhibits PKC and cell proliferation. For further information, see Hanai *et al.* (1988a,b), Song *et al.*, *J. Biol. Chem.* **266,** 10174 (1991), Zhou *et al.* (1994), and Sjoberg *et al.* (1995).

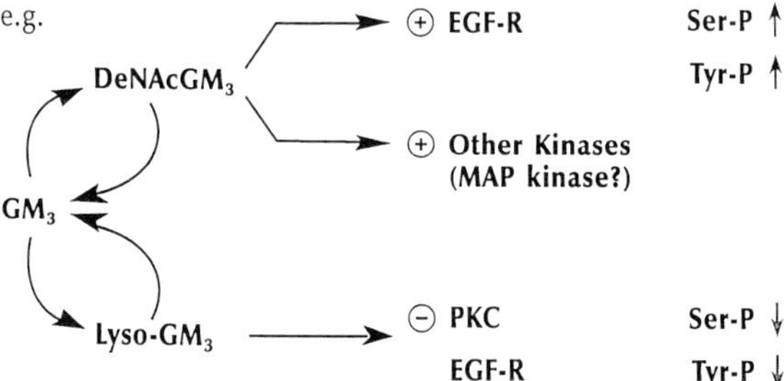

FIGURE 6   Direct catabolites of gangliosides function as nonspecific second messengers. De-*N*-acetyl-GM3 and lyso-GM3 (see Fig. 5) are not further degraded, but rather reconverted to GM3 by *N*-acetylation or *N*-fatty acylation. De-*N*-acetyl-GM3 promotes activity of EGF receptor kinase and other kinases. Lyso-GM3 inhibits PKC and EGF receptor kinase. Both of these catabolites act as second messengers.

the -L-threo-, -L-erythro-, or -D-threo- forms (Igarashi *et al.*, 1989a). DMS also strongly promotes EGF RK activity (Igarashi *et al.*, 1990) and induces *in vitro* phosphorylation of multiple unidentified proteins in A431 cells (T. Megidish, Y. Igarashi, K. Takio, K. Titani, and S. Hakomori, unpublished data). Therefore, DMS derived from Sph inhibits PKC and also promotes multiple DMS-dependent kinases, including EGF RK, as summarized in Figure 7. This figure also shows the importance of ceramide (Cer) as a possible second messenger. When leukemic cell lines are stimulated by TNF, sphingomyelinase is activated, leading to release of Cer, which has been reported to activate the transcription factor NF-κB (Schütze *et al.*, 1992). Similar studies also showed enhancement of Cer level through activation of sphingomyelin pathways (Kolesnick and Golde, 1994).

Another important Sph metabolite that modulates cellular function is Sph-1-phosphate (Sph-1-P), the initial catabolite that is subsequently degraded to phosphoethanolamine and palmitaldehyde. Sph-1-P strongly inhibits cell motility. Among various SLs tested, only Sph-1-P inhibits cell motility strongly at a 10- to 100-nM level. Sph, DMS, and N,N,N-trimethyl-Sph (TMS) had minimal effects on cell motility (Sadahira *et al.*, 1992) (Fig. 8). The molecular mechanism behind the motility-inhibitory effect of Sph-1-P is still unclear, and the target molecule is not clearly identified. Actin-nucleation, as measured by actin filament formation using labeled actin subunit and membrane components, is strongly inhibited by Sph-1-P. Thus, actin polymerization is inhibited and greatly delayed by addition of Sph-1-P.

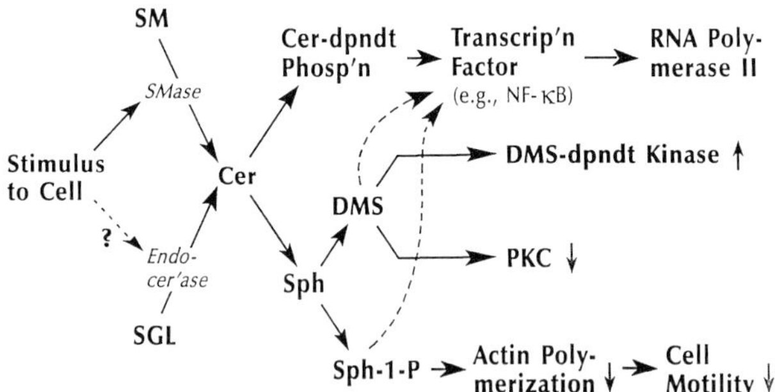

FIGURE 7   Proposed role of SL derivatives Cer, Sph, DMS, and Sph-1-P, in control of signal transduction. Cer is derived from sphingomyelin by action of sphingomyelinase or from GSLs by action of endoceramidase. An increased Cer level is observed as a result of physiological cell stimulation by IL-1 or TNF-α. Cer in turn activates "Cer-dependent kinase," which activates transcription factors to initiate transcription through activation of specific RNA polymerase II. Sph (yielded from Cer by action of ceramidase) is immediately converted to (i) DMS, which strongly inhibits PKC and strongly promotes "DMS-dependent kinase"; and (ii) Sph-1-P, which is further degraded into palmitaldehyde and phosphoethanolamine by a well-established catabolic pathway. Sph-1-P in nM concentrations strongly inhibits cell motility, possibly through inhibition of actin polymerization.

V. GM3 as a Modulator of Integrin Receptor Function ______

GM3 has been shown to act as a modulator of integrin receptor function and integrin-dependent cell signaling and adhesion. Liposomes consisting of fixed quantities of phosphatidylcholine, cholesterol, and $\alpha5\beta1$ receptor, and varying quantities of GM3 or other GSLs, showed remarkable differences in fibronectin-dependent adhesion (Fig. 9). Detailed adhesion activity of such liposomes as a function of GM3 concentration is shown in Figure 10. We observed remarkable GM3-dependent enhancement of $\alpha5\beta1$ function at 0.2–0.5 nmol and inhibition of $\alpha5\beta1$ at higher quantities in $\alpha5\beta1$ liposomes. It is possible that organization of various components in "podo-

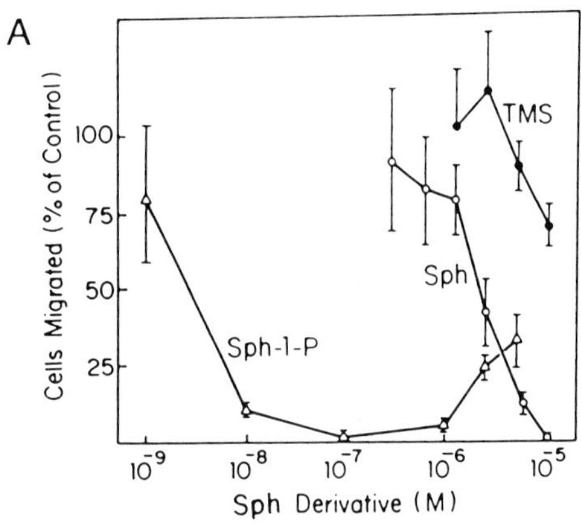

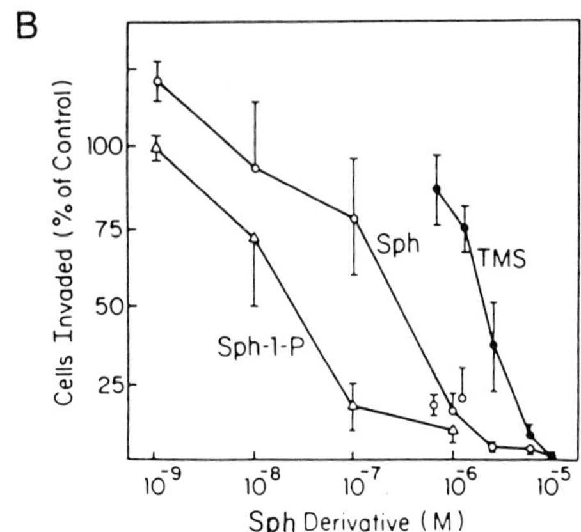

FIGURE 8   Effect of Sph-1-P, Sph, and TMS on B16 melanoma cell motility. (A) Chemotactic motility assay through polycarbonate filter coated with 1 μg Matrigel. Sph-1-P produced significant inhibition at much lower concentrations (10–100 nM) than Sph or TMS. (B) Chemoinvasion assay through filter coated with 10 μg Matrigel. Again, Sph-1-P had a strong inhibitory effect at a much lower concentration (100 nM) than Sph and TMS. TMS was effective at 10 μM. [Reproduced by permission from Sadahira *et al.* (1992).]

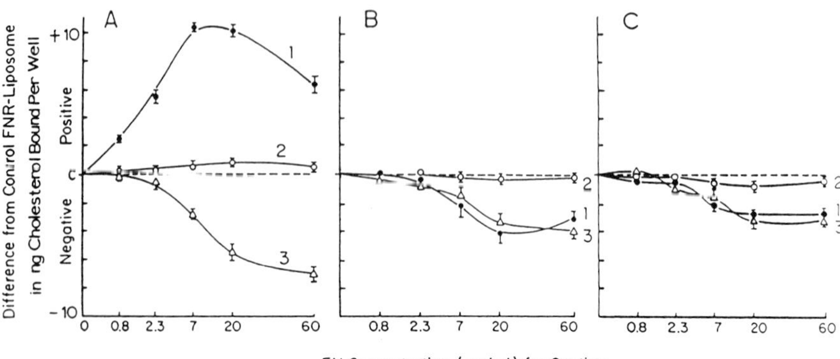

FIGURE 9 FN binding by $\alpha 5\beta 1$ integrin liposomes containing various quantities of GSLs. Purified $\alpha 5\beta 1$ integrin receptor was incorporated in liposomes. Liposomes of constant composition in terms of PC, [^{14}C]cholesterol, and $\alpha 5\beta 1$, with various quantities of GM3, LacCer, or GlcCer, were prepared, and aliquots of ≈ 1.25 µg cholesterol per well were added to Pro-bind plates previously coated by a 1 : 3 sequence dilution of FN (abscissa). Specific binding was quantitated by scintillation counting and expressed as ng cholesterol bound. (A) $\alpha 5\beta 1$ Liposomes (55 µg PC, 33 µg [^{14}C]cholesterol, 5 µg $\alpha 5\beta 1$ receptor) containing 2.2 µg GM3 (2 nM) ($\triangle$) (curve 3), 0.44 µg (0.4 nM) ($\bullet$) (curve 1), and 0.088 µg (0.08 nM) ($\bigcirc$) (curve 2) were compared with control FN receptor liposomes containing no GM3 (dashed line, C) in terms of binding activity. (B) All experimental conditions and symbols are as described previously in A except LacCer was used instead of GM3. (C) Conditions and symbols as described previously in A except GlcCer was used instead of GM3. [Reproduced by permission from Zheng *et al.* (1993).]

somes," i.e., talin, vinculin, tensin, P60*src,* and P125FAK, is maintained by an appropriate concentration of GM3, and actin filament formation is affected by Sph-1-P. Signal transduction through such organization, mediated by Ca^{2+} influx, would be greatly influenced by GM3 as well as by Sph-1-P.

VI. Plasmalopsychosine as Signaling Molecule in Neuronal Cells

Nudelman *et al.* (1992) discovered a novel signaling GSL, termed *plasmalopsychosine,* in 1992. This compound is the conjugate of plasmal through 4,6- or 3,4-cyclic acetal linkage to β-Gal residue of psychosine (Nudelman *et al.,* 1992). Plasmalopsychosine is found exclusively in white matter of brain, but not in gray matter. In contrast to all other complex lipids, plasmalopsychosine has a unique structure with two aliphatic tails oriented in opposite directions (Fig. 11). It strongly induces differentiation in neuronal cells leading to neurite outgrowth. Neuritogenic differentiation of most neuronal cells is induced by nerve growth factor (NGF), but rarely by GSLs alone. However, plasmalopsychosine by itself, without NGF, induces neuritogenic differentiation. In other words, it mimics NGF activity. When PC12 cells were incubated with plasmalopsychosine, tyrosine kinase associated with NGF receptor

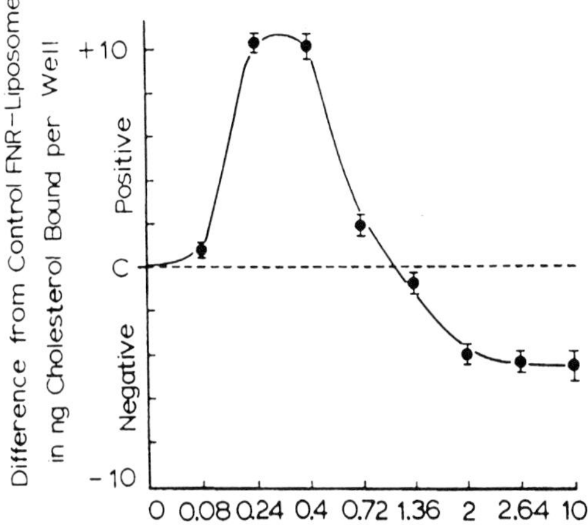

FIGURE 10   Adhesion of FN receptor liposomes containing various concentrations of GM3 to plastic plates coated with a constant concentration of FN. Each well of Pro-bind assay plates was coated with 12.5 μg/ml FN in PBS and saturated with 1% BSA in PBS. Binding assays were performed using PC/[^{14}C]cholesterol/FN receptor liposomes containing various concentrations of GM3. Abscissa: GM3 quantity (nmol) contained in liposomes. Ordinate: difference of adhesion (expressed as ng cholesterol bound per well) of GM3-containing liposome vs. control liposome. Value for control (i.e., FN receptor liposome without GM3) was 26 ng cholesterol bound per well. [Reproduced by permission from Zheng *et al.* (1993).]

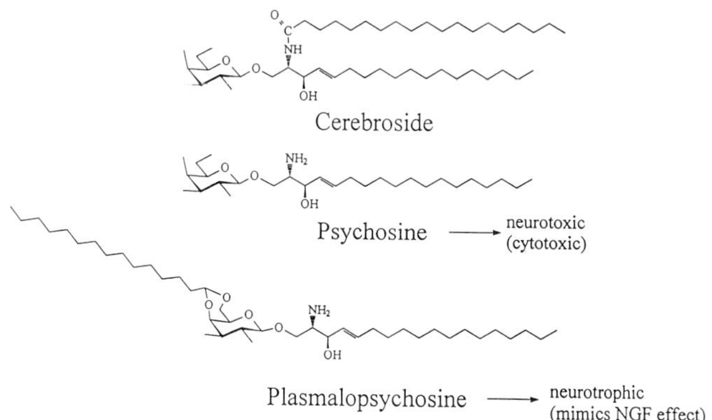

FIGURE 11   Structure of plasmalopsychosine compared to psychosine and cerebroside. Fatty aldehyde (myristyl or palmityl) conjugated to β-galactosyl residue of psychosine through 4,6-cyclic acetal is a unique feature of plasmalopsychosine. Two aliphatic chains are oriented in opposite directions. Plasmalopsychosine is neurotrophic. In contrast, psychosine is highly neurotoxic.

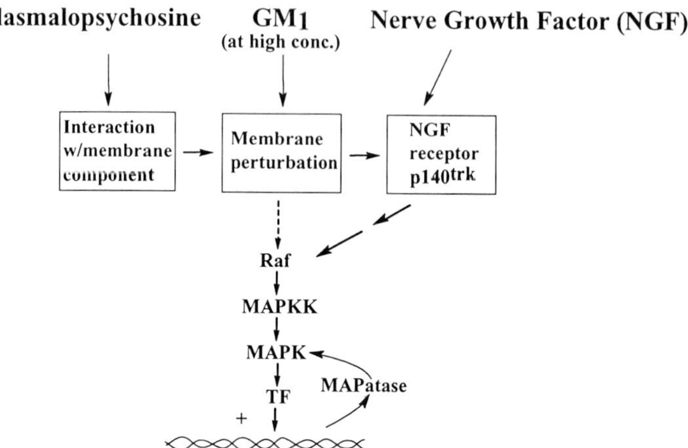

FIGURE 12   Possible mechanism of neurotrophic effect of plasmalopsychosine. Experiments described in the text clearly indicate that plasmalopsychosine activates NGF receptor kinase (p140[trk]) and MAPK, and consequently induces neuritogenesis in PC12 cells. The effect of plasmalopsychosine mimics that of NGF and is observed only in NGF-susceptible cells (i.e., those possessing NGF receptor). Plasmalopsychosine and NGF share a common signaling pathway leading to activation of MAPK. We postulate that plasmalopsychosine initially interacts with an unknown membrane component, leading to membrane perturbation with activates p140[trk]. Plasmalopsychosine does not bind to NGF receptor directly because binding of ^{125}I-labeled NGF to its receptor is not affected by plasmalopsychosine (Sakakura *et al.*, 1996).

(P140[trk]) was strongly activated. No other GSL is capable of activating P140[trk]. This effect is the same as when PC12 cells are incubated with NGF. MAP kinase (MAPK) is also strongly and immediately activated when PC12 cells are incubated with plasmalopsychosine and the activity is sustained for a long period. This enhancement of MAPK occurred within a few minutes of incubation, similar to the effect of NGF. No other GSL or growth factor showed such activation of MAPK. The effect of plasmalopsychosine may not directly rely on the activation of NGF receptor and P140[trk] because binding of [125I]NGF to its receptor was not inhibited by plasmalopsychosine. However, plasmalopsychosine enhanced P140[trk] activity to the same extent as NGF, leading to strong sustained enhancement of MAPK. Thus, plasmalopsychosine may have an independent receptor that interacts with the lipid bilayer and causes membrane perturbation, activating tyrosine phosphorylation of P140[trk] (Fig. 12; Sakakura *et al.*, 1996).

VII. Application of SLs in "Orthosignaling" Therapy of Inflammation

The possible application and utilization of signal-affecting GSLs and SLs are discussed. Many diseases are caused by and progress through a series of aberrant signal transductions; examples are tumor progression and harmful inflammatory response following thrombosis, as often seen in heart attack and stroke. Tissue damage caused by acute inflammatory response is less affected by circulatory disturbances resulting in ischemia than by neutrophil migration mediated by P- and E-selectin. In particular, the earliest cell adhesion response and neutrophil recruitment are mediated by P-selectin and result in O_2^- production and consequent "reperfusion injury." DMS is a stereospecific inhibitor of PKC and inhibits the secretory response of platelets. Therefore, TMS, the analog of DMS, was synthesized. TMS, in contrast to Sph and DMS, is completely water soluble. TMS inhibits PKC and phosphorylation of a 47-kDa protein that is considered to be the direct substrate of PKC in platelets and is required for the secretory response of α-granules. TMS therefore blocks the secretory response and expression of P-selectin at the surface of platelets (Fig. 13). Similarly, TMS inhibits translocation of Weibel–Pallade bodies of endothelial cells, possibly through PKC inhibition (Fig. 14). Surface expression of P-selectin is efficiently inhibited by 5–10 μM TMs for platelets and 1–3 μM TMS for endothelial cells.

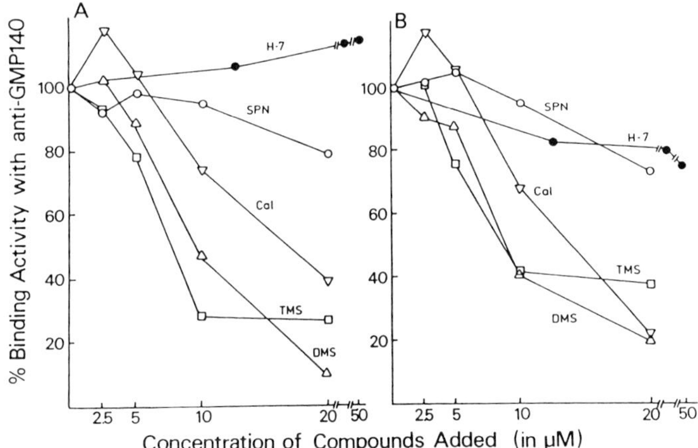

FIGURE 13 Effect of various compounds on thrombin- (A) and PMA- (B) induced P-selectin expression detected by flow cytometry. Platelet suspension (25 μl) was mixed with 225 μl of Tyrode's buffer, pH 7.2, containing Sph derivatives or PKC inhibitors H-7 or Calphostin-C, as indicated in the figure. For example, for preparation of 0, 2.5, 5, 10, and 20 μM solution, 2 mM TMS stock solution was diluted 0-, 12.5-, 25-, 50-, or 100-fold, respectively, with Tyrode's buffer. The cell suspensions were incubated at 37°C for 5 min, then stimulated by addition of 10 μl of thrombin (final concentration 1 U/ml) or PMA (final concentration 10^{-7} M), followed by incubation at 37°C for 10 min. After fixation, the platelets were stained by mAb AC1.2 and analyzed by flow cytometry. The mean fluorescence intensity of resting platelets (incubated in the absence of inhibitor and activator) was subtracted from the value of each activated platelet sample. Addition of ethanol (0.5 and 1.0%) alone, which was used for preparing 10 or 20 μM DMS or TMS in Tyrode's buffer, had no effect. Results represent the average of three similar experiments using different platelet sources. [Reproduced with permission from Handa *et al.*, *Biochemistry* **30**, 11682–11686 (1991). Copyright 1991 American Chemical Society.]

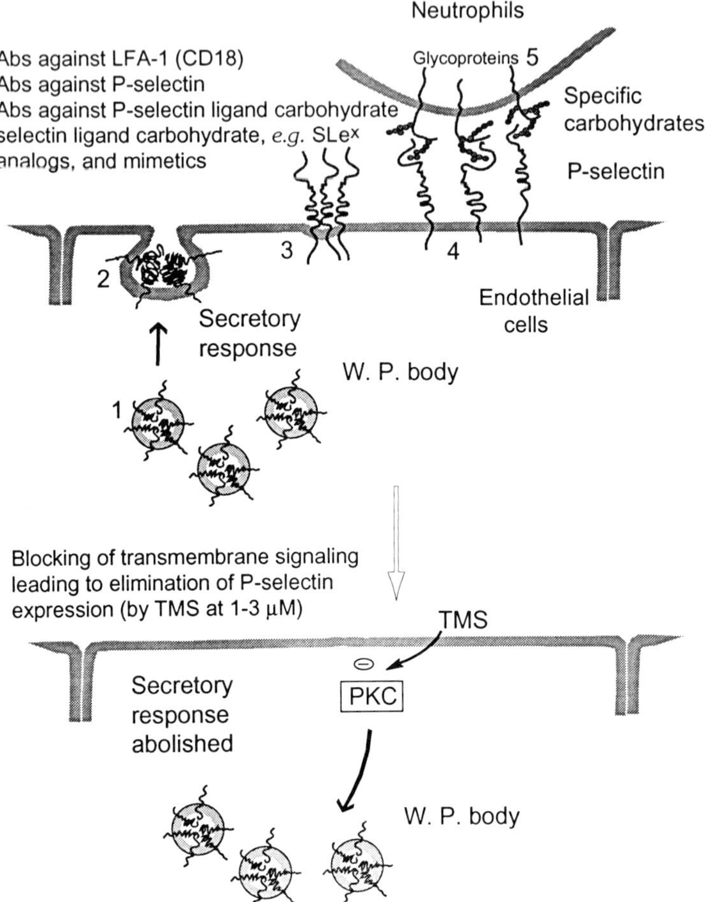

FIGURE 14   As an initial mechanism of acute inflammatory response, P-selectin packed in Weibel–Pallade bodies (W. P. bodies) (1 in top panel) in cytoplasm of ECs is translocated to the luminal surface (2) through a complex transmembrane signaling via EC surface stimuli such as O_2^-, histamine, thrombin, etc. The bodies are thereby exposed to the cell surface ("secretory response"), and P-selectin is expressed (3). Neutrophils are recognized by P-selectin through their specific P-selectin binding ligand (PSGL-1) (5). This process is essential for initiation of acute inflammatory responses associated with thrombosis, heart attack, stroke, and wounding. Antibodies directed to P-selectin, or P-selectin ligand carbohydrates such as SLe^x analogs, can be used to inhibit neutrophil recognition by P-selectin at the EC surface. This process is efficiently blocked by administration of TMS, which inhibits PKC and abolishes consequent secretory responses.

Thus, TMS may be useful for inhibiting neutrophil accumulation associated with inflammatory lesions, the cause of reperfusion injury. Experiments along this line were performed in collaboration with Alan Lefer of Jefferson Medical College, Philadelphia. Neutrophil adhesion to reperfused coronary artery (left anterior descending artery) was strongly inhibited by 3 μM TMS, compared to control coronary artery endothelium (left circumferent artery). Clearly, administration of TMS strongly suppressed neutrophil adhesion to endothelial cells. Cat coronary arteries were blocked for 90 min followed by reperfusion, and the area of injury/necrosis was examined (with methylene blue vital staining and triphenyl tetrazolium chloride staining). In animals treated with 0.5 mg/kg TMS, the necrotic area was greatly reduced (Muro-hara *et al.*, 1995).

A similar experiment using a rabbit ear model was performed in collaboration with Nicholas Vedder, University of Washington and Harborview Hospital, Seattle. The ear was cut, leaving the central artery and vein intact. The central artery was blocked for 90 min, then the ear was sutured. During this time, the animal was infused with 0.5 mg/kg TMS. TMS was administered continuously for 1 week. During this period, ears of control animals (not treated with TMS) showed strong inflammation, and in many cases developed necrosis such that the ear dropped off. In TMS-treated animals, the ear remained intact and inflammation was minimal. Volumetry of ear (by immersion in water and measurement of displaced water) showed that TMS-treated animals had much smaller ear volume (i.e., less inflammation) than control animals. Results of both these studies clearly demonstrate the ability of TMS to suppress reperfusion injury. DMS and TMS inhibit tumor cell-dependent activation of platelets and tumor cell proliferation through inhibition of PKC (Endo *et al.*, 1991; Okoshi *et al.*, 1991) and are thus capable of inhibiting tumor metastasis (Park *et al.*, 1995).

Acknowledgment

I thank Dr. Stephen Anderson for scientific editing and preparation of the manuscript.

References

Bremer, E. G., and Hakomori, S. (1982). GM$_3$ ganglioside induces hamster fibroblast growth inhibition in chemically-defined medium: Ganglioside may regulate growth factor receptor function. *Biochem. Biophys. Res. Commun.* **106**, 711–718.

Bremer, E., Schlessinger, J., and Hakomori, S. (1986). Ganglioside-mediated modulation of cell growth: Specific effects of GM$_3$ on tyrosine phosphorylation of the epidermal growth factor receptor. *J. Biol. Chem.* **261**, 2434–2440.

Endo, K., Igarashi, Y., Nisar, M., Zhou, Q., and Hakomori, S. (1991). Cell membrane signaling as target in cancer therapy: Inhibitory effect of *N,N*-dimethyl and *N,N,N*-trimethyl sphin-gosine derivatives on *in vitro* and *in vivo* growth of human tumor cells in nude mice. *Cancer Res.* **51**, 1613–1618.

Hakomori, S. (1984). Tumor associated carbohydrate antigens. *Annu. Rev. Immunol.* **2**, 103–126.

Hanai, N., Dohi, T., Nores, G. A., and Hakomori, S. (1988a). A novel ganglioside, de-*N*-acetyl-GM₃ (II³NeuNH₂LacCer), acting as a strong promoter for epidermal growth factor receptor kinase and as a stimulator for cell growth. *J. Biol. Chem.* **263**, 6296–6301.

Hanai, N., Nores, G. A., MacLeod, C., Torres-Mendez, C.-R., and Hakomori, S. (1988b). Ganglioside-mediated modulation of cell growth: Specific effects of GM₃ and lyso-GM₃ in tyrosine phosphorylation of the epidermal growth factor receptor. *J. Biol. Chem.* **263**, 10915–10921.

Hannun, Y. A., and Bell, R. M. (1987). Lysosphingolipids inhibit protein kinase C: Implications for the sphingolipidoses. *Science* **235**, 670–674.

Hannun, Y. A., Loomis, C. R., Merrill, A. H., Jr., and Bell, R. M. (1986). Sphingosine inhibition of protein kinase C activity and of phorbol dibutyrate binding *in vitro* and in human platelets. *J. Biol. Chem.* **261**, 12604–12609.

Hirabayashi, Y., Hanaoka, A., Matsumoto, M., Matsubara, T., Tagawa, M., Wakabayashi, S., and Taniguchi, M. (1985). Syngeneic monoclonal antibody against melanoma antigen with interspecies cross-reactivity recognizes GM₃, a prominent ganglioside of B16 melanoma. *J. Biol. Chem.* **260**, 13328–13333.

Igarashi, Y., and Hakomori, S. (1989). Enzymatic synthesis of *N,N*-dimethyl-sphingosine: Demonstration of the sphingosine:*N*-methyltransferase in mouse brain. *Biochem. Biophys., Res. Commun.* **164**, 1411–1416.

Igarashi, Y., Hakomori, S., Toyokuni, T., Dean, B., Fujita, S., Sugimoto, M., Ogawa, T., El-Ghendy, K., and Racker, E. (1989a). Effect of chemically well-defined sphingosine and its *N*-methyl derivatives on protein kinase C and src kinase activities. *Biochemistry* **28**, 6796–6800.

Igarashi, Y., Nojiri, H., Hanai, N., and Hakomori, S. (1989b). Gangliosides that modulate membrane protein function. *Methods Enzymol.* **179**, 521–541.

Igarashi, Y., Kitamura, K., Toyokuni, T., Dean, B., Fenderson, B. A., Ogawa, T., and Hakomori, S. (1990). A specific enhancing effect of *N,N*-dimethylsphingosine on epidermal growth factor receptor autophosphorylation: Demonstration of its endogenous occurrence (and the virtual absence of unsubstituted sphingosine) in human epidermoid carcinoma A431 cells. *J. Biol. Chem.* **265**, 5385–5389.

Kojima, N., and Hakomori, S. (1989). Specific interaction between gangliotriaosylceramide (Gg3) and sialosyllactosylceramide (GM3) as a basis for specific cellular recognition between lymphoma and melanoma cells. *J. Biol. Chem.* **264**, 20159–20162.

Kojima, N., and Hakomori, S. (1991). Cell adhesion, spreading, and motility of G_M3-expressing cells based on glycolipid–glycolipid interaction. *J. Biol. Chem.* **266**, 17552–17558.

Kolesnick, R. N., and Golde, D. W. (1994). The sphingomyelin pathway in tumor necrosis factor and interleukin-1 signaling. *Cell* **77**, 325–328.

Merrill, A. H., Jr., Nimkar, S., Menaldino, D., Hannun, Y. A., Loomis, C. R., Bell, R. M., Tyagi, S. R., Lambeth, J. D., Stevens, V. L., Hunter, R., and Liotta, D. C. (1989). Structural requirements for long-chain (sphingoid) base inhibition of protein kinase C in vitro and for the cellular effects of these compounds. *Biochemistry* **28**, 3138–3145.

Murohara, T., Buerke, M., Margiotta, J., Ruan, F., Igarashi, Y., Hakomori, S., and Lefer, A. M. (1995). Myocardial and endothelial protection by TMS in ischemia-reperfusion injury. *Am. J. Physiol.* **269** (*Heart Circ. Physiol.* 38), H504–H514.

Nojiri, H., Stroud, M. R., and Hakomori, S. (1991). A specific type of ganglioside as a modulator of insulin-dependent cell growth and insulin receptor tyrosine kinase activity: Possible association of ganglioside-induced inhibition of insulin receptor function and monocytic differentiation induction in HL60 cells. *J. Biol. Chem.* **266**, 4531–4537.

Nores, G. A., Dohi, T., Taniguchi, M., and Hakomori, S. (1987). Density-dependent recognition of cell surface GM₃ by a certain anti-melanoma antibody, and GM₃ lactone as a possible immunogen: Requirements for tumor-associated antigen and immunogen. *J. Immunol.* **139**, 3171–3176.

Nudelman, E. D., Levery, S. B., Igarashi, Y., and Hakomori, S. (1992). Plasmalopsychosine, a novel plasmal (fatty aldehyde) conjugate of psychosine with cyclic acetal linkage: Isolation and characterization from human brain white matter. *J. Biol. Chem.* **267**, 11007–11016.

Okoshi, H., Hakomori, S., Nisar, M., Zhou, Q., Kimura, S., Tashiro, K., and Igarashi, Y. (1991). Cell membrane signaling as target in cancer therapy II: Inhibitory effect of *N,N,N*-trimethylsphingosine on metastatic potential of murine B16 melanoma cell line through blocking of tumor cell-dependent platelet aggregation. *Cancer Res.* **51**, 6019–6024.

Park, Y. S., Ruan, F., Hakomori, S., and Igarashi, Y. (1995). Cooperative inhibitory effect of *N,N,N*-trimethylsphingosine and sphingosine-1-phosphate, coincorporated in liposomes, on B16 melanoma cell metastasis: Cell membrane signaling as a target in cancer therapy IV. *Int. J. Oncol.* **7**, 487–494.

Sadahira, Y., Ruan, F., Hakomori, S., and Igarashi, Y. (1992). Sphingosine 1-phosphate, a specific endogenous signaling molecule controlling cell motility and tumor cell invasiveness. *Proc. Natl. Acad. Sci. U.S.A.* **89**, 9686–9690.

Sakakura, C., Igarashi, Y., Anand, J. K., Sadozai, K. K., and Hakomori, S. (1996). Plasmalopsychosine of human brain mimics the effect of nerve growth factor by activating its receptor kinase and mitogen-activated protein kinase in PC12 cells: Induction of neurite outgrowth and prevention of apoptosis. *J. Biol. Chem.* **271**, 946–952.

Schütze, S., Potthoff, K., Machleidt, T., Berkovic, D., Wiegmann, K., and Krönke, M. (1992). TNF activates NF-κB by phosphatidylcholine-specific phospholi pase C-induced "acidic" sphingomyelin breakdown. *Cell* **71**, 765–776.

Sjoberg, E. R., Chammas, R., Ozawa, H., Kawashima, I., Khoo, K.-H., Morris, H. R., Dell, A., Tai, T., and Varki, A. (1995). Expression of de-*N*-acetyl-gangliosides in human melanoma cells is induced by genistein or nocodazole. *J. Biol. Chem.* **270**, 2921–2930.

Tsuruoka, T., Tsuji, T., Nojiri, H., Holmes, E. H., and Hakomori, S. (1993). Selection of a mutant cell line based on differential expression of glycosphingolipid, utilizing anti-lactosylceramide antibody and complement. *J. Biol. Chem.* **268**, 2211–2216.

Ullrich, A., and Schlessinger, J. (1990). Signal transduction by receptors with tyrosine kinase activity. *Cell* **61**, 203–212.

Weis, F. M. B., and Davis, R. J. (1990). Regulation of epidermal growth factor receptor signal transduction: Role of gangliosides. *J. Biol. Chem.* **265**, 12059–12066.

Yamakawa, T., and Suzuki, S. (1951). The chemistry of lipids of posthemolytic residue or stroma of erythrocytes: I. Concerning the ether-insoluble lipids of lyophilized horse blood serum. *J. Biochem. (Tokyo)* **38**, 199–212.

Zheng, M., Fang, H., Tsuruoka, T., Tsuji, T., Sasaki, T., and Hakomori, S. (1993). Regulatory role of G_{M3} ganglioside in $\alpha 5\beta 1$ integrin receptor for fibronectin-mediated adhesion of FUA169 cells. *J. Biol. Chem.* **268**, 2217–2222.

Zhou, Q., Hakomori, S., Kitamura, K., and Igarashi, Y. (1994). G_{M3} directly inhibits tyrosine phosphorylation and de-*N*-acetyl-G_{M3} directly enhances serine phosphorylation of epidermal growth factor receptor, independently of receptor–receptor interaction. *J. Biol. Chem.* **269**, 1959–1965.

Tatsuya Haga
Kazuko Haga
Kimihiko Kameyama
Hirofumi Tsuga

Department of Biochemistry
Institute for Brain Research
Faculty of Medicine
University of Tokyo
Hongo, Tokyo 113, Japan

G Protein-Coupled Receptor Kinase: Phosphorylation of Muscarinic Receptors and Facilitation of Receptor Sequestration

Various kinds of hormones and autacoids, almost all kinds of neuro-transmitters, and a variety of external signals including odorants and light, interact with and activate receptors on cell membranes, which are coupled with GTP-binding regulatory proteins (G proteins). Agonist-bound receptors interact with G proteins composed of an $\alpha\beta\gamma$-trimer and facilitate the exchange of GDP and GTP thereby causing dissociation of $\alpha_{GDP}\beta\gamma$ into α_{GTP}- and $\beta\gamma$-subunits. Dissociated α_{GTP}-subunits interact with and activate or inhibit effectors such adenylyl cyclase, phospholipase $C\beta$, or ion channels (Gilman, 1987). $\beta\gamma$-Subunits have also been shown to interact with effectors (Clapham and Neer, 1993).

Agonist-bound receptors are known to induce another reaction—phosphorylation of the receptors. The phosphorylation is considered to cause desensitization of the reaction induced by agonist-bound receptors. Three different processes of desensitization are known—uncoupling of receptors from G proteins, sequestration (or internalization) of receptors from the cell surface, and decrease in the numbers of receptors (down-regulation) (Lefkowitz *et al.*, 1990).

Kinases that phosphorylate G protein-coupled receptors in an agonist-dependent manner are known as G protein-coupled receptor kinases (GRKs). At present, six different kinds of GRKs have been cloned. They constitute a subgroup of serine/threonine kinases and have catalytic domains in the central part of each molecule. Prototypes of GRKs are rhodopsin kinase (GRK1) and β-adrenergic receptor kinases (βARK1 and -2, or GRK2 and -3) that phosphorylate rhodopsin and β-adrenergic receptors in a light- or an agonist-dependent manner, respectively. GRKs are subdivided into three groups, GRK1, GRK2/GRK3, and GRK4/GRK5/GRK6, based on the homology of amino acid sequences. GRK2 and -3 are larger by 100–200 amino acid residues at the C-terminal domain than the other GRKs. The C-terminal domains are the sites that interact with G protein $\beta\gamma$-subunits. Several reviews are available on GRKs (Palczewski and Benovic, 1991; Lefkowitz, 1993; T. Haga *et al.*, 1994).

In this chapter, we describe phosphorylation of muscarinic acetylcholine receptor m_2 subtypes (m_2 receptors) by GRKs belonging to the GRK2/GRK3 subgroup. We focus on synergistic activation of GRK2 by G protein $\beta\gamma$-subunits and agonist-bound receptors and on facilitation of m_2 receptor sequestration that results from phosphorylation.

I. Activation of GRK2 by G Protein $\beta\gamma$-Subunits

A. Dual Regulation by G Proteins of the Agonist-Dependent Phosphorylation of Muscarinic Receptors

We began our experiments by questioning whether there is a kinase that phosphorylates muscarinic receptors in an agonist-dependent manner, because extensive studies had been carried out on phosphorylation of β-adrenergic receptors and rhodopsin, but very little was known about the agonist-dependent phosphorylation of receptors coupled to G_i, G_o, or G_q. We identified such an activity in porcine brain extracts and partially purified the relevant kinase. The partially purified kinase, which will be termed *mAChR kinase,* was found to phosphorylate muscarinic receptors purified from porcine brain (largely m_1 subtype) or from atria (m_2 subtype) (Haga and Haga, 1989). The atrial m_2 receptors were better substrates than brain receptors and have been used in most of the following experiments. The mAChR kinase showed essentially the same enzymatic properties as those reported for β-adrenergic receptor kinases, which include inhibition by heparin or salts and independence of second messengers or Ca^{2+} ion.

We found that the agonist-dependent phosphorylation of m_2 receptors by mAChR kinase is affected by G proteins in a dual manner (Fig. 1) (Haga and Haga, 1990). Purified m_2 receptors were reconstituted with different

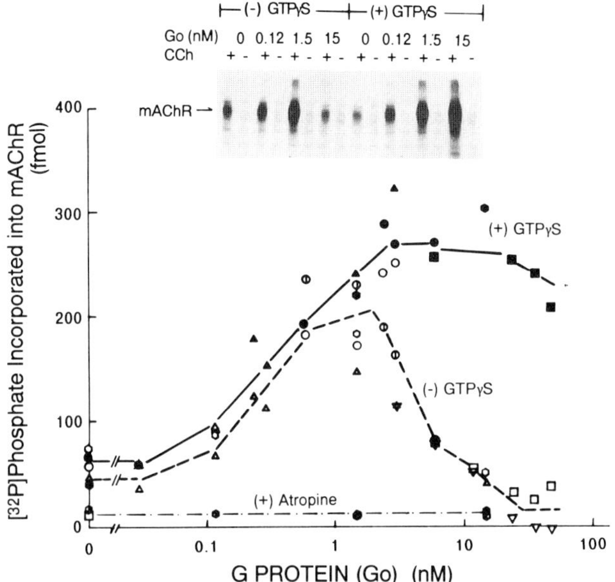

FIGURE I   Effect of G protein concentration on phosphorylation of m_2 receptors by mAChR kinase. Atrial mAChRs were reconstituted with different concentrations of G_o and then subjected to phosphorylation by mAChR kinase in the presence (closed symbols) or absence (open symbols) of 10 μM GTPγS and in the presence of 1 mM carbamylcholine or 10 μM atropine (broken line), followed by SDS–PAGE and autoradiography or counting of the receptor band. The concentration of receptors is 1.1 nM. Different symbols represent different sets of experiments. [From K. Haga and T. Haga (1990). *FEBS Lett.* **268**, 43–47.]

concentrations of purified G proteins, G_o, in lipid vesicles and then subjected to phosphorylation by mAChR kinase in the presence of the muscarinic agonist, carbamylcholine, and in the presence or absence of the nonhydrolyzable guanine nucleotide GTPγS. The phosphorylation in the absence of GTPγS increased with G_o concentration up to 2 nM and then decreased with a further increase in G_o concentration, no phosphorylation being observed in the presence of more than 30 nM G_o. However, phosphorylation in the presence of GTPγS increased up to 3 nM G_o and remained essentially the same upon further addition of G_o. The presence of carbamylcholine was necessary for phosphorylation, irrespective of the presence or absence of G proteins, and very little phosphorylation was observed in the presence of the muscarinic antagonist, atropine.

B. Inhibition of Agonist-Dependent Phosphorylation of Muscarinic Receptors by Guanine Nucleotide-Free G Protein $\alpha\beta\gamma$-Trimer

The inhibition of agonist-dependent phosphorylation of m_2 receptors in the absence of guanine nucleotides was observed with three different kinds of G proteins, G_o, G_{i1}, and G_{i2}. Both α- and $\beta\gamma$-subunits were necessary for inhibition. This inhibition was restored by addition of either GTP or GDP as well as by GTPγS. The m_2 receptor is known to interact with G_{i1}, G_{i2}, or G_o in reconstituted vesicles of purified proteins, and the interaction can be detected as high-affinity agonist binding. The high-affinity state is converted to the low-affinity state, which corresponds to that of m_2 receptors free from G proteins, by addition of guanine nucleotides such as GTPγS, GTP, or GDP (Haga *et al.*, 1986). The presence of both α- and $\beta\gamma$-subunits of G proteins is necessary for m_2 receptors to show a high affinity for a given agonist (Florio and Sternweis, 1989). The high-affinity agonist binding is thought to represent the formation of a ternary complex of agonist, receptor, and guanine nucleotide-free G protein trimer (Berstein and Haga, 1990). Both the inhibition of phosphorylation and the high-affinity agonist binding were observed in the presence of essentially the same concentrations of G proteins. In addition, effective concentrations of guanine nucleotides were also essentially the same for suppression of both the inhibition of phosphorylation and the high-affinity agonist binding. These results indicate that the two phenomena are based on the same event—the formation of a ternary complex of agonist, receptor, and guanine nucleotide-free G protein. A similar inhibition by G proteins has been reported for the phosphorylation of rhodopsin by rhodopsin kinase (Kelleher and Johnson, 1988).

C. Stimulation by G Protein $\beta\gamma$-Subunits of the Phosphorylation of m$_2$, β-Adrenergic Receptors, and Rhodopsin

The stimulation of an agonist-dependent phosphorylation of m_2 receptors by mAChR kinase was reproduced by G protein $\beta\gamma$-subunits (Haga and Haga, 1990). The stimulation by $\beta\gamma$-subunits increased with $\beta\gamma$ concentration and was not affected by the presence or absence of guanine nucleotides. The α-subunit alone did not affect the phosphorylation, although the presence of both α- and $\beta\gamma$-subunits at high concentrations caused inhibition of the phosphorylation in the absence of guanine nucleotides.

When we found the stimulatory effect of G protein $\beta\gamma$-subunits on the phosphorylation of muscarinic receptors, no effect of $\beta\gamma$-subunits on the phosphorylation of β-adrenergic receptors and rhodopsin had been reported. Therefore, we tried to examine whether the stimulation of $\beta\gamma$-subunits was specific for phosphorylation of only muscarinic receptors. Rhodopsin and β-adrenergic receptors were partially purified and subjected to phosphorylation by mAChR kinase. Both rhodopsin and β-adrenergic receptors were found to be phosphorylated by mAChR kinase in a light- or an agonist-dependent manner (Haga and Haga, 1992; Kameyama *et al.*, 1993). Figure 2 shows the effect of three different species of $\beta\gamma$-subunits on the phosphorylation of mAChRs, β-adrenergic receptors, and rhodopsin by mAChR kinase. Two kinds of $\beta\gamma$-subunits, $\beta\gamma$I and $\beta\gamma$II, which were purified from bovine

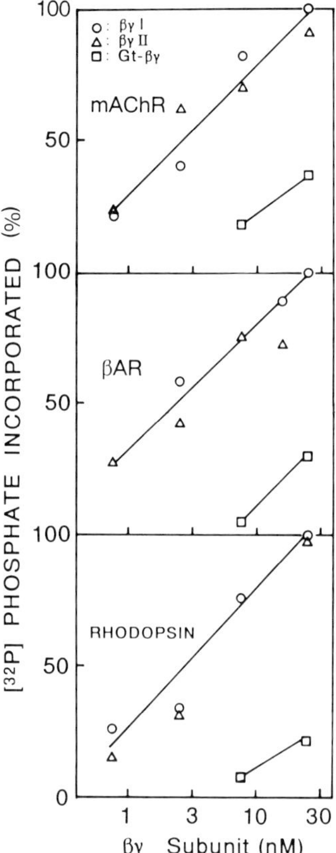

FIGURE 2 Effects of different species of $\beta\gamma$-subunits on the phosphorylation of m$_2$, β_2, and rhodopsin by mAChR kinase. $\beta\gamma$I and -II were purified from bovine brain and G$_t$-$\beta\gamma$ from bovine retina. Phosphorylation was carried out in the presence of 1 mM carbamylcholine or 10 μM isoproterenol or under fluorescent light. [From K. Kameyama *et al.* (1993). *J. Biol. Chem.* **268**, 7753–7758.]

brain, had the same stimulatory effects on the phosphorylation of the three substrates. However, the $\beta\gamma$-subunit of transducin, which was purified from bovine retina, had much smaller effects on the phosphorylation of all three substrates including rhodopsin (Kameyama *et al.*, 1993). This finding, that the effect of $\beta\gamma$-subunits is similar for the three different substrates, suggests that the target of $\beta\gamma$-subunits may not be the substrates but the kinase. Therefore, we examined whether the stimulation by $\beta\gamma$-subunits is a specific property of mAChR kinase or is shared by rhodopsin kinase and other GRKs.

D. Target of G Protein $\beta\gamma$-Subunits

We have partially purified rhodopsin kinase and examined the effect of $\beta\gamma$-subunits on the phosphorylation of rhodopsin and m$_2$ receptors by rhodopsin kinase and mAChR kinase. Phosphorylation of rhodopsin by rhodopsin kinase was stimulated by light but not affected by the presence or absence of $\beta\gamma$-subunits, in contrast to the light- and $\beta\gamma$-dependent phosphorylation of rhodopsin by mAChR kinase (Fig. 3) (Haga and Haga, 1992). The m_2 receptor was not phosphorylated by rhodopsin kinase, irrespective of the presence or absence of agonists and $\beta\gamma$-subunits, under our experimental conditions.

cDNAs for GRK2 and rhodopsin kinase were cloned by Benovic *et al.* (Benovic *et al.*, 1989; Lorenz *et al.*, 1991). They are similar to each other, but GRK2 is longer than rhodopsin kinase and has an extra peptide of about 130 amino acids in the C terminus.

We expressed GRK2 in COS7 cells and examined the effect of $\beta\gamma$-subunits on GRK2 by using cell extract as an enzyme source. Both m_2 receptors and rhodopsin were found to be phosphorylated by GRK2 and their phosphorylation was stimulated by G protein $\beta\gamma$-subunits (Fig. 4) (Kameyama *et al.*, 1993). These results show that the stimulatory effect of $\beta\gamma$-subunits depends on the species of kinases and suggest that the target of $\beta\gamma$-subunits may be the kinase.

Assuming that the C terminus may be responsible for activation by $\beta\gamma$-subunits, we constructed a mutant lacking a part of this sequence in GRK2 as shown in Figure 4. As we expected, m_2 receptors and rhodopsin were phosphorylated by the mutant in an agonist- or a light-dependent manner, but the phosphorylation was not stimulated by G protein $\beta\gamma$-subunits. This result suggests that $\beta\gamma$-subunits interact with the C-terminal portion of GRK2 and activate it.

Lefkowitz and colleagues have confirmed the activation by $\beta\gamma$-subunits of GRK2 by using β-adrenergic receptors as a substrate (Pitcher *et al.*, 1992)

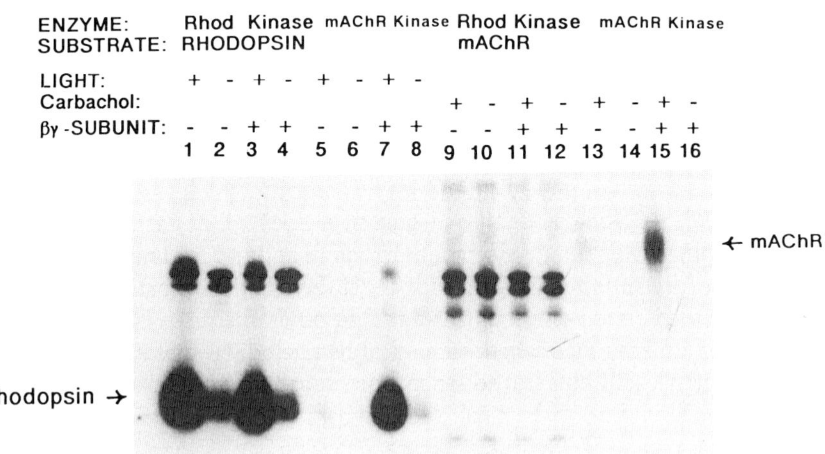

FIGURE 3   Effects of $\beta\gamma$-subunits, light, and carbamylcholine on the phosphorylation of rhodopsin and m_2 receptors by rhodopsin kinase and mAChR kinase. [From K. Haga and T. Haga (1992). *J. Biol. Chem.* **267**, 2222–2227.]

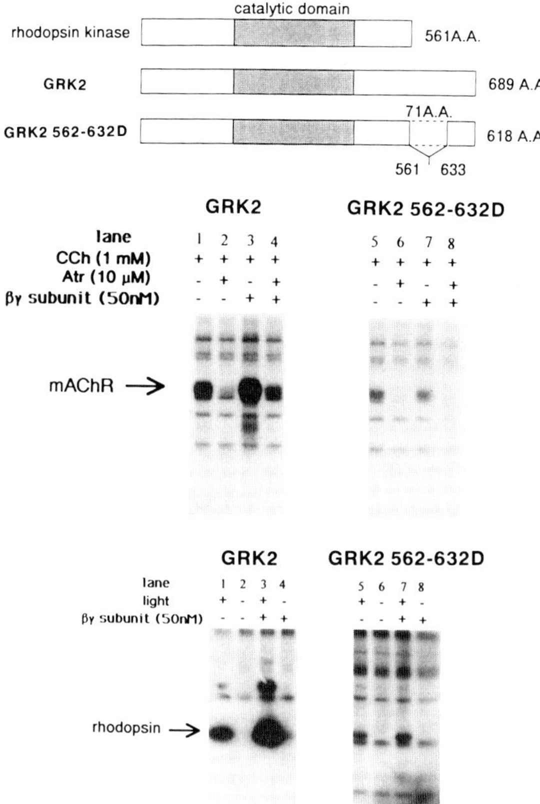

FIGURE 4 Phosphorylation of m_2 receptors and rhodopsin by GRK2 and a mutated form of GRK2 with a deletion from 562 to 632 residues. [From Kameyama *et al.* (1993). *J. Biol. Chem.* **268**, 7753–7758.]

and also reported that the peptides corresponding to the C-terminal portion of GRK2 interact with $\beta\gamma$-subunits (Koch *et al.*, 1993). Hosey and colleagues (Richardson *et al.*, 1993) have shown that $\beta\gamma$-subunits stimulate the phosphorylation of m_2 receptors by GRK3 as well as by GRK2. The m_2 receptor is also phosphorylated by GRK5 but the stimulatory effect of $\beta\gamma$-subunits is not observed (Kunapuli *et al.*, 1994). This is consistent with the assignment that the target of $\beta\gamma$-subunits is the C terminus of GRK2 or GRK3 because GRK5 does not have an equivalent C-terminal domain.

II. Synergistic Activation of GRK2 by $\beta\gamma$-Subunits and Agonist-Bound Receptors

A. Phosphorylation Sites in m_2 Receptors

Phosphorylation sites in m_2 receptors by mAChR kinase have been located in the central part of the third intracellular loop (I3) by protein chemical methods (Nakata *et al.*, 1994), although exact serine and threonine residues have not been identified yet. The number of phosphorylated amino acids was estimated to be 10 or 11 per receptor, with three or four threonine residues and seven or eight serine residues. This location was confirmed by examining the phosphorylation of an I3-deleted m_2 mutant and a glutathione S-transferase (GST) fusion protein of the peptide corresponding to the phosphorylation sites (I3-GST) (Fig. 5). The I3-deleted mutant and I3-GST were expressed in and purified from Sf9 cells and *Escherichia coli,* respectively. As was expected, I3-GST was phosphorylated by mAChR kinase or GRK2, which was expressed in and purified from Sf9 cells, but the I3-deleted m_2 mutant was not phosphorylated by either kinase.

B. Substrate Specificity of GRK2

GRK2 phosphorylates β-adrenergic receptors (β_2-subtype), rhodopsin, m_2 receptors, and substance P receptor (Kwatra *et al.*, 1993), which are linked to G_s, G_t, G_i/G_o, and G_q, respectively. In addition, muscarinic m_3 (Debburman *et al.*, 1995), α_2-adrenergic (α_{2A} and α_{2B} but not α_{2C}) (Kurose and Lefkowitz, 1994), thrombin (Ishii *et al.*, 1994), adenosine A_1 (Malek *et al.*, 1993), and N-formyl peptide (Prossnitz *et al.*, 1995) receptors are also reported to be phosphorylated by GRK2. Phosphorylation sites are located in the C-terminal tail for β_2, thrombin, and N-formyl peptide receptors, and are in the central part of the third intracellular loop for m_2 and α_2 receptors. Phosphorylated residues in α_2 receptors have been identified as serine residues in the sequence EESSSS, present in the third intracellular loop (Eason *et al.*, 1995). Phosphorylated residues in β_2, m_2, and other receptors have not been identified. It is not likely that there is a strict consensus sequence for the phosphorylation because there are no repetitive sequences found within the multiple serine and threonine residues phosphor-

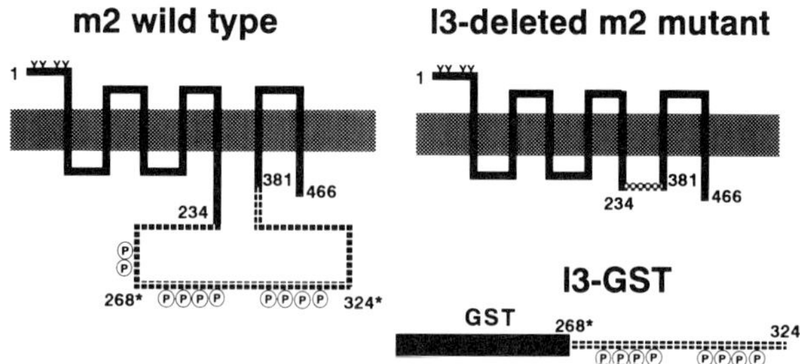

FIGURE 5 Molecular models for m_2 receptor, the third intracellular loop-detected m_2 mutant, and a fusion protein of the third intracellular loop peptide with glutathione S-transferase. The number of phosphorylation sites was estimated to be seven or eight serine residues and three or four threonine residues (Nakata *et al.*, 1994). Phosphorylation sites are depicted as Ⓟ.

ylated in these receptors. The presence of acidic residues next to serine or threonine residues appears to be the only requisite for phosphorylation (Onorato *et al.*, 1991).

These results indicate that the substrate specificity of GRK2 is not strict. On the other hand, the substrate specificity of GRK2 is strict in a sense that the substrates of GRK2 are restricted to only G protein-coupled receptors that are activated by light or agonist. The simplest explanation for these apparently contradictory facts is to assume that the phosphorylation sites in receptors are concealed at resting states and are exposed by activation of receptors. If this assumption is correct, peptides containing phosphoryla-tion sites would be good substrates because the serine and threonine residues in the peptides are expected to be exposed.

I3-GST, however, was found to be a much poorer substrate compared with m_2 receptors and the ratio of V_{max} to K_m was 1000-fold lower (K. Haga *et al.*, 1994). Synthetic peptides without GST are also poor substrates. Thus, these results do not support the assumption that agonist stimulates receptor phosphorylation by exposing the phosphorylation sites.

C. Activation by Mastoparan of the Phosphorylation of I3-GST

The phosphorylation of I3-GST by mAChR kinase or GRK2 was found to be synergistically stimulated by mastoparan and G protein $\beta\gamma$-subunits (T. Haga *et al.*, 1994). Mastoparan is a peptide in bee venom and is known to mimic the function of agonist-bound receptor and activates G proteins (Higashijima *et al.*, 1988). Figure 6 shows the phosphorylation of various concentrations of I3-GST by mAChR kinase in the absence or presence of

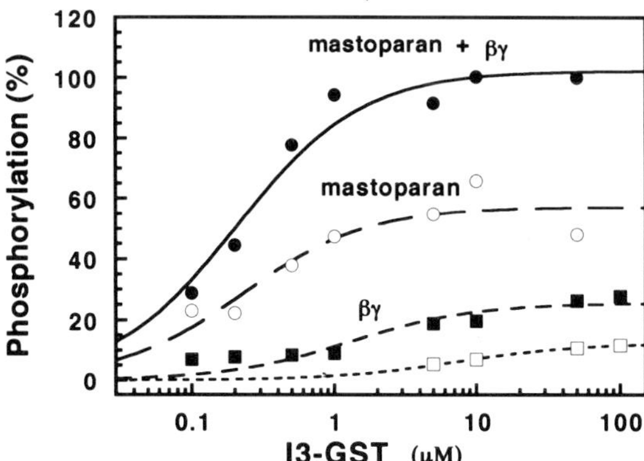

FIGURE 6 Synergistic stimulation by mastoparan and $\beta\gamma$-subunits of phosphorylation of I3-GST by mAChR kinase. Concentrations of mastoparan and $\beta\gamma$-subunits are 10 μM and 6.6 nM, respectively. Curves were fitted to the Michaelis–Menten equation. [From K. Haga *et al.* (1994). *J. Biol. Chem.* **269**, 12594–12599.]

mastoparan and G protein $\beta\gamma$-subunits. An increase in V_{max} and a decrease in K_m are observed in the presence of mastoparan alone and further changes are observed in the presence of both mastoparan and $\beta\gamma$-subunits. Stimulation of the phosphorylation of I3-GST was observed at 10–100 μM of mastoparan, and an apparent Hill coefficient for the activation was estimated to be 1.7, suggesting multiple binding sites for mastoparan. Addition of $\beta\gamma$-subunits decreased the effective concentrations of mastoparan, indicating that the interaction of $\beta\gamma$-subunits with GRK2 increases the affinity of GRK2 for mastoparan. The phosphorylation of m_2 receptors was stimulated to a limited extent by the increase of mastoparan concentrations up to 10 μM and then inhibited with the further increase of mastoparan concentration, in contrast with the increase of the phosphorylation of I3-GST with the increase of concentration of mastoparan up to 1 mM (K. Haga *et al.*, 1994).

The ratio of V_{max} to K_m for the phosphorylation of I3-GST in the presence of mastoparan is only 30-fold lower than the ratio for phosphorylation of m_2 receptors, in contrast to a 1000-fold difference in the absence of mastoparan. These results indicate that the presence of both I3-GST and mastoparan is necessary for I3-GST to be as good a substrate as the agonist-bound m_2 receptor and suggest the possibility that the agonist-bound m_2 receptor may play dual roles as a substrate and an activator.

Pitcher *et al.* (1992) have proposed that $\beta\gamma$-subunits stimulate the phosphorylation of receptors by facilitating the translocation of GRK2 from cytoplasm to membranes but not by activating GRK2, because they could not detect the stimulatory effect of $\beta\gamma$-subunits on the phosphorylation of synthetic peptides. However, we could observe stimulatory effects of $\beta\gamma$-subunits on the phosphorylation of I3-GST or synthetic peptides, and the effect was particularly clear in the presence of mastoparan. It is possible that they failed to detect the effect of $\beta\gamma$-subunits because the phosphorylation of peptides was performed in the absence of activators such as mastoparan or agonist-bound receptors. I3-GST and synthetic peptides are soluble substrates, and hence the translocation of GRK2 from soluble to vesicle fractions does not explain the stimulatory effect of $\beta\gamma$-subunits on the phosphorylation of I3-GST or peptides. We propose that the $\beta\gamma$-subunit not only translocates but also activates mAChR kinase or GRK2 probably by increasing their affinity for mastoparan or agonist-bound receptors.

D. Synergistic Activation by Agonist-Bound Receptors and G Protein $\beta\gamma$-Subunits

We have tested the assumption that the agonist-bound receptor activates GRK2 by examining the phosphorylation of I3-GST in the presence of m_2 receptors. As shown in Figure 7, the phosphorylation of I3-GST as well as m_2 receptors was stimulated by carbamylcholine in the presence of both m_2 receptors and G protein $\beta\gamma$-subunits (Kameyama *et al.*, 1994). Phosphoryla-

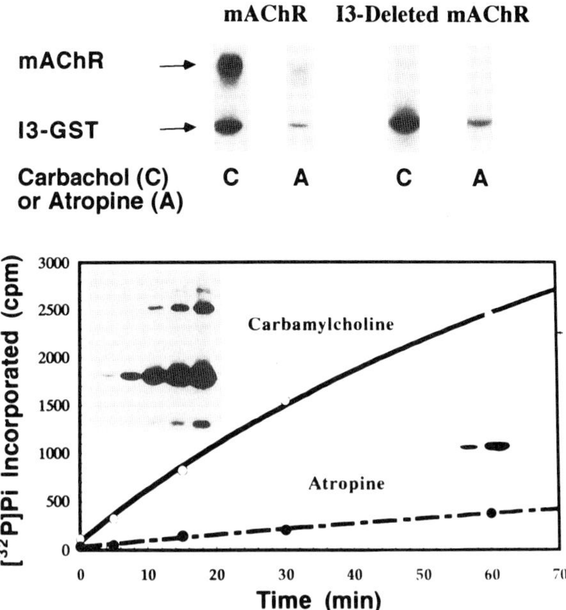

FIGURE 7   Effect of carbamylcholine on the phosphorylation of I3-GST in the presence of m_2 or I3-deleted m_2 receptors. The time course was measured in the presence of I3-deleted m_2 receptors. [From K. Kameyama *et al.* (1994). *Eur. J. Biochem.* **226**, 267–276.]

tion was hardly observed in the absence of either $\beta\gamma$-subunits or m_2 receptors or in the presence of atropine. Because I3-GST is not thought to interact with carbamylcholine, $\beta\gamma$-subunits, or m_2 receptors, these results indicate that the carbamylcholine-bound m_2 receptor together with $\beta\gamma$-subunits interact with and activate GRK2, which then interacts with and phosphorylates I3-GST.

The stimulation by carbamylcholine of the phosphorylation of I3-GST was also observed in the presence of the I3-deleted m_2 mutant (Fig. 7). The m_2 mutant is thought to interact with GRK2 in domains distinct from the phosphorylation sites because the mutant does not contain the phosphorylation sites. Thus, we may conclude that the agonist-bound m_2 receptors interact with GRK2 at two different sites—the phosphorylation sites and the activation sites.

The activation sites are expected to be the domain that undergoes the conformational change by agonist binding. Intracellular regions adjacent to transmembrane segments are possible sites. Therefore, we have examined the effects of synthetic peptides corresponding to these regions on phosphorylation of m_2 receptors and I3-GST. Peptides corresponding to the second intracellular loop, the C terminus of the third intracellular loop, and the N terminus of the C-terminal tail were found to stimulate the phosphorylation of both m_2 receptors and I3-GST (K. Haga *et al.*, 1994). These regions overlap with the domains that are assumed to be involved in the interaction with G proteins. It is reasonable to speculate that the binding of agonist to receptors induces conformational changes in these regions, and that G proteins and GRK2 competitively interact with these regions. The α-subunit, but not $\beta\gamma$-subunits, of G proteins is assumed to be competitive with GRK2 because GRK2 may interact with both $\beta\gamma$-subunits and agonist-bound receptors. This assumption is consistent with the finding that agonist-dependent phosphorylation of m_2 receptors is inhibited under conditions in which a ternary complex of agonist, receptor, and guanine nucleotide-free $\alpha\beta\gamma$-trimer is formed. The α-subunit alone does not inhibit phosphorylation nor form a complex with agonist-bound receptors. It is tempting to speculate that $\beta\gamma$-subunits facilitate the binding of the α-subunit as well as GRK2 with agonist-bound receptors (aR) and that either an aR–α–$\beta\gamma$ or an aR–GRK2–$\beta\gamma$ ternary complex is formed in which an interaction between aR and $\beta\gamma$ is also assumed.

III. Sequestration of Phosphorylated m_2 Receptors

A. Interaction of Phosphorylated m_2 Receptors with G Proteins

We have examined the effect of phosphorylation of m_2 receptors on their interaction with G proteins. We have reconstituted phosphorylated or nonphosphorylated m_2 receptors with the G protein, G_o, in crude lipid vesicles and examined the displacement by carbamylcholine of [^{3}H]QNB binding and the effect of carbamylcholine on [^{35}S]GTPγS binding (Fig. 8) (Kameyama *et al.*, 1994). The GTP-sensitive high-affinity agonist binding and the stimulation by carbamylcholine of [^{35}S]GTPγS binding were observed for both phosphorylated and nonphosphorylated receptors, and virtually no difference was found between the two receptors. There was a tendency for the effective concentrations of carbamylcholine to be slightly higher for phosphorylated receptors than for nonphosphorylated receptors but the difference was not statistically significant. These results indicate that the phosphorylation of m_2 receptors by GRK2 does not affect their interaction with muscarinic ligands and G proteins. However, Richardson *et al.* (Richardson and Hosey, 1992; Richardson *et al.*, 1992) have reported that the ability of m_2 receptors to interact with G proteins was partially impaired by their phosphorylation by GRK2 under similar experimental conditions. The reason for the discrepancy is not currently known. It is possible that the interaction of pure m_2 receptors and pure G proteins is not impaired by phosphorylation of m_2 receptors but is impaired in the presence of third components like β-arrestin. Interaction of β-arrestin with

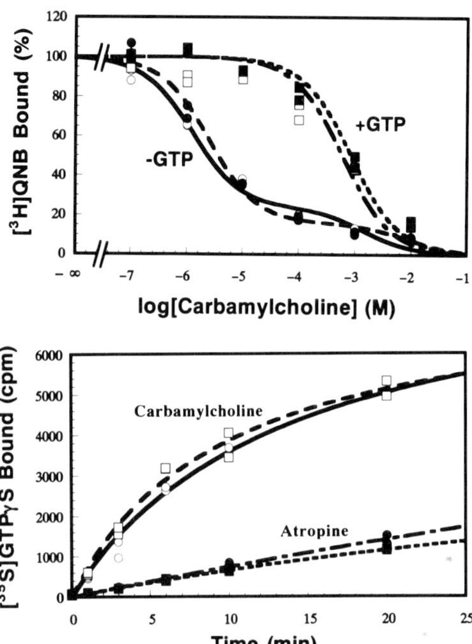

FIGURE 8 Interactions of phosphorylated or nonphosphorylated m_2 receptors with G protein G_o. Nonphosphorylated (circles and solid lines) or fully phosphorylated (squares and dotted lines) m_2 receptors were reconstituted with G_o and then subjected to [³H]QNB binding in the presence or absence of 0.1 mM GTP or to [³⁵S]GTPγS binding for indicated times in the presence of 1 mM carbamylcholine or 10 μM atropine. [From Kameyama *et al.* (1994). *Eur. J. Biochem.* **226**, 267–276.]

m_2 receptors in an agonist- and phosphorylation-dependent manner has been reported (Gurevich *et al.*, 1993), although its functional and physiological roles have not yet been elucidated.

The I3-deleted m_2 receptors are also capable of interacting with muscarinic ligands and G proteins G_o and G_{i1} as was assessed by guanine nucleotide-sensitive high-affinity agonist binding and carbamylcholine-stimulated [³⁵S]GTPγS binding of reconstituted vesicles containing purified m_2 mutants and G proteins. These results indicate that the central part of the third intracellular loop is not necessary for the interaction of m_2 receptors with muscarinic agonists and G proteins. This conclusion is consistent with the findings that the ligand binding sites of muscarinic receptors are located in the transmembrane domain and the sites for interaction with G proteins are located in the intracellular regions adjacent to transmembrane segments (Spalding *et al.*, 1994; Wess, 1993; Moro *et al.*, 1994). This conclusion is also consistent with the fact that the third intracellular loops are short for many G protein-coupled receptors including receptors, which are linked to pertussis toxin-sensitive G proteins (G_i and G_o), such as adenosine A_1 and opioid receptors. Thus, the question is raised: what is the physiological function(s) of the presence of the long third intracellular loop and its phosphorylation.

B. Phosphorylation by GRK2 of m$_2$ Receptors *in Vivo*

We have expressed m$_2$ receptors in cultured cells and examined if they are phosphorylated by GRK2 *in vivo* as well as *in vitro*. COS7 cells were transfected by expression vectors encoding m$_2$ receptors in combination with vectors encoding GRK2 or a mutant of GRK2 (DN-GRK2), which has a tryptophan substituted for a lysine at residue 220 of the catalytic domain, and which consequently does not have kinase activity. As shown in Figure 9, m$_2$ receptors were phosphorylated *in vivo* in the presence of carbamylcholine but hardly phosphorylated in its absence (Tsuga *et al.*, 1994). The agonist-dependent phosphorylation was further increased by coexpression of GRK2 and suppressed by coexpression of DN-GRK2. This result indicates that the DN-GRK2 functions as a dominant-negative mutant and that m$_2$ receptors are phosphorylated by an endogenous GRK2 or a related kinase(s).

C. Sequestration of m$_2$ Receptors

Sequestration or internalization of m$_2$ receptors was assessed as loss of binding sites of a hydrophilic ligand [^{3}H]NMS from the cell surface. In the presence of 10^{-6} M carbamylcholine, virtually no sequestration was observed for control cells expressing m$_2$ receptor alone or cells expressing m$_2$ receptors plus DN-GRK2, but approximately 40% of m$_2$ receptors were sequestered with a half-life of 20 min from cells expressing GRK2 (Fig. 9) (Tsuga *et al.*, 1994). In the presence of 10^{-4} M carbamylcholine, 30–40% of m$_2$ receptors were sequestered from control cells expressing m$_2$ receptor alone as well as from cells expressing m$_2$ receptor plus GRK2, but the proportion of sequestered receptors was 20–30% and apparently less for cells expressing DN-GRK2 (Fig. 9). Thus, GRK2 facilitates the sequestration of m$_2$ receptors at low concentrations of carbamylcholine and DN-GRK2 acts as a dominant-negative mutant of GRK2 attenuating the sequestration in the presence of carbamylcholine. These results suggest that the sequestration of m$_2$ receptors is facilitated by phosphorylation of m$_2$ receptors by GRK2 or an endogenous kinase that is the same as or similar to GRK2.

This is in sharp contrast to results on phosphorylation of β-adrenergic receptors by GRK2. Several lines of evidence have indicated that phosphorylation of β-adrenergic receptors by GRK2 is not linked with their sequestration (Lohse *et al.*, 1990; Kong *et al.*, 1994). It is likely that the phosphorylation by GRKs has different consequences for different receptors.

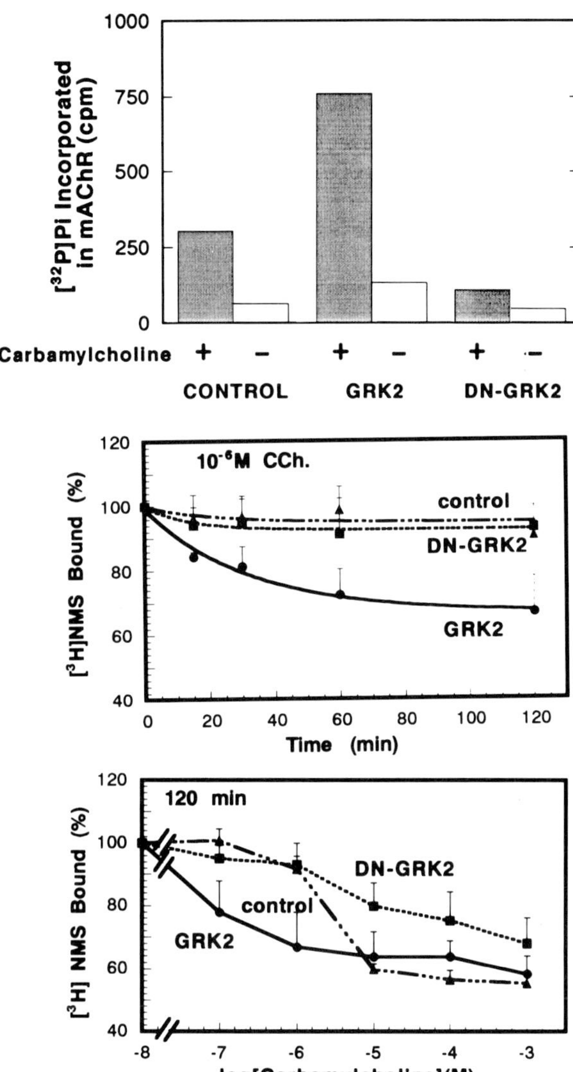

FIGURE 9 Effects of coexpression of GRK2 or dominant-negative GRK2 (DN-GRK2) on phosphorylation and sequestration of m_2 receptors. The m_2 receptor was transiently expressed in COS-7 cells alone or together with GRK2 or DN-GRK2 and then subjected to phosphorylation *in vivo* and to sequestration in the presence of carbamylcholine. The phosphorylated m_2 receptors with tagged *myc* were precipitated with anti-*myc* antibody and then subjected to SDS–PAGE, autoradiography, and counting. Sequestration of m_2 receptors induced by carbamylcholine was assessed as the loss of [^{3}H]NMS binding sites from cell surface. [From H. Tsuga *et al.* (1994). *J. Biol. Chem.* **269**, 32522–32527.]

IV. Conclusion

We may conclude that (i) GRK2, a G protein-coupled receptor kinase, is synergistically activated by G protein $\beta\gamma$-subunits and agonist-bound receptors; (ii) agonist-bound receptors serve as both substrates and activators of GRK2 by interacting with GRK2 at two different sites; (iii) phosphorylation of muscarinic receptor m_2 subtypes by GRK2 does not affect their interaction with G proteins but does facilitate their sequestration; and (iv) phosphorylation by GRK2 may cause different effects depending on the species of receptor.

These conclusions are summarized in a scheme shown in Figure 10. The binding of agonist to receptors induces conformational change in receptors, and the agonist–receptor complex interacts with G protein trimer $\alpha_{GDP}\beta\gamma$, resulting in the release of GDP and the formation of agonist, receptor, and $\alpha\beta\gamma$ complex. This agonist–receptor–$\alpha\beta\gamma$ complex represents a state that has a high affinity for agonist and is broken down by addition of guanine nucleotides. The complex formed in the presence of Mg has a much lower affinity for GDP than for GTP, thereby facilitating the binding of GTP (Shiozaki and Haga, 1992). Upon binding of GTP, α_{GTP} dissociates from both $\beta\gamma$-subunits and the receptor and then interacts with effectors. No

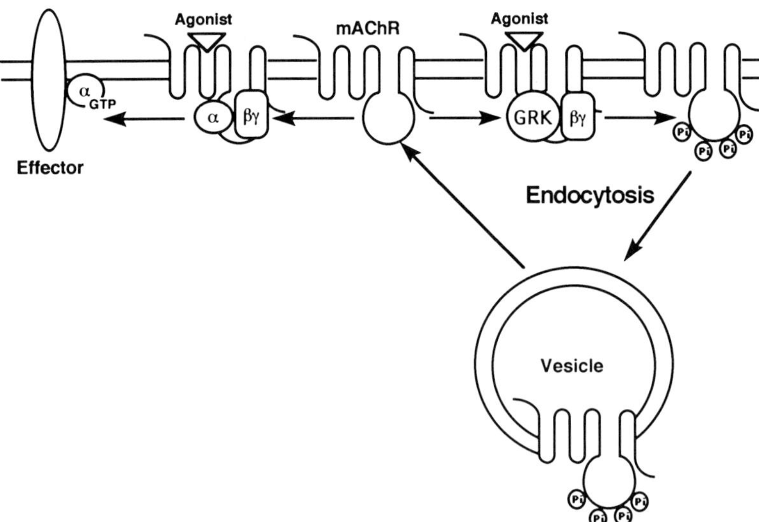

FIGURE 10   A scheme for interaction of agonist-bound receptor with $\beta\gamma$- and α-subunits or with $\beta\gamma$-subunits and GRK2 and for facilitation of endocytosis by phosphorylation. In this scheme, we assume that (i) agonist-bound receptors serve as both substrates and activators of GRK2, (ii) GRK2 interacts with $\beta\gamma$ and agonist-bound receptors and is synergistically activated by both, (iii) GRK2 competes with α-subunits for interaction with $\beta\gamma$-subunits and agonist-bound receptors, and (iv) phosphorylated receptors are internalized.

direct evidence is available for the interaction between m$_2$ receptors and G protein $\beta\gamma$-subunits, but rhodopsin has been shown to directly interact with $\beta\gamma$-subunits of transducin (Kelleher and Johnson, 1988; Phillips and Cerione, 1992). We assume that $\beta\gamma$-subunits take two forms—one bound to and the other dissociated from the receptor—and there is an equilibrium between the two forms. GRK2 is expected to bind with both agonist-bound receptors and $\beta\gamma$-subunits in the same manner as the α-subunit does, and GRK and α-subunit compete with each other. GRK2 is activated by both $\beta\gamma$-subunits and agonist-bound m$_2$ receptors resulting in the phosphorylation of the central part of the receptor. The phosphorylation of m$_2$ receptors is assumed to initiate internalization of phosphorylated m$_2$ receptors, but the mechanism of the initiation of internalization remains to be elucidated.

Acknowledgments

This work was supported in part by grants from the Ministry of Education, Science, and Culture of Japan, the Mitsubishi Foundation, Japan Research Foundation for Clinical Pharmacology, and Yamada Science Foundation.

References

Benovic, J. L., Deblasi, A., Stone, W. C., Caron, M. G., and Lefkowitz, R. J. (1989). Beta-adrenergic receptor kinase: Primary structure delineates a multigene family. *Science* **246**, 235–240.

Berstein, B., and Haga, T. (1990). Molecular aspects of muscarinic receptors. *In* "Current Aspects of the Neurosciences" (N. N. Osborne, ed.), pp. 245–284. London: Macmillan.

Clapham, D. E., and Neer, E. J. (1993). New roles for G-protein beta-gamma-dimers in transmembrane signalling. *Nature* **365**, 403–406.

Debburman, S. K., Kunapuli, P. K., Benovic, J. L., and Hosey, M. M. (1995). Agonist-dependent phosphorylation of human muscarinic receptors in *Spodoptera frugiperda* insect cell membranes by G protein-coupled receptor kinases. *Mol. Pharmacol.* **47**, 224–233.

Eason, M. G., Moreira, S. P., and Liggett, S. B. (1995). Four consecutive serines in the third intracellular loop are the sites for beta-adrenergic receptor kinase-mediated phosphorylation and desensitization of the alpha2A-adrenergic receptor. *J. Biol. Chem.* **270**, 4681–4688.

Florio, V. A., and Sternweis, P. C. (1989). Mechanisms of muscarinic receptor action on G$_o$ in reconstituted phospholipid vesicles. *J. Biol. Chem.* **264**, 3909–3915.

Gilman, A. G. (1987). G proteins: Transducers of receptor-generated signals. *Annu. Rev. Biochem.* **56**, 615–649.

Gurevich, V. V., Richardson, R. M., Kim, C. M., Hosey, M. M., and Benovic, J. L. (1993). Binding of wild type and chimeric arrestins to the m$_2$ muscarinic cholinergic receptor. *J. Biol. Chem.* **268**, 16879–16882.

Haga, K., and Haga, T. (1989). Agonist-dependent phosphorylation of cerebral and atrial muscarinic receptors: Blockade of the phosphorylation by GTP-binding regulatory proteins and its reversal by guanine nucleotides. *Biomed. Res.* **10**(4), 293–299.

Haga, K., and Haga, T. (1990). Dual regulation by G proteins of agonist-dependent phosphorylation of muscarinic acetylcholine receptors. *FEBS Lett.* **268**, 43–47.

Haga, K., and Haga, T. (1992). Activation by G protein beta gamma subunits of agonist- or light-dependent phosphorylation of muscarinic acetylcholine receptors and rhodopsin. *J. Biol. Chem.* **267**, 2222–2227.

Haga, K., Haga, T., and Ichiyama, A. (1986). Reconstitution of the muscarinic acetylcholine receptor. Guanine nucleotide-sensitive high affinity binding of agonists to purified muscarinic receptors reconstituted with GTP-binding proteins (G_i and G_o). *J. Biol. Chem.* **261**, 10133–10140.

Haga, K., Kameyama, K., and Haga, T. (1994). Synergistic activation of a G protein-coupled receptor kinase by G protein beta gamma subunits and mastoparan or related peptides. *J. Biol. Chem.* **269**, 12594–12599.

Haga, T., Haga, K., and Kameyama, K. (1994). G protein-coupled receptor kinases. *J. Neurochem.* **63**, 400–412.

Higashijima, T., Uzu, S., Nakajima, T., and Ross, E. M. (1988). Mastoparan, a peptide toxin from wasp venom, mimics receptors by activating GTP-binding regulatory proteins (G proteins). *J. Biol. Chem.* **263**, 6491–6494.

Ishii, K., Chen, J., Ishii, M., Koch, W. J., Freedman, N. J., Lefkowitz, R. J., and Coughlin, S. R. (1994). Inhibition of thrombin receptor signaling by a G-protein coupled receptor kinase. Functional specificity among G-protein coupled receptor kinases. *J. Biol. Chem.* **269**, 1125–1130.

Kameyama, K., Haga, K., Haga, T., Moro, O., and Sadée, W. (1994). Activation of G protein and G protein-coupled receptor kinase (βARK1) by a muscarinic receptor m_2 mutant lacking phosphorylation sites. *Eur. J. Biochem.* **226**, 267–276.

Kameyama, K., Haga, K., Haga, T., Kotani, K., Katada, T., and Fukada, Y. (1993). Activation by G protein beta-gamma subunits of beta adrenergic and muscarinic receptor kinase. *J. Biol. Chem.* **268**, 7753–7758.

Kelleher, D. J., and Johnson, G. L. (1988). Transducin inhibition of light-dependent rhodopsin phosphorylation: Evidence for beta gamma subunit interaction with rhodopsin. *Mol. Pharmacol.* **34**, 452–460.

Koch, W. J., Inglese, J., Stone, W. C., and Lefkowitz, R. J. (1993). The binding site for the beta gamma subunits of heterotrimeric G proteins on the beta-adrenergic receptor kinase. *J. Biol. Chem.* **268**, 8256–8260.

Kong, G., Penn, R., and Benovic, J. L. (1994). A β-adrenergic receptor kinase dominant negative mutant attenuates desensitization of the β_2-adrenergic receptor. *J. Biol. Chem.* **269**, 13084–13087.

Kunapuli, P., Onorato, J. J., Hosey, M. M., and Benovic, J. L. (1994). Expression, purification, and characterization of the G protein-coupled receptor kinase GRK5. *J. Biol. Chem.* **269**, 1099–1105.

Kurose, H., and Lefkowitz, R. J. (1994). Differential desensitization and phosphorylation of three cloned and transfected alpha 2-adrenergic receptor subtypes. *J. Biol. Chem.* **269**, 10093–10099.

Kwatra, M. M., Schwinn, D. A., Schreurs, J., Blank, J. L., Kim, C. M., Benovic, J. L., Krause, J. E., Caron, M. G., and Lefkowitz, R. J. (1993). The substance P receptor, which couples to Gq/11, is a substrate of beta-adrenergic receptor kinase 1 and 2. *J. Biol. Chem.* **268**, 9161–9164.

Lefkowitz, R. J. (1993). G protein-coupled receptor kinases. *Cell* **74**, 409–412.

Lefkowitz, R. J., Hausdorff, W. P., and Caron, M. G. (1990). Role of phosphorylation in desensitization of the beta-adrenoceptor. *Trends Pharmacol. Sci.* **11**, 190–194.

Lohse, M. J., Benovic, J. L., Caron, M. G., and Lefkowitz, R. J. (1990). Multiple pathways of rapid beta 2-adrenergic receptor desensitization. Delineation with specific inhibitors. *J. Biol. Chem.* **265**, 3202–3211.

Lorenz, W., Inglese, J., Palczewski, K., Onorato, J. J., Caron, M. G., and Lefkowitz, R. J. (1991). The receptor kinase family: Primary structure of rhodopsin kinase reveals similarities to the beta-adrenergic receptor kinase. *Proc. Natl. Acad. Sci. U.S.A.* **88**, 8715–8719.

Malek, D., Muench, G., and Palm, D. (1993). Two sites in the third inner loop of the dopamine D_2 receptor are involved in functional G protein-mediated coupling to adenylate cyclase. *FEBS Lett.* **325,** 215–219.

Moro, O., Shockley, M. S., Lameh, J., and Sadée, W. (1994). Overlapping multi-site domains of the muscarinic cholinergic hm1 receptor involved in signal transduction and sequestration. *J. Biol. Chem.* **269,** 6651–6655.

Nakata, H., Kameyama, K., Haga, K., and Haga, T. (1994). Location of agonist-dependent-phosphorylation sites in the third intracellular loop of muscarinic acetylcholine receptors (m_2 subtype). *Eur. J. Biochem.* **220,** 29–36.

Onorato, J. J., Palczewski, K., Regan, J. W., Caron, M. G., Lefkowitz, R. J., and Benovic, J. L. (1991). Role of acidic amino acids in peptide substrates of the beta-adrenergic receptor kinase and rhodopsin kinase. *Biochemistry* **30,** 5118–5125.

Palczewski, K., and Benovic, J. L. (1991). G-protein-coupled receptor kinases. *Trends Biochem. Sci.* **16,** 387–391.

Phillips, W. J., and Cerione, R. A. (1992). Rhodopsin/transducin interactions. I. Characterization of the binding of the transducin beta and gamma subunit complex to rhodopsin using fluorescence spectroscopy. *J. Biol. Chem.* **267,** 17032–17039.

Pitcher, J. A., Inglese, J., Higgins, J. B., Arriza, J. L., Casey, P. J., Kim, C., Benovic, J. L., Kwatra, M. M., Caron, M. G., and Lefkowitz, R. J. (1992). Role of beta-gamma subunits of G proteins in targeting the beta-adrenergic receptor kinase to membrane-bound receptors. *Science* **257,** 1264–1267.

Prossnitz, E. R., Kim, C. M., Benovic, J. L., and Ye, R. D. (1995). Phosphorylation of the N-formyl peptide receptor carboxyl terminus by the G protein-coupled receptor kinase, GRK2. *J. Biol. Chem.* **270,** 1130–1137.

Richardson, R. M., and Hosey, M. M. (1992). Agonist-induced phosphorylation and desensitization of human m_2 muscarinic cholinergic receptors in Sf9 insect cells. *J. Biol. Chem.* **267,** 22249–22255.

Richardson, R. M., Kim, C., Benovic, J. L., and Hosey, M. M. (1993). Phosphorylation and desensitization of human m_2 muscarinic cholinergic receptors by two isoforms of the beta-adrenergic receptor kinase. *J. Biol. Chem.* **268,** 13650–13656.

Shiozaki, K., and Haga, T. (1992). Effects of magnesium on the interaction of atrial muscarinic acetylcholine receptors and GTP-binding regulatory proteins. *Biochemistry* **31,** 10634–10642.

Spalding, T. A., Birdsall, N. J. M., Curtis, C. A. M., and Hulme, E. C. (1994). Acetylcholine mustard labels the binding site aspartate in muscarinic acetylcholine receptors. *J. Biol. Chem.* **269,** 4092–4097.

Tsuga, H., Kameyama, K., Haga, T., Kurose, H., and Nagao, T. (1994). Sequestration of muscarinic acetylcholine receptor m_2 subtypes: Facilitation by G protein-coupled receptor kinase (GRK2) and attenuation by a dominant-negative mutant of GRK2. *J. Biol. Chem.* **269,** 32522–32527.

Wess, J. (1993). Molecular basis of muscarinic acetylcholine receptor function. *Trends Pharmacol. Sci.* **14,** 308–313.

Hiroyoshi Hidaka
Hisayuki Yokokura

Department of Pharmacology
Nagoya University School of Medicine
Nagoya, 466, Japan

Molecular and Cellular Pharmacology of a Calcium/ Calmodulin-Dependent Protein Kinase II (CaM Kinase II) Inhibitor, KN-62, and Proposal of CaM Kinase Phosphorylation Cascades

The autonomic nervous system plays a central role in maintaining homeostasis under resting as well as stressful conditions. Neurohumoral substances released in response to environmental stimuli induce different changes in different types of cells. This process, commonly referred to as signal transduction, has been extensively investigated for many years. cAMP was regarded as a prototype second messenger (Rall and Sutherland, 1958; Sutherland and Rall, 1958) responsible for signal transduction in many cells. Soon it was discovered that protein kinase A (PKA) was involved in the cAMP-mediated signal transduction (Walsh *et al.*, 1968). Based on such background, there is now a large list of second messengers and protein kinases associated in numerous physiological processes involving cell differentiation and growth to death.

It is known that Ca^{2+} is an important intracellular second messenger. Ca^{2+} binds to a variety of proteins and alters their functions. Calmodulin (CaM) is a ubiquitous Ca^{2+}-binding protein and the Ca^{2+}/CaM complex

Advances in Pharmacology, Volume 36

interacts with target proteins, resulting in their functional changes (Wolff and Brostrom, 1979; Tanaka and Hidaka, 1980). Evidence has elucidated that Ca^{2+} activates diverse intracellular signaling pathways via protein phosphorylation. In this respect, it is believed that the family of Ca^{2+}/CaM-dependent protein kinases (CaM kinases) is important for Ca^{2+}-mediated protein phosphorylation. The prototype of this family is smooth muscle myosin light-chain kinase (MLCK), which phosphorylates myosin regulatory light chain in a Ca^{2+}/CaM-dependent manner resulting in the activation of actomyosin ATPase required for muscle fiber contraction (Walsh and Hartshone, 1982). A CaM kinase distinct from MLCK was identified in a rabbit brain extract (Endo and Hidaka, 1980) and many CaM kinases have subsequently been discovered. They are classified into two groups (Table I)—CaM kinase I/V, II, and IV (-Gr) assigned to multifunctional enzymes and MLCK, phosphorylase kinase, and CaM kinase III assigned to specific enzymes—in which the terms *multifunctional* and *specific* are referred to as having the ability to phosphorylate multiple endogenous proteins and a specific protein, respectively. These findings are reviewed by Schulman (1993) and Nairn and Picciotto (1994).

The multifunctional CaM kinases, as well as MLCK, have been found to be regulated in an intrasteric manner by their own pseudosubstrates (Cruzalegui and Means, 1993). The enzymes are kept in a basal state by the interaction of catalytic domain with pseudosubstrate within a molecule. The multifunctional CaM kinases have different tissue distributions and substrate specificities, suggesting that each of the enzymes is involved in different types of Ca^{2+}-mediated processes. Despite these advances, the precise physiological role of these kinases remains unclear.

Since the discovery of a CaM antagonist, W-7 (N-(6-aminohexyl)-5-chloro-1-naphthalenesulfonamide), by Hidaka *et al.* in 1978, we have syn-

TABLE I Ca^{2+}/Calmodulin-Dependent Protein Kinases (CaM Kinases) and the Classification of Them into Two Groups— Specific and Multifunctional Enzymes

Ca^{2+}/calmodulin-dependent protein kinases	
Specific	*Multifunctional*
Phosphorylase kinase	CaM kinase I[c]
MLCK[a]	CaM kinase II
CaM kinase III[b]	CaM kinase IV[d]

[a] Myosin light-chain kinase.
[b] Elongation factor-2 (EF-2) kinase.
[c] Related enzymes include CaM kinases Ia, Ib, and V.
[d] It appears to be identical to CaM kinase-Gr.

thesized and screened more than 3000 chemical compounds for their inhibitory effects on the protein kinases PKA, protein kinase G (PKG), protein kinase C (PKC), smooth muscle MLCK, CaM kinases I, II, and V, and casein kinase I and II. Of these, H-series compounds have proved to be relatively specific protein kinase inhibitors. Methods of the synthesis of H-series compounds and their applications were first described in *Biochemistry* (Hidaka *et al.*, 1984), which has so far been cited 2007 times.

We have succeeded in synthesizing the potent, specific inhibitor of CaM kinase II, KN-62 (1-[N,O-bis(5-isoquinolinesulfonyl)-N-methyl-L-tyrosyl)-4-phenylpiperazine) (Tokumitsu *et al.*, 1990). CaM kinase II is distributed in a wide range of tissues, but the brain contains the highest amount (e.g., 2% of total proteins in rat hippocampus). The enzyme is activated by Ca^{2+}/CaM and autophosphorylated, accompanying the generation of Ca^{2+}/CaM-independent activity. This constitutively active form is believed to prolong the effects of a transient increase in intracellular Ca^{2+} concentration. It has been accepted that CaM kinase II is highly involved in neuronal activities such as neurotransmitter synthesis and release, neuronal plasticity, and memory (Schulman, 1988). In this chapter, we review previous studies with KN-62 and propose the involvement of CaM kinase II in various aspects of cell function.

A CaM kinase V activator was identified in a crude CaM kinase V preparation and partially purified (Mochizuki *et al.*, 1993a). The activator phosphorylated the highly purified CaM kinase V in a Ca^{2+}/CaM-dependent manner and this phosphorylation enhanced the Ca^{2+}/CaM-dependent CaM kinase V activity. The activator may be a novel CaM kinase, which has also been found to activate the recombinant CaM kinase I and constitute—together with CaM kinase I/V—a CaM kinase phosphorylation cascade. Since the proposal of the potential cascade (Mochizuki *et al.*, 1993a), CaM kinases Ia and IV activators have subsequently been purified and characterized (Lee and Edelman, 1994; Okuno *et al.*, 1994; Tokumitsu *et al.*, 1994). In this chapter, the new group of CaM kinases and potential CaM kinase phosphorylation cascades are also described.

I. Molecular Pharmacology: CaM Antagonists to Protein Kinase Inhibitors

A. Development of H-Series Compounds

In 1978, we succeeded in synthesizing W-7, a naphthalenesulfonamide derivative (Hidaka *et al.*, 1978), which can inhibit a protein activator-stimulated cyclic nucleotide phosphodiesterase (PDE). The use of W-7 has elucidated the involvement of CaM in platelet aggregation and secretion (Nishikawa *et al.*, 1980), vascular contraction (Hidaka *et al.*, 1980), the degranulation of neutrophils (Naccache *et al.*, 1980), and cell proliferation (Hidaka *et al.*, 1981). W-7 has also been useful in studying the activation mechanism of Ca^{2+}/CaM-dependent enzymes. W-7 has become the prototype of H-series compounds and we have synthesized many naphthalenesulfonamide derivatives to date (Fig. 1).

During the screening of the derivatives for CaM antagonists, we have uncovered the structure and function relationship—the shorter alkyl chain derivatives ($n = 2–5$) of W-7 ($n = 6$) have more inhibitory effects on protein kinases such as smooth muscle MLCK, PKA, and PKC (Inagaki *et al.*, 1986). They include A-3 (*N*-aminoethyl)-5-chloro-1-naphthalenesulfonamide) and ML-9 (1-(5-chloronaphthalenesulfonyl)-1-H-hexahydro-1,4-diazepine), which inhibit the enzymes competitively with respect to ATP without interacting with Ca^{2+}/CaM. ML-9, which has the alkyl chain replaced by a diazepine group, proved to be a specific MLCK inhibitor with a K_i value of 3.8 μM (Saitoh *et al.*, 1987). SC-9 (*N*-(6-phenylhexyl)-5-chloro-1-naphthalenesulfonamide) has a phenyl group added to the alkyl chain ($n = 6$) and is a Ca^{2+}-dependent activator of PKC (Ito *et al.*, 1986). SC-9 appears to resemble phosphatidylserine, another cofactor of PKC, in activating the enzyme. These observations have encouraged us to synthesize new protein kinase inhibitors.

When the naphthalene ring was replaced by an isoquinoline ring in the derivatives with a shorter alkyl chain, the isoquinolinesulfonamide derivatives were no longer CaM antagonists but potent protein kinase inhibitors

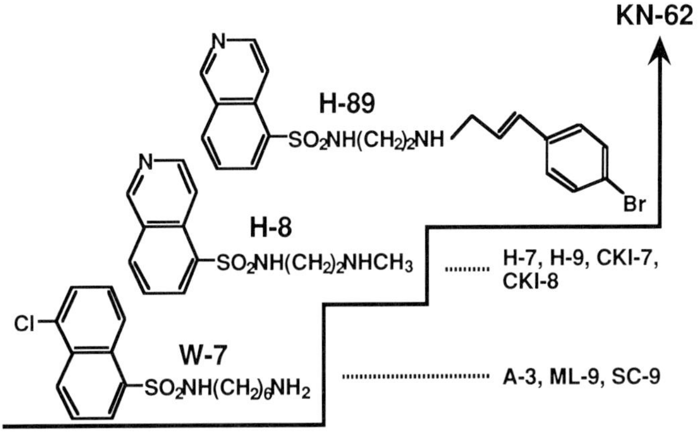

FIGURE 1   Development of H-series compounds. W-7, a naphthalenesulfonamide derivative, was identified as a CaM antagonist. Protein kinase inhibitors, including A-3, ML-9, and SC-9, were subsequently developed within its derivatives. The sequential introductions of an isoquinoline ring and a cinnamyl group have given birth to two sets of protein kinase inhibitors represented by H-8 and H-89. H7, H-9, CKI-7, and CKI-9 isoquinolinesulfonamide derivatives are related compounds of H-8. The potency and specificity of each compound were described in the text. The most recent compound, KN-62, has been synthesized on the course to developing H-series compounds.

(Hidaka *et al.*, 1984). The introduction of the isoquinoline ring into H-series compounds has produced another set of protein kinase inhibitors, H-8 (*N*-[2-(methylamino)ethyl]-5-isoquinolinesulfonamide), H-7 (1-(5-isoquinolinesulfonyl)-2-methylpiperazine), and CKI-7 (*N*-(2-aminoethyl)-5-chloro-isoquinoline-8-sulfonamide). H-8 proved to be a potent protein kinase inhibitor relatively specific to PKA and PKG (Hidaka *et al.*, 1984). This inhibition was competitive with respect to the phosphate donor, ATP, and noncompetitive with respect to the phosphate acceptors, substrate proteins. H-8 slightly resembles ATP in structure, raising the possibility that the compound binds to the ATP binding site of the enzymes. H-8 bound to the enzymes with a stoichiometry of 1:1 and had unique features that differ from those of ATP analogs (Hagiwara *et al.*, 1987): (i) H-8 specifically inhibited cyclic nucleotide-dependent protein kinases but not other protein kinases, (ii) the binding constants of H-8 to the enzymes were much lower than those of ATP, (iii) the binding of H-8 to the enzymes was independent of Mg^{2+}, and (iv) the binding subsite of H-8 was related to a phosphate binding site at the ATP binding site of the enzymes and differed slightly from that of ATP.

H-7 is a relatively specific PKC inhibitor and this inhibition was competitive with respect to ATP with a K_i value of 6.0 μM (Hidaka *et al.*, 1984). H-7 inhibited PKC by its direct interaction with PKC, but did not inhibit the cofactors of the enzyme, Ca^{2+} or phospholipid. H-7 has been widely used to elucidate the physiological roles of PKC in specifically inhibiting enzyme-mediated protein phosphorylation. H-9 is closely related to H-7 in structure and has a free amino group. H-9 was used in subsequent studies as an affinity ligand and PKC was highly purified by affinity chromatography (Ohno *et al.*, 1987). Of the isoquinolinesulfonamide derivatives, we found the casein kinase I inhibitor CKI-7 (*N*-(2-aminoethyl)-isoquinoline-8-sulfonamide) (Chijiwa *et al.*, 1989). Except for CKI-7, there is no potent, specific casein kinase I inhibitor available and we must depend on this compound to discover the biological significance of the enzyme.

An additional set of H-series compounds were produced by the addition of a cinnamyl group to the isoquinoline ring. They include H-89 (*N*-[2-(*p*-bromocinnamylamino)ethyl]-5-isoquinolinesulfonamide), which is the most potent, highly specific PKA inhibitor. H-89 inhibited PKA competitively with respect to ATP with a K_i value of 0.05 μM, whereas the compound was a relatively poor inhibitor of PKG, PKC, MLCK, and casein kinase I with K_i values of 0.5, 30, 30, and 40 μM, respectively (Chijiwa *et al.*, 1990). H-89 will aid as a pharmacological tool to clarify the physiological significance of PKA-mediated protein phosphorylation.

B. KN-62, a CaM Kinase II Inhibitor

CaM regulates a variety of proteins in response to an increase of intracellular Ca^{2+} concentration. These proteins range from membrane-associated proteins to protein kinases, including type I PDE, Ca^{2+}-pump, adenylate cyclase, calcineurin, "constitutive" nitric oxide synthase, 90-kDa heat shock protein, cytoskeletal proteins such as adducin, protein 4.1, and caldesmon, and the family of CaM kinases. Many CaM antagonists including W-7 are now available, but specific compounds are required to affect each of the target proteins in the Ca^{2+}/CaM signaling pathway. In this respect, we have succeeded in synthesizing the CaM kinase II inhibitor, KN-62 (Fig. 2). KN-62 possesses such a large (M_r 721.9 kDa) and complex structure that we cannot regard it as a isoquinolinesulfonamide derivative at first glance. However, it consists of isoquinolinesulfonamide and piperazine groups, which are frequently used in H-series compounds. KN-62 inhibited the phosphorylation of chicken gizzard myosin regulatory light chain by CaM kinase II without affecting the activity of PKA, PKC, or MLCK in doses up to 100 μM (Fig. 3; Tokumitsu $et\ al.$, 1990). KN-62 also inhibited the Ca^{2+}/ CaM-dependent autophosphorylation of both α- (M_r 50 kDa) and β- (M_r 60 kDa) subunits of CaM kinase II, but did not affect the Ca^{2+}/CaM-independent activity of previously autophosphorylated enzymes. In addition, KN-62 was proven to bind to CaM kinase II, but not Ca^{2+}/CaM, by affinity chromatography using the compound coupled to Sepharose, consistent with the results of kinetic analyses that KN-62 inhibited CaM kinase II competitively with respect to Ca^{2+}/CaM with a K_i value of 0.8 μM. Of H-series protein kinase inhibitors, KN-62 is unique because the compound binds to the Ca^{2+}/CaM- but not the ATP-binding domain of CaM kinase II. The developmental history originating from the prototype, W-7 (CaM antagonist), to the most recent derivative, KN-62 (CaM kinase II inhibitor), may promise the development of specific inhibitors of any protein kinase by H-series compounds. Treatment with KN-62 suppressed the A_{23187}-stimulated autophosphorylation of CaM kinase II in PC12D pheochromocytoma cells (Tokumitsu $et\ al.$, 1990). These results suggest that KN-62 may be useful for approaching the physiological roles of CaM kinase II. Generally,

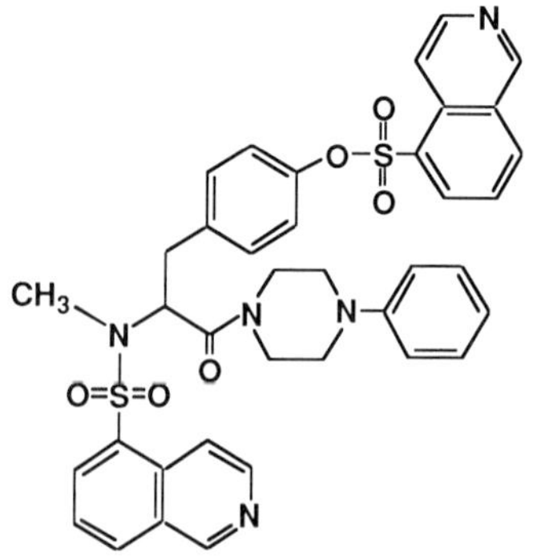

FIGURE 2   Chemical structure of KN-62—1-[N,O-bis(5-isoquinolinesulfonyl)-N-methyl-L-tyrosyl]-4-phenylpiperazine.

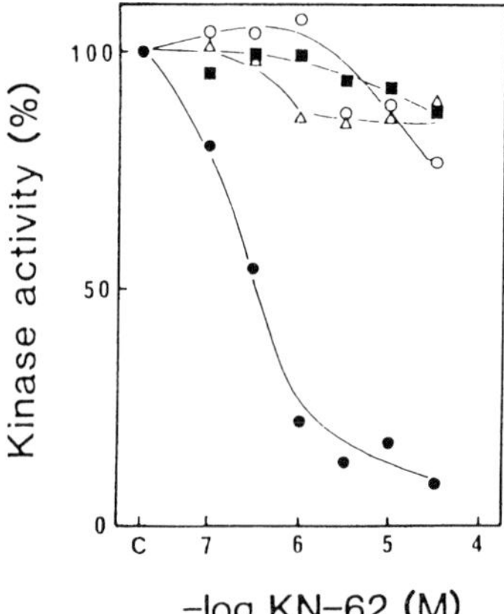

FIGURE 3 Effects of KN-62 on PKA, PKC, smooth muscle MLCK, and CaM kinase II. PKA (○), PKC (△), smooth muscle MLCK (■), and CaM kinase II (●) activities were measured in the presence of $0–10^{-4.5}$ M KN-62. For determining PKA and PKC activities, 20 μg histone IIB and IIIS were used as a substrate protein, respectively. Smooth muscle MLCK and CaM kinase II activities were determined in the presence of 0.1 μM CaM by using myosin 20-kDa light chain. Protein kinase activities were expressed as a percentage of control activity in the absence of KN-62. C, the absence of KN-62.

it is accepted that protein kinase inhibitors should meet the following three criteria: (i) direct binding to the protein kinase, (ii) specificity to the protein kinase, and (iii) cell membrane permeability.

KN-62 has not yet fulfilled the criteria except for its direct binding to CaM kinase II. We are now examining the remaining two criteria of specificity and membrane permeability. Currently, there are no other pharmacological agents besides KN-62 that are available to explore the physiological role of CaM kinase II. It is anticipated that this agent, as other agents, will be of great promise.

II. Cellular Pharmacology of the Specific CaM Kinase II Inhibitor, KN-62

A. Involvement of CaM Kinase II in Neuronal Activities

Tyrosine hydroxylase (TH) regulates the formation of L-3,4-dihydroxyphenylalanine (DOPA) from L-tyrosine, which is the rate-limiting step of catecholamine biosynthesis. There is convincing evidence that TH itself is regulated by undergoing phosphorylation. TH has been found to be *in vitro* phosphorylated by PKA, PKG, PKC, and CaM kinase II. However, which kinases phosphorylate TH mainly *in vivo* is still not clear. Treatment with

10 μM KN-62 suppressed 56 mM K$^+$-stimulated TH phosphorylation in rat pheochromocytoma PC12h cells to a basal level without affecting that stimulated by 10 μM forskolin (Ishii *et al.*, 1991). In addition, the treatment with KN-62 inhibited a high K$^+$- but not forskolin-stimulated increase in *in situ* DOPA formation. KN-62 did not affect rat-purified TH in doses up to 30 μM. These results suggest that CaM kinase II phosphorylates TH in response to neuronal depolarization with a resultant stimulation of neurotransmitter synthesis. TH has been found to be phosphorylated primarily at Ser19 and slightly at Ser40 by CaM kinase II (Campbell *et al.*, 1986). Protein 14-3-3 is required for a two or three times increase in V_{max} of previously phosphorylated TH by CaM kinase II (Ichimura *et al.*, 1987).

The long-term potentiation (LTP) of synaptic efficacy in hippocampus is regarded as the experimental model of learning and memory. Two types of LTP with different regulatory mechanisms exist: one observed at Schaffer commissural-fiber CA1 pyramidal cell synapses, and the other at mossy-fiber CA3 pyramidal cell synapses. The induction of LTP in the CA1 region is completely dependent on the N-methyl-D-aspartate (NMDA) type of glutamate receptors (GluRs) and needs the activation of PKC or CaM kinase II (Malinow *et al.*, 1989). KN-62 strongly suppressed the CA1 region LTP at 3 μM in rat hippocampal slices when applied prior to, not immediately after, a tetanic stimulation (Ito *et al.*, 1991). In contrast, KN-62 did not affect the generation of LTP in the CA3 region. Thus, CaM kinase II may be critical for the initial phase of LTP in the CA1 region. 2-Amino-5-phosphonovalerate (AP5), an NMDA antagonist, resembled KN-62 in its effects on LTP in the CA1 and CA3 regions. It was proven using *Xenopus* oocytes microinjected with NMDA receptor mRNA that KN-62 does not affect the activity of the receptors (Ito *et al.*, 1991). Considering the enrichment of CaM kinase II in postsynaptic neurons, the enzyme—together with NMDA receptors—may operate in the generation of LTP in the CA1 region. In visual cortex, long-term depression (LTD), as well as LTP, is regarded as the basis for the modifiability of neurons during the critical period of postnatal development. After the injection of KN-62 into layer two or three neurons of the visual cortex slice of young rats, a tetanic stimulation of white matter induced LTD at tetanized synapses, but excitatory postsynaptic potentials (EPSP) at nontetanized synapses remained unchanged (Funauchi *et al.*, 1992). The majority of cells not treated with KN-62 showed LTP. These observations suggest that postsynaptic CaM kinase II functions in the induction of both LTP and LTD in visual cortex at a developing period.

Regarding the presynaptic and postsynaptic participation in an increased synaptic efficacy during the expression of LTP, presynaptic terminals have been shown to increase the release of glutamate and postsynaptic GluRs to enhance responsiveness to glutamate. The enhanced postsynaptic responsiveness presumably involves non-NMDA GluRs because non-NMDA antagonists, but not NMDA antagonists, block the generation of LTP

(Madison *et al.*, 1991). A variety of ion channels have been shown to be regulated by protein phosphorylation. Treatment with glutamate/glycine (Glu/Gly) enhanced the phosphorylation of α-amino-3-hydroxy-5-methyl-4-isoxazoleproprionate (AMPA)-type GluRs in cultured rat hippocampal neurons and the addition of KN-62 at 10 μM, as well as AP5 at 100 μM, blocked this phosphorylation (Tan *et al.*, 1994). The treatment with Glu/Gly gave a twofold increase in the ratio of Ca^{2+}/CaM-independent activity to Ca^{2+}/CaM-dependent activity of CaM kinase II in which the enzyme was more phosphorylated. Ca^{2+} influx through NMDA GluRs activates CaM kinase II and this activated enzyme may in turn regulate AMPA-type GluRs via their phosphorylation. This potential mechanism may contribute to the postsynaptic component of LTP in hippocampus.

Survival of different types of neurons is enhanced by depolarizing culture conditions. Grown in serum-containing medium, rat cerebellar granule cells require depolarizing conditions, such as 25 mM K$^+$ on the second or third day, for their survival. In the absence of excess K$^+$, a large proportion of the cells die relatively abruptly by the end of first week. Similarly, NMDA enhanced the survival of the cells at 140 μM (Hack *et al.*, 1993). The cell survival-promoting action of high K$^+$ was inhibited by KN-62 with an IC$_{50}$ value of 2.5 μM, but H-7 (50 μM), polymixin B (50 μM), nor disialoganglioside GD1a (100 μM) had any effect (Hack *et al.*, 1993). In addition, KN-62 suppressed the survival dependence on NMDA in a dose-dependent manner. KN-62 did not affect the $^{45}Ca^{2+}$ influx induced by either high K$^+$ or NMDA plus Mg^{2+}-free medium in doses up to 10 μM. These results suggest the involvement of CaM kinase II in neuronal survival.

Concerning the pathophysiological aspects, paired helical filaments (PHF) of neurofibrillary tangles seen in brain from patients with Alzheimer's disease contain hyperphosphorylated forms of τ. This aberrant protein phosphorylation is implicated in the peculiar secretion and processing of amyloid peptides. τ Phosphorylation in human temporal slices, which were obtained from resections for intractable hippocampal epilepsy, has been shown to increase by treatment with okadaic acid (a protein phosphatase inhibitor) in a dose-dependent manner, but treatment with NMDA, kainate, and quisqualate had no effect (Harris *et al.*, 1993). The most phosphorylated forms of τ (PHF-like τ) by okadaic acid were dephosphorylated by calcineurin. KN-62, as well as staurosporine, inhibited the formation of PHF-like τ. τ phosphorylation may be regulated by protein kinases, such as CaM kinase II and PKC, or okadaic acid-sensitive protein phosphatases, of which dysfunctions may be involved in the pathogenesis of Alzheimer's disease.

B. Involvement of CaM Kinase II in Cardiovascular Regulations

An important step in smooth muscle contraction is the phosphorylation of myosin regulatory light chain. However, the extent of its phosphorylation is determined by the ratio of activity of MLCK to myosin light-chain phosphatase, of which a cDNA has been isolated (Shimizu *et al.*, 1994). Purified MLCK has been shown to be phosphorylated by CaM kinase II at Ser512 (site A) around its Ca^{2+}/CaM binding domain (Ikebe and Reardon, 1990). This phosphorylation in turn reduces its affinity for Ca^{2+}/CaM. MLCK was phosphorylated at site A in cultured bovine tracheal smooth muscle cells by treatment with the Ca^{2+} ionophore, ionomycin (Tansey *et al.*, 1992). This phosphorylation was inhibited by KN-62 but not by the structural analog, KN-04, which has no effect on CaM kinase II (Tansey *et al.*, 1992). Addition of KN-62 suppressed Ca^{2+}-stimulated MLCK phosphorylation in β-escin-permeabilized bovine tracheal smooth muscle cells (Tansey *et al.*, 1994). In the permeable cells, Ca^{2+} concentrations required for the half-maximal phosphorylation of MLCK and myosin light chain were 500 and 250 nM, respectively. In the presence of KN-62 (1.5 μM), the Ca^{2+} concentration required for the half-maximal phosphorylation of myosin light chain decreased to 170 nM. These results suggest that Ca^{2+}/CaM available for MLCK activation is limiting and that CaM kinase II phosphorylates MLCK and modulates its activity in smooth muscle.

Intracellular Ca^{2+} plays a central role in cardiac muscle contraction. CaM kinase II is distributed in cardiac myocytes and has recently been purified from rat heart, which phosphorylated troponin I, glycogen synthase, and phospholamban. KN-62, but not KN-04, reduced the spontaneous beating rate of fetal mouse ventricular myocytes in a dose-dependent manner accompanying a decrease in the maximal value of relative fluo-3 fluorescence intensity, an indicator of Ca^{2+} concentration, at the systolic phase (Okazaki *et al.*, 1994). It is believed that the phosphorylation of phospholamban by CaM kinase II releases the inhibition by phospholamban of Ca^{2+} pump in sarcoplasmic reticulum. The physiological significance of troponin I phosphorylation remains unclear. However, these phosphorylations may play some roles in cardiac contraction by altering Ca^{2+} mobilization or actomyosin contractility.

C. Involvement of CaM Kinase II in Secretory Regulations

Inherent in cell secretion are prominent morphological changes including the translocation of intracellular tubulovesicles to plasma membrane. This translocation depends on the dynamic reorganization of cytoskeletal components. The significance of CaM kinase II in the regulation of synaptic neurotransmission has been recognized. The enzyme phosphorylates cytoskeletal proteins, such as synapsin I, τ, microtubule-associated protein-2 (MAP-2), and caldesmon, resulting in the alteration of their functions. It is of great interest to study the physiological roles of CaM kinase II in various secretory cells. Polypeptides (M_r 60 and 61 kDa) have been identified as CaM kinase

II in rat gastric parietal cells because they were predominantly phosphory-lated in a cytosolic fraction and both bound to $[^{125}I]$CaM in a Ca^{2+}-dependent manner (Mayer *et al.*, 1994). The polypeptide-containing fraction had the ability to Ca^{2+}/CaM-dependently phosphorylate the specific peptide sub-strate for CaM kinase II, autocamtide II. It was revealed by using restriction mapping following the amplification of cDNA with RT-PCR that γ- and δ-subunits of CaM kinase II, but not α- and β-subunits, are expressed in the parietal cells (Mayer *et al.*, 1994). KN-62 inhibited the autocamtide-2 phosphorylation in a dose-dependent manner with an IC_{50} value of 2 μM, suggesting that KN-62 is a potent inhibitor of all the subunits of CaM kinase II—α, β, γ, and δ. The acid secretion in parietal cells is stimulated by histamine or carbachol. The histamine-stimulated secretion is mediated by an increase in intracellular cAMP, whereas the carbachol-stimulated secretion is achieved via IP_3-induced Ca^{2+} release from intracellular Ca^{2+} stores. The acid secretion in parietal cells is often assessed by indirect measurement of the uptake of $[^{14}C]$aminopyrine. KN-62 did not significantly inhibit the histamine-stimulated aminopyrine uptake in isolated rabbit parietal cells, but it inhibited the carbachol-stimulated aminopyrine uptake in a dose-dependent manner with an IC_{50} value of 20 μM (Tsunoda *et al.*, 1992). It was also shown by using fura-2/AM, a fluorescent dye, that KN-62 did not affect a carbachol-stimulated increase in cytoplasmic-free Ca^{2+} con-centration or its basal level. These results suggest that CaM kinase II plays an important role, at least in part, in the acid secretion in gastric parietal cells.

In pancreatic β-cells, glucose is considered to be a primary physiological stimulus for insulin secretion, which is dependent on extracellular Ca^{2+}. The metabolism of glucose within β-cells results in a transient increase in the ratio of ATP to ADP, leading to the depolarization and opening of voltage-dependent Ca^{2+} channels. However, the regulatory mechanism by which an increase in cytoplasmic Ca^{2+} concentration is translated into the insulin secretion is still largely unknown. A polypeptide (M_r 50 kDa) in a crude extract of rat pancreatic β-cells immunoreacted with an polyclonal antibody against rat brain CaM kinase II (Wenham *et al.*, 1992). Alloxan, a potent diabetogenic agent, inhibited both the glucose-induced insulin secretion and CaM kinase II activity, which suggested the involvement of the enzyme in insulin secretion (Colca *et al.*, 1983). However, the agent also has a potent inhibitory effect on glucokinase in β-cells (Lenzen *et al.*, 1988). In this respect, KN-62 seems to be a useful tool for elucidating the roles of CaM kinase II in insulin secretion. Treatment with KN-62 suppressed 28 mM glucose-induced insulin secretion in isolated rat pancreatic islets in a dose-dependent manner with the half-maximal effect achieved at a concentration of 1.5 $\pm$ 0.5 μM without affecting its basal secretion (Wenham *et al.*, 1992). KN-62 partially inhibited the secretion (67 $\pm$ 11%) at a concentration of 10 μM, whereas KN-04 had only a modest effect at 10 μM. Insulin secretion is also activated by high K^+, glucose plus carbachol or forskolin, and TPA. KN-62 suppressed the 40 mM K^+-induced insulin secretion in isolated rat pancreatic islets by $59 \pm 11\%$ at 10 μM. KN-62 further suppressed the carbachol enhancement of 8 mM glucose-stimulated insulin secretion by 53% at 10 μM. It is believed that carbachol increases intracellular cAMP concentration, which in turn cooperates with glucose to elevate intracellular Ca^{2+} concentration. By contrast, insulin secretion stimulated by 100 nM TPA, a direct activator of PKC, was not affected by KN-62 at 5 μM. Similar experiments using isolated rat pancreatic β-cells, not islets, showed that KN-62 inhibited insulin secretion induced by 10 mM glucose, 20 mM K^+, or 10 mM glucose plus 5 μM forskolin at slightly lower concentrations (Niki *et al.*, 1993). Taken together, Ca^{2+} plays a crucial role in insulin secretion and CaM kinase II may be involved in the Ca^{2+}-mediated insulin secretion.

In pancreatic acinar cells, acetylcholine or cholecystokinin (CCK) stimulates phosphoinositide turnover to produce diacylglycerol (DG) and inositol 1,4,5-triphosphate (IP_3) with the subsequent activation of PKC (by DG) and Ca^{2+} mobilization (by IP_3). On the other hand, secretin or vasoactive intestinal peptide (VIP) stimulates adenylyl cyclase to produce cAMP with the subsequent activation of PKA. KN-62 had no effect on the maximal amylase secretion induced by 1 μM carbachol or 100 pM CCK octapeptide (CCK8) in isolated rat pancreatic acini, but dose-dependently inhibited amylase secretion induced by 300 nM carbachol or 30 pM CCK8 by 27.0 ± 2.7 and $27.2 \pm 4.6\%$, respectively, at the highest concentration tested (30 μM) (Ishiguro *et al.*, 1992). KN-04, however, had no effect on the submaximal amylase secretion. Ca^{2+} (1 μM) stimulated amylase secretion in streptolysin O-permeabilized pancreatic acini, which was enhanced by cAMP, TPA, or GTPγS. KN-62 had no effect on the Ca^{2+}-stimulated amylase secretion but slightly inhibited the enhancement of Ca^{2+}-stimulated amylase secretion by cAMP, TPA, or GTPγS. It is difficult to draw a firm conclusion from these findings about the involvement of CaM kinase II in amylase secretion in pancreatic acini.

The regulatory mechanism of CCK secretion in STC-1 cells, an intestinal CCK-secretory cell line, has been studied (Prpic *et al.*, 1994). CCK secretion was stimulated by 20 mM L-phenylalanine or 100 μM IBMX, a phosphodiesterase inhibitor. Diltiazem (10 μM), an L-type Ca^{2+} channel blocker, inhibited the CCK secretion stimulated by L-phenylalanine or IBMX by 30.5 ± 2.3 and $22.6 \pm 4.5\%$, respectively. KN-62 did not affect the L-phenylalanine-stimulated CCK secretion but inhibited the IBMX-stimulated CCK secretion by $32 \pm 5.7\%$ at 1.4 μM. 8-Br–cAMP, an unhydrolyzable analog of cAMP, stimulated CCK secretion and this CCK secretion was similarly inhibited by KN-62. It was revealed by using fura-2/AM that L-phenylalanine or IBMX induced a slow, linear increase in cytoplasmic Ca^{2+} concentration. Notably, KN-62 completely suppressed the IBMX-induced Ca^{2+} increase to a basal level without affecting the basal concentration of cytoplasmic Ca^{2+} concentration. These observations should not be interpreted to mean that KN-62 acts as a typical Ca^{2+} channel blocker in secretory cells. Each secretagogue, L-phenylalanine or IBMX, elicits the diltiazem-sensitive response and only the response to IBMX is further sensitive to KN-62. cAMP may activate L-type Ca^{2+} channels through a KN-62-sensitive related pathway.

A number of hormones have been found to stimulate the secretion of atrial natriuretic factor (ANF) from atrial myocytes. Endothelin (ET) is a most potent secretagogue of the ANF secretion. ET stimulates the hydrolysis of phosphoinositides to produce DG and IP_3. PKC may be involved in the ET-stimulated ANF secretion. ET-stimulated ANF secretion by fivefold greater than PDBu in primary rat atrial myocyte cultures (Irons *et al.*, 1992). This indicates that PKC activation alone does not fully account for the effect of ET on the ANF secretion. Treatment of the cultures with PDBu at 100 nM for 24 hr resulted in the down-regulation of PKC activity and a >50% decrease in the ET-stimulated ANF secretion. H-7, but not HA-1077 (a specific PKA inhibitor), inhibited the ET-stimulated ANF secretion in a dose-dependent manner by 55% at 5 μM. H-7 produced only a partial inhibition and did not reduce the ET-stimulated ANF secretion to a basal level even up to 100 μM. Interestingly, BAY K 8644, an L-type Ca^{2+} channel agonist, stimulated ANF secretion without affecting the phosphorylation of p80, indicating the existence of a PKC-independent signaling pathway of the ANF secretion. KN-62, but not KN-04, gave a 50–60% inhibition of the BAY K 8644-stimulated secretion at 1 μM, but this inhibition did not significantly increase with increases in drug concentration to 10 μM. KN-62 inhibited the H-7-insensitive ANF secretion and the combination of H-7 (25 μM) and KN-62 (1 μM) showed complete inhibition of the ET-stimulated ANF secretion. These results suggest that the full effect of ET on ANF secretion in atrial myocytes requires both PKC and CaM kinase activities.

D. Involvement of CaM Kinase II in Regulation of Gene Expression

Changes in intracellular Ca^{2+} concentration regulate the expression of target genes by affecting the phosphorylation state of specific transcription factors. Many Ca^{2+}-mediated changes in gene expression are attributed to the phosphorylation of transcription factors by PKC. However, changes in intracellular Ca^{2+} concentration can also activate Ca^{2+}/CaM-dependent protein kinases. CaM kinase II may either directly or indirectly alter the phosphorylation state of transcription factors and activate the expression of target genes. In transient transfection experiments, the DNA sequence motif, 5'-AAATGTAGTCTTATGCAATACACTTGTAGTCTTGCAACA-

3′, renders a reporter gene responsive to a constitutively active mutant of the α-subunit of CaM kinase II in GC cells, a human pituitary cell line (Kapiloff, 1991). Critical positions within the CaM kinase II responsive element (CaMRE) coincide with binding sites for the transcription factors, C/EBPs (Ryden and Beemon, 1989). Purified CaM kinase II has been shown not to phosphorylate bacterially expressed recombinant C/EBPα, but has been shown to phosphorylate recombinant C/EBPβ as well as the cAMP-responsive element-binding protein, CREB, which is the only other transcription factor known to be phosphorylated by CaM kinase II (Wegner *et al.*, 1992). CaM kinase II phosphorylated C/EBPβ at Ser276 within a leucine zipper dimerization domain, which is replaced by Asp336 and Ala245 in C/EBPα and A/EBPδ, respectively. Treatment with A_{23187} enhanced the phosphorylation of C/EBPβ in GC cells and KN-62 suppressed this increase to a basal level without affecting the expression of $[^{35}S]$methionine-labeled C/EBPβ. These observations indicate the existence of a signal transduction pathway distinct from the classical PKC pathway by which changes in intracellular Ca^{2+} concentration are translated to alterations of gene expression. If this is the case, we may have to focus on the significance of CaM kinase II in the regulation of Ca^{2+}-mediated events at the gene level.

Immediate early genes (IEGs) are rapidly and transiently transcribed in various cells in response to each various extracellular stimuli. Several IEGs, including the c-*fos* gene, encode transcription factors and control secondary programs of gene expression, which may ultimately determine cell phenotypes. The segregation of Ca^{2+} channels into distinct subcellular regions in cells serves to generate highly localized Ca^{2+} signals, thereby resulting in specific cell activities. In neurons, the activation of either of two types of Ca^{2+} channels, NMDA receptor/ion channels or L-type voltage-dependent Ca^{2+} channels, induces the expression of c-*fos* gene and other IEGs. Treatment with glutamate or KCl induced an increase in the autophosphorylation of CaM kinase II in rat hippocampal neurons and KN-62 significantly inhibited this autophosphorylation (Bading *et al.*, 1993). Interestingly, KN-62 did not largely affect the induction of c-*fos* expression by glutamate but inhibited the KCl-induced expression of c-*fos* mRNA in a dose-dependent manner (Bading *et al.*, 1993). These experiments suggest that Ca^{2+} signals evoked by the activation of either NMDA receptors or L-type Ca^{2+} channels are propagated to the nucleus through two pharmacologically distinguishable pathways. CaM kinase II may be involved in the L-type Ca^{2+} channel-associated pathway but not the NMDA receptor-associated pathway.

Rat glucagon gene is a single-copy gene expressed in a highly cell-specific manner mainly in the α-cells of pancreatic islets as well as in intestinal L cells and a few neurons in brain. The principal hormonal function of pancreatic glucagon is to stimulate the hepatic production of glucose, balancing the effects of insulin. To maintain blood glucose levels within narrow limits, the secretion and production of pancreatic glucagon (α-cells) and insulin

(β-cells) are influenced by food consumption and fasting. Glucose and other energy sources, as well as hormones and neurotransmitters, act on pancreatic islet cells and many of these signals seem to be mediated by second messengers including Ca^{2+}. An increase in cytosolic Ca^{2+} concentration is considered to be a principal signal that regulates pancreatic hormone secretion. However, it remains to be determined whether alterations in the Ca^{2+} concentration are signals for glucagon gene transcription. Treatment with KN-62 (1 μM), but not KN-04 (1 μM), inhibited high K^+-stimulated glucagon gene transcription by 60% in HIT cells, a rat pancreatic islet cell line (Schwaninger *et al.*, 1993). However, 8-Br–cAMP-stimulated glucagon gene transcription was not affected by KN-62. Glucagon gene sequence contains three control elements (G_1, G_2, and G_3) that mediate cell-specific expression and a CRE that confers cAMP responsiveness (Knepel *et al.*, 1990, 1991). HIT cells transiently transfected with the only thymidine kinase promoter of herpes simplex virus in the luciferase vector, pT81Luc, did not respond to high K^+, but insertion of the glucagon CRE cassette (5xCRET81Luc) led to the induction of reporter gene in response to the depolarization (Schwaninger *et al.*, 1993). Treatment with KN-62 (1 μM) completely inhibited the high K^+-stimulated 5xCRET81Luc transcription, indicating that the CRE-mediated gene induction by K^+ depolarization depends on CaM kinase II.

E. Miscellaneous

In K562 cells, a human myeloid leukemia cell line, a polypeptide (M_r 65 kDa) immunoreacted with a polyclonal antibody of rat CaM kinase II (Minami *et al.*, 1994). The polypeptide, immunoprecipitated by the antibody, had the ability to phosphorylate syntide-2 and be autophosphorylated, both in a Ca^{2+}/CaM-dependent manner. The immunoprecipitated polypeptide gave a myosin light-chain phosphorylation activity on renatured gel and KN-62 inhibited this phosphorylation in a dose-dependent manner with an IC_{50} value of 1.8 μM. These results indicate that the polypeptide in K562 cells is closely related to CaM kinase II. KN-62 suppressed K562 cell growth by 37% at 10 μM and in a dose-dependent manner (Minami *et al.*, 1994). Flow cytometric analyses showed that KN-62 inhibits the cell cycle progression at S phase. CaM kinase II or its related enzyme(s) may play an important role in cell cycle progression.

In cancer chemotherapy, intrinsic or acquired resistance to anticancer agents is a problem. It is of great importance to elucidate the mechanism of this resistance and develop pharmacological agents to overcome the drug-resistance problem. A serous cystadenocarcinoma cell line, NOS3, was obtained from a patient with ovarian cancer (Obata *et al.*, 1995). An adriamycin-resistant cell line, NOS3AR, has been established by the sequential selection of NOS3 cells in media containing increasing concentrations of adriamycin. A polypeptide related to CaM kinase II was partially purified from NOS3AR cells on the basis of its ability to phosphorylate syntide-2 in a Ca^{2+}/CaM-dependent manner and immunoreact with a polyclonal antibody of rat CaM kinase II (Obata *et al.*, 1995). KN-62 dose-dependently inhibited this phosphorylation up to 60% at 10 μM, the maximal concentration tested. Interestingly, KN-62 did not directly affect NOS3AR cell growth but dose-dependently reduced the concentration of adriamycin required for a 50% inhibition of the growth (Obata *et al.*, 1995). By addition of KN-62 (10 μM), the IC_{50} value decreased from 38.1 to 0.19 μM. The adriamycin resistance is in part caused by an increase in glutathione *S*-transferase (GST) activity. Treatment with KN-62 did not affect the GST activity in NOS3AR cells. It is generally accepted that P-glycoprotein functions as an energy-dependent pump of pharmacological agents. Indeed, a gene encoding P-glycoprotein was detected by Northern blotting in NOS3AR cells but not in NOS3 cells (Obata *et al.*, 1995). Notably, the content of adriamycin in NOS3AR cells was one-third less compared with that in NOS3 cells and increased approximately by threefold by treatment with 10 μM KN-62. Recently, it has been demonstrated that P-glycoprotein is regulated by undergoing phosphorylation by PKA and PKC. By analogy, the resistance of NOS3AR cells to adriamycin is ascribed to pumping out of the agent by P-glycoprotein, which may be regulated by CaM kinase II phosphorylation.

Intercellular adhesion molecule-1 (ICAM-1) is an important cell surface adhesion molecule whose significance has been recognized in the immune system, particularly in cell–cell contact-mediated immune reactions. ICAM-1 expression is regulated by cytokines in a variety of cells including cancer cells. The cytokine-induced enhancement of ICAM-1 expression increases the vulnerability of tumor cells to tumor-infiltrating lymphocyte-, monocyte-, and natural killer-mediated lysis. The involvement of PKA or PKC in the regulation of ICAM-1 expression is still controversial. Treatment with A_{23187} or thapsigargin, a Ca^{2+}–ATPase inhibitor, has been found to enhance ICAM-1 expression in SK-N-SH cells, a human neuroblastoma cell line (Bouillon and Audette, 1994). Treatment with γ-interferon (IFN-γ) and retinoic acid also enhanced ICAM-1 expression in SK-N-SH cells. The effect of either Ca^{2+}-mobilizing agent, A_{23187} or thapsigargin, on ICAM-1 expression was additive to that of IFN-γ but not to retinoic acid. KN-62 inhibited the retinoic acid-stimulated ICAM-1 expression by 87% at 10 μM, whereas KN-62 did not give any significant effects on the IFN-γ-stimulated ICAM-1 expression. Constitutive CaM kinase II activity was substantially increased by treatment with retinoic acid, but not by IFN-γ, and this increase was inhibited by KN-62 in a dose-dependent manner (Bouillon and Audette, 1994). These observations suggest that CaM kinase II is involved in the retinoic acid- but not IFN-γ-stimulated ICAM-1 expression.

III. KN-62 Is Not a Specific Inhibitor of CaM Kinase II but Is a Group-Specific Inhibitor of CaM Kinases ___________

KN-62 was proposed to be a specific inhibitor of CaM kinase II because the compound could selectively inhibit the enzyme among a limited number of protein kinases such as PKA, PKC, and smooth muscle MLCK (Tokumitsu *et al.*, 1990). KN-62 differed from other H-series compounds in that KN-62 competed with CaM, but not ATP, on CaM kinase II. In addition, KN-62 was proven to bind to CaM kinase II, but not Ca^{2+}/CaM, by affinity chromatography using the compound coupled to Sepharose 4B and had no effect on the Ca^{2+}/CaM-independent activity of previously autophosphorylated enzyme. These findings suggest that KN-62 occupies the Ca^{2+}/CaM binding domain of CaM kinase II, thereby inhibits its Ca^{2+}/CaM-dependent activation. There is additional evidence that CaM kinases I and IV may be involved in various aspects of cell function (Miyano *et al.*, 1992; Picciotto *et al.*, 1993). It remains to be demonstrated whether KN-62 affects the multifunctional CaM kinases.

CaM kinase I was purified from bovine brain based on its ability to phosphorylate site I of the neuronal protein, synapsin I (Nairn and Greengard, 1987). The enzyme is widely distributed and the highest amounts are found in brain. A cDNA encoding CaM kinase I was isolated from rat brain (Picciotto *et al.*, 1993); subsequently, another cDNA was isolated from fetal rat lung (Cho *et al.*, 1994). The differences between the cDNA sequences have been shown to be derived from some sequencing errors (Yokokura *et al.*, 1995). The deduced amino acid sequence predicts a protein of M_r -42 kDa (Cho *et al.*, 1994). Several enzymes related to CaM kinase I including CaM kinases Ia, Ib, and V, have been purified and characterized. CaM kinases Ia (M_r 42 kDa) and Ib (M_r 39 kDa) were purified from rat brain based on their ability to phosphorylate site I peptide (DeRemer *et al.*, 1992a,b). CaM kinase V (M_r 41 kDa) was purified from rat brain using syntide-2 as a substrate from our laboratory (Mochizuki *et al.*, 1993b). Syntide-2, derived from the amino acid sequence of glycogen synthase around a phosphorylation site for CaM kinase II, has frequently been used as a peptide substrate for CaM kinase II (Hashimoto and Soderling, 1987). The amino acid sequence of syntide-2 conforms well to the consensus amino acid sequence for phosphorylation by CaM kinase I. Based on a high identity of amino acid sequence, CaM kinase V appears closely related to CaM kinase I. Synapsin I is a synaptic vesicle-associated, actin-binding protein. Its phosphorylation by CaM kinase I has not yet been shown to affect the interaction of synapsin I with synaptic vesicles or actin. Purified CaM kinase I, as well as PKA, phosphorylated the transcription factor CREB at Ser133 (Sheng *et al.*, 1991). The Ser133 was phosphorylated in response to an increase in intracellular Ca^{2+} or cAMP concentration (Sheng *et al.*, 1991). CaM kinase I appears to be predominantly located in cytosol, although

CREB is restricted to the nucleus. Whether the Ser133 phosphorylation in CREB by CaM kinase I is a physiological event has not yet been determined.

CaM kinase IV was originally identified from isolation of a cDNA clone encoding a Ca^{2+}/CaM binding domain of the enzyme (Sikela and Hahn, 1987). Subsequently, CaM kinase-Gr enriched in cerebellar granule cells was purified and characterized (Ohmstede *et al.*, 1989). Based on their deduced amino acid sequences, it appears that they are identical. CaM kinase IV is a monomer of M_r 65–67 kDa and expressed predominantly in brain with particularly high levels found in cerebellar granule cells and slightly high levels in nonneuronal tissues such as thymus, spleen, and testis (Ohmstede *et al.*, 1989). Interestingly, calspermin, a testis-specific protein, shares the Ca^{2+}/ CaM binding and COOH-terminal noncatalytic domains of CaM kinase IV (Ohmstede *et al.*, 1991). It is believed that the calspermin-identical portion of CaM kinase IV determines its unique localization to the nucleus. In neurons, CaM kinase IV has been shown to be localized to the nucleus as well as axons and terminals (Jensen *et al.*, 1991). The consensus amino acid sequence of substrates for CaM kinase IV is the same as that for CaM kinase II, which means that their substrates are overlapping. However, distinct sets of physiological substrates for the two enzymes may exist because they show different cellular and subcellular distributions. Transfection experiments have revealed that CaM kinase IV, but not CaM kinase II, may be important in Ca^{2+}-mediated transcriptional activation through the phosphorylation of Ser133 in CREB (Enslen *et al.*, 1994). CaM kinase II was shown to phosphorylate a second site, Ser142, in CREB besides Ser133, and this second site phosphorylation has been established to give negative effects on its activity (Enslen *et al.*, 1994).

In general, the Ca^{2+}/CaM binding domains of proteins are not highly conserved at the level of primary structure but are conserved at the level of secondary structure. They are all predicted to form an amphipathic α-helix (O'Neil and DeGrado, 1990). Ca^{2+} binds to CaM and this Ca^{2+} binding exposes hydrophobic region(s) on the protein (Tanaka and Hidaka, 1980). This activated CaM folds around the helical peptide with hydrophobic bonds formed between the hydrophobic region on CaM and key hydrophobic residues on the peptide (Ikura *et al.*, 1992). In addition, extensive contacts between charged residues are important. These critical residues are conserved at key positions in the Ca^{2+}/CaM binding domain of CaM kinases I, II, and IV and smooth muscle MLCK (Fig. 4; Yokokura *et al.*, 1995). Despite such common structures, KN-62 can inhibit CaM kinase II but not smooth muscle MLCK, the difference of which remains to be studied. Notably, CaM kinases II and IV, in particular, resemble each other in the Ca^{2+}/CaM binding domain (Fig. 4).

One report has revealed that KN-62 can inhibit CaM kinase IV, expressed in the baculovirus/Sf9 cell system, with an IC_{50} value of 3 μM (Enslen *et al.*, 1994). However, the recombinant CaM kinase IV may be more highly

skMLCK	[571] LKKYLMKRR**W**KKN**F**IA**V**SAANR**F**KKISS [598]
smMLCK	[791] MKKYMARRKWQKTGHAVRAIG**R**LSSMAM [818]
CaM kinase I	[294] IKKNFAKSKWKQAFNRTAVVRHMRKLQLGHQP [325]
CaM kinase II	[290] LKKFNARRKLKGAILTTMLATRNFSGGK [317]
CaM kinase IV	[310] QKKLQEFNARRKLKAAVKAVVASSRLGSASS [340]

FIGURE 4   Comparison of the Ca^{2+}/CaM binding domain of CaM kinases. Trp^{580}, Phe^{584}, Val^{587}, and Phe^{593} of skeletal muscle MLCK (skMLCK) as well as Arg^{812} of smooth muscle MLCK (smMLCK) are important for binding to Ca^{2+}/CaM (positions indicated by asterisks). Conserved amino acid residues of Ca^{2+}/CaM binding domains were aligned with these key positions. Amino acid sequences known to be directly involved in binding Ca^{2+}/CaM with structural studies (Ikura *et al.*, 1992; Meador *et al.*, 1992, 1993) are underlined. Ca^{2+}/CaM binding domains: skMLCK (Ikura *et al.*, 1992), smMLCK (Meador *et al.*, 1992), CaM kinase I (Picciotto *et al.*, 1993), CaM kinase II (Meador *et al.*, 1993), and CaM kinase IV (Sikera *et al.*, 1987).

sensitive to KN-62 than the native enzyme, which remains to be studied. In parallel experiments, KN-62 inhibited CaM kinase II, purified from rat brain, with almost the same IC_{50} value. In addition, KN-62 has been shown to inhibit CaM kinase V, purified from rat brain, competitively with respect to Ca^{2+}/CaM with a K_i value of 0.8 μM (Mochizuki *et al.*, 1993b). CaM kinases I and V are closely related and it is quite possible that CaM kinase I is similarly inhibited by KN-62. These results indicate that KN-62 is not a specific inhibitor of CaM kinase II, necessitating reconsideration of the use of the compound. As described under Section II, KN-62 has been, over the past 4 years, useful in elucidating the involvement of CaM kinase II in various aspects of cell function. In fact, KN-62 has been shown to inhibit multifunctional CaM kinases IV and V in addition to CaM kinase II with almost the same potency (Tokumitsu *et al.*, 1990; Mochizuki *et al.*, 1993b; Enslen *et al.*, 1994). Of the many experiments with KN-62, any unfavorable nonspecific effects of KN-62 have not been proven. It is quite likely that KN-62 binds to CaM kinases at their Ca^{2+}/CaM binding domains and accordingly inhibits them. KN-62 may be used as a group-specific inhibitor of CaM kinases II, IV, and V and other potential CaM kinases.

IV. A New Group of CaM Kinases and CaM Kinase Phosphorylation Cascades

To purify CaM kinase II, a rat brain extract was directly applied onto a Sepharose 4B column without preapplication onto an ion-exchange column and two peaks of Ca^{2+}/CaM-dependent syntide-2 phosphorylation activity were observed (Mochizuki *et al.*, 1993b). The two peaks have been shown to be derived from CaM kinase II (first flow peak) and CaM kinase V (second flow peak). CaM kinase V has been purified to near homogeneity as determined by SDS–PAGE gel with silver staining and its partial amino acid sequence has been determined (Mochizuki *et al.*, 1993b). Gel filtration with a TSK gel G3000SW column has been used as the final step of its purification. CaM kinase V activity, as well as its autophosphorylation, largely decreased when gel filtration was repeated. In a fraction of the second gel filtration, an activator with the ability to phosphorylate the highly purified CaM kinase V and increase its syntide-2 phosphorylation activity (by 10-fold), both in a Ca^{2+}/CaM-dependent manner, has been identified. The most probable interpretation of these results may be the existence of a CaM kinase V kinase and a CaM kinase phosphorylation cascade (Mochizuki *et al.*, 1993a). This possibility is consistent with our recent finding that the activator fraction contained a polypeptide (M_r 64 kDa) undergoing Ca^{2+}/CaM-dependent autophosphorylation in renatured gel methods. It is known that many CaM kinases are autophosphorylated in a Ca^{2+}/CaM-dependent manner. We have recently succeeded in purifying the CaM kinase V activator. Isolation of a cDNA of this activator is in progress. The activator has also been shown to activate recombinant CaM kinase I, expressed in bacteria, accompanying phosphorylations mainly at Thr177 in CaM kinase I (the autophosphorylation site) and slightly at other site(s) not identified (H. Hidaka and H. Yokokura, unpublished data). Autophosphorylation occurs in CaM kinase I and up-regulates the enzyme. The activator may rapidly phosphorylate the autophosphorylation site of CaM kinase (I/V) and this phosphorylation may in turn enhance the enzyme activity. Notably, KN-62 has been shown not to affect recombinant CaM kinase I in concentrations up to 10 μM (H. Hidaka and H. Yokokura, unpublished data). This finding raises three possibilities: (i) CaM kinase V is closely related to CaM kinase I but they may be different entities, thereby showing different sensitivities to KN-62; (ii) recombinant CaM kinase I may differ from native CaM kinase I/V in structural and functional aspects, including their sensitivities to KN-62; and (iii) the preparation of CaM kinase V used may be contaminated by traces of its activator: KN-62 may not directly affect CaM kinase I/V, but may indirectly inhibit them through inhibition of the activator.

It is important to examine these possibilities because they include several interests such as the growth of CaM kinase family and the specificity of KN-62 to CaM kinases. For the answer to possibility (i), isolation of a cDNA encoding CaM kinase V is under investigation. A CaM kinase phosphorylation cascade exists and may be affected by KN-62, a group-specific inhibitor of CaM kinases.

By analogy, during the purification of CaM kinase Ia, low recoveries of site I peptide phosphorylation activity were often obtained at the CaM–Sepharose chromatography step (DeRemer *et al.*, 1992a,b). Addition of an aliquot of the CaM–Sepharose column wash fraction significantly enhanced the activity of the purified CaM kinase Ia. Based on its ability to activate and phosphorylate CaM kinase Ia in the presence of both Ca^{2+}/CaM and MgATP, CaM kinase Ia activator (M_r ~52 kDa) has been purified (Lee and

Edelman, 1994). Note that this activator is of low affinity for Ca^{2+}/CaM because it passed the CaM–Sepharose column under the standard condition and did not bind to $[^{125}I]$ CaM with blotting procedures. The activator, distinct from many CaM kinases, has not been shown to be autophosphorylated (Lee and Edelman, 1994). The CaM kinase Ia activator is presumably a CaM kinase, but it is apparently distinct from the CaM kinase V activator.

From recent experiments concerning the regulation of CaM kinase IV, the existence of its activators has been implicated. It was demonstrated that the activation of T cell antigen receptor–CD3 complex in Jurkat cells produces a 17-fold increase in CaM kinase IV activity (Hanissian *et al.*, 1993). Interestingly, one report indicated that recombinant CaM kinase IV, expressed in bacteria, is almost completely inactive, and that the addition of a brain extract highly stimulates the enzyme (Okuno and Fujisawa, 1993). Two CaM kinase IV kinases, which can phosphorylate recombinant CaM kinase IV, expressed in bacteria or the baculovirus/Sf9 cell system and concomitantly activate its activity, both in a Ca^{2+}/CaM-dependent manner, have been purified from rat brain and characterized (Okuno *et al.*, 1994; Tokumitsu *et al.*, 1994). One kinase is a polypeptide of M_r 66 kDa on SDS–PAGE gel and phosphorylates CaM kinase IV exclusively at Ser residue(s) (Okuno *et al.*, 1994). Of a number of substrates, syntide-2 in addition to CaM kinase IV has been found to be a fairly good substrate for the CaM kinase IV kinase. In this respect, the CaM kinase V kinase may be distinct from the CaM kinase IV kinase because the former has been found to phosphorylate syntide-2 very weakly (H. Hidaka and H. Yokokura, unpublished data). The other CaM kinase IV kinase (M_r 68 kDa) has been shown to activate recombinant CaM kinase IV, expressed using the baculovirus/Sf9 cell system, with a 10-fold decrease in K_m for syntide-2 but little effect on K_m for ATP or V_{max} (Tokumitsu *et al.*, 1994). Distinct from the previous CaM kinase IV kinase, this 68-kDa enzyme has been found to phosphorylate CaM kinase IV, resulting in a large increase (by >20-fold) in Ca^{2+}/CaM-independent CaM kinase IV activity (Tokumitsu *et al.*, 1994).

These kinases appear to form a new group of CaM kinases with the ability to activate multifunctional CaM kinases. By analogy with mitogen-activated protein (MAP) kinase phosphorylation cascade, CaM kinase phosphorylation cascades should be studied (Fig. 5).

V. Conclusions

Since the 1980s, we have synthesized a large number of chemical compounds, known as H-series compounds, including W-7 (CaM antagonist) and H-89 (PKA inhibitor), H-7 (PKC inhibitor), and KN-62 (CaM kinase II inhibitor). They have been useful in understanding second messenger (cAMP or Ca^{2+})-mediated processes in various aspects of cell function. In

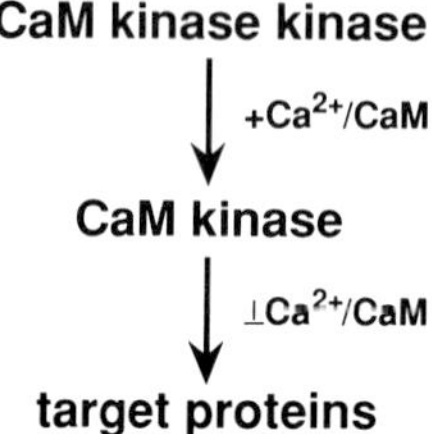

FIGURE 5   CaM kinase phosphorylation cascades. CaM kinase kinase phosphorylates CaM kinase in a Ca^{2+}/CaM-dependent manner resulting in a stimulation of the CaM kinase-catalyzed, Ca^{2+}/CaM-independent or -dependent phosphorylation of target proteins.

addition, W-7 and its close derivatives have been important in uncovering the structural changes [exposure of hydrophobic region(s)] in CaM on its binding to Ca^{2+}. Therefore, H-series componds have significantly contributed to the establishment of modern "molecular and cellular pharmacology." It is of great interest to know which protein kinases phosphorylate target proteins in response to an increase in intracellular Ca^{2+} concentration. Despite extensive studies, the classical isozymes of PKC (α, β, and γ) are not fully responsible for the Ca^{2+}-mediated protein phosphorylation. Therefore, attention should be paid to multifunctional Ca^{2+}/CaM-dependent protein kinases including CaM kinases I/V, II, and IV. From previous studies that used KN-62, many aspects of cell function, including neurotransmitter synthesis and release, LTP and LTD, smooth muscle contraction, insulin secre-

TABLE II   Inhibition of CaM Kinases by KN-62

CaM kinase	Inhibition
CaM kinase II	$K_i = 0.9\ \mu M$[a]
CaM kinase IV	$IC_{50} = \leq 3\ \mu M$[b]
CaM kinase V	$K_i = 0.8\ \mu M$[c]
r. CaM kinase I	NE[d]
CaM kinase III	NE[e]

Note. Inhibitory effects of KN-62 on CaM kinases were shown as a K_i or IC_{50} value. KN-62 inhibited the enzymes competitively with respect to Ca^{2+}/CaM, but not ATP. KN-62 gave no effect (NE) on a bacterially expressed recombinant CaM kinase I or CaM kinase III.

[a] Tokumitsu *et al.* (1990).
[b] Enslen *et al.* (1994).
[c] Mochizuki *et al.* (1993b).
[d] H. Yokokura *et al.*, unpublished data.
[e] A. C. Nairn *et al.*, unpublished data.

tion, and gene expression, have been shown to involve protein phosphorylation by CaM kinase II. Recently, it has been revealed that KN-62 can significantly inhibit CaM kinases IV and V in addition to CaM kinase II (Table II). KN-62 should not be used as a specific inhibitor of CaM kinase II, but should be used as a group-specific inhibitor of CaM kinases II, IV, and V, and other potential CaM kinases. Increasing members of the CaM kinase family have been identified, including an activator that activates CaM kinase V via its phosphorylation. Other activators have subsequently been identified and purified based on their ability to phosphorylate CaM kinases Ia or IV and stimulate their activity in a Ca^{2+}/CaM-dependent manner. The potential CaM kinase phosphorylation cascade remains to be proven a physiological event. However, these activators may be classified putatively into the third group of CaM kinases because they activate another group of multifunctional CaM kinases. KN-62 has been shown to be a group-specific inhibitor of CaM kinases including CaM kinase II. Our goal is to develop inhibitors specific to each member of the CaM kinase family and elucidate their physiological roles in various aspects of cell function.

References

Bading, H., Ginty, D. D., and Greenberg, M. E. (1993). Regulation of gene expression in hippocampal neurons by distinct calcium signaling pathways. *Science* **260**, 181–186.

Bouillon, M., and Audette, M. (1994). Retinoic acid-stimulated intercellular adhesion molecule-1 expression on SK-N-SH cells: Calcium/calmodulin-dependent pathway. *Cancer Res.* **54**, 4144–4149.

Campbell, D. G., Hardie, D. G., and Vulliet, P. R. (1986). Identification of four phosphorylation sites in the N-terminal regions of tyrosine hydroxylase. *J. Biol. Chem.* **261**, 10489–10492.

Chijiwa, T., Hagiwara, M., and Hidaka, H. (1989). A newly synthesized selective casein kinase I inhibitor, *N*-(2-aminoethyl)-5-chloroisoquinoline-8-sulfonamide, and affinity purification of casein kinase I from bovine testis. *J. Biol. Chem.* **264**, 4924–4927.

Chijiwa, T., Mishima, A., Hagiwara, M., Sano, M., Hayashi, K., Inoue, T., Naito, K., Toshioka, T., and Hidaka, H. (1990). Inhibition of forskolin-induced neurite outgrowth and protein phosphorylation by a newly synthesized selective inhibitor of cyclic AMP-dependent protein kinase, *N*-[2-(*p*-bromocinnamylamino)ethyl]-5-isoquinolinesulfonamide (H-89), of PC12D pheochromocytoma cells. *J. Biol. Chem.* **265**, 5267–5272.

Cho, F. C., Phillips, K. S., Bofucki, B., and Weaver, T. E. (1994). Characterization of a rat cDNA clone encoding calcium/calmodulin-dependent protein kinase I. *Biochem. Biophys. Acta* **1224**, 156–160.

Colca, J. R., Kotogal, N., Brooks, C. L., Lacy, P. E., Landt, M., and McDaniel, M. L. (1983). Alloxan inhibition of a Ca^{2+}- and calmodulin-dependent protein kinase activity in pancreatic islets. *J. Biol. Chem.* **258**, 7260–7263.

Cruzalegui, F. H., and Means, A. R. (1993). Biochemical characterization of the multifunctional Ca^{2+}/calmodulin-dependent protein kinase type IV expressed in insect cells. *J. Biol. Chem.* **268**, 1993.

DeRemer, M. F., Saeli, R. J., Brautigan, D. L., and Edelman, A. R. (1992b). Ca^{2+}-calmodulin dependent protein kinase Ia and Ib from rat brain: II. Enzymatic characteristics and regulation of activities by phosphorylation and dephosphorylation. *J. Biol. Chem.* **267**, 13466–13471.

DeRemer, M. F., Saeli, R. J., and Edelman, A. R. (1992a). Ca^{2+}-calmodulin dependent protein kinase Ia and Ib from rat brain: I. Identification, purification, and structural comparisons. *J. Biol. Chem.* **267**, 13460–13465.

Endo, T., and Hidaka, H. (1980). Ca^{2+}-calmodulin dependent phosphorylation of myelin isolated from rabbit brain. *Biochem. Biophys. Res. Commun.* **97**, 553–558.

Enslen, H., Sun, P., Brickey, D., Soderling, S. H., Klamo, E., and Soderling, T. R. (1994). Characterization of Ca^{2+}/calmodulin-dependent protein kinase IV: Role in transcriptional regulation. *J. Biol. Chem.* **269**, 15520–15527.

Funauchi, M., Tsumoto, T., Nishigori, A., Yoshimura, Y., and Hidaka, H. (1992). Long-term depression is induced in Ca^{2+}/calmodulin kinase-inhibited visual cortex neurons. *NeuroReport* **3**, 173–176.

Hack, N., Hidaka, H., Wakefield, M. J., and Balazs, R. (1993). Promotion of granule cell survival by high K^+ or excitatory amino acid treatment and Ca^{2+}/calmodulin-dependent protein kinase activity. *Neuroscience* **57**, 9–20.

Hagiwara, M., Inagaki, M., and Hidaka, H. (1987). Specific binding of a novel compound, N-[2-(methylamino)ethyl]-5-isoquinolinesulfonamide (H-8) to the active site of cAMP-dependent protein kinase. *Mol. Pharmacol.* **31**, 523–528.

Hanissian, S. H., Frangatis, M., Bland, M. M., Jawahar, S., and Chatila, T. A. (1993). Expression of a Ca^{2+}/calmodulin-dependent protein kinase, CaM kinase-Gr, in human T lymphocytes. Regulation of kinase activity by T cell receptor signaling. *J. Biol. Chem.* **268**, 20055–20063.

Harris, K. A., Oyler, G. A., Doolittle, G. M., Vincent, I., Lehman, R. A. W., Kincaid, R. L., and Billingsley, M. L. (1993). Okadaic acid induces hyperphosphorylated forms of tau protein in human brain slices. *Ann. Neurol.* **33**, 77–87.

Hashimoto, Y., and Soderling T. R. (1987). Calcium/calmodulin-dependent protein kinase II and calcium/phospholipid-dependent protein kinase activities in rat tissues assayed with a synthetic peptide. *Arch. Biochem. Biophys.* **252**, 418–425.

Hidaka, H., Asano, T., Iwadare, S., Matsumoto, I., Totsuka, T., and Aoki, N. (1978). A novel vascular relaxing agent, N-(6-aminohexyl)-5-chloro-1-naphthalenesulfonamide which affects vascular smooth muscle actomyosin. *J. Pharmacol. Exp. Ther.* **207**, 8–15.

Hidaka, H., Inagaki, M., Kawamoto, S., and Sasaki, Y. (1984). Isoquinolinesulfonamides, novel and potent inhibitors of cyclic nucleotide dependent protein kinase and protein kinase C. *Biochemistry* **23**, 5036–5941.

Hidaka, H., Sasaki, Y., Tanaka, T., Endo, T., Ohno, S., Fujii, Y., and Nagata, T. (1981). N-(6-aminohexyl)-5-chloro-1-naphthalenesulfonamide, a calmodulin antagonist, inhibits cell proliferation. *Proc. Natl. Acad. Sci. U.S.A.* **78**, 4354–4357.

Hidaka, H., Yamaki, T., Naka, M., Tanaka, T., Hayashi, H., and Kobayashi, R. (1980). Calcium-regulated modulator protein interacting agents inhibit smooth muscle calcium-stimulated protein kinase and ATPase. *Mol. Pharmacol.* **17**, 66–72.

Ichimura, T., Isobe, T., Okuyama, T., Yamauchi, T., and Fujisawa, H. (1987). Brain 14-3-3 protein is an activator protein that activates tryptophan 5-monooxygenase and tyrosine 3-monooxygenase in the presence of Ca^{2+}, calmodulin-dependent protein kinase II. *FEBS Lett.* **219**, 79–82.

Ikebe, M., and Reardon, S. (1990). Phosphorylation of smooth muscle myosin light chain kinase by smooth muscle Ca^{2+}/calmodulin-dependent multifunctional protein kinase. *J. Biol. Chem.* **265**, 8975–8978.

Ikura, M., Clore, G. M., Gronenborn, A. M., Zhu, G., Klee, C. B., and Bax, A. (1992). Solution structure of a calmodulin-target peptide complex by multidimensional NMR. *Science* **256**, 632–638.

Inagaki, M., Kawamoto, S., Ito, H., Saitoh, M., Hagiwara, M., Takahashi, J., and Hidaka, H. (1986). Naphthalenesulfonamides as calmodulin antagonists and protein kinase inhibitors. *Mol. Pharmacol.* **29**, 577–581.

Irons, C. E., Sei, C. A., Hidaka, H., and Glembotski, C. C. (1992). Protein kinase C and calmodulin kinase are required for endothelin-stimulated atrial natriuretic factor secretion from primary atrial myocytes. *J. Biol. Chem.* **267**, 5211–5216.

Ishiguro, H., Hayakawa, T., Kondo, T., Shibata, T., Kitagawa, M., Sakai, Y., Sobajima, H., Nakane, Y., Tanikawa, M., and Hidaka, H. (1992). Effects of calmodulin inhibitors on amylase secretion from rat pancreatic acini. *Digestion* **53**, 162–170.

Ishii, A., Kiuchi, K., Kobayashi, R., Sumi, M., Hidaka, H., and Nagatsu, T. (1991). A selective Ca^{2+}/calmodulin-dependent protein kinase II inhibitor, KN-62, inhibits the enhanced phosphorylation and the activation of tyrosine hydroxylase by 56 mM K^+ in rat pheochromocytoma PC12h cells. *Biochem. Biophys. Res. Commun.* **176**, 1051–1056.

Ito, I., Hidaka, H., and Sugiyama, H. (1991). Effects of KN-62, a specific inhibitor of calcium/calmodulin-dependent protein kinase II, on long-term potentiation in the rat hippocampus. *Neurosci. Lett.* **121**, 119–121.

Ito, M., Tanaka, T., Inagaki, M., Nakanishi, K., and Hidaka, H. (1986). N-(6-phenylhexyl)-5-chloro-1-naphthalenesulfonamide, a novel activator of protein kinase C. *Biochemistry* **25**, 4179–4184.

Jensen, K. F., Ohmstede, C.-A., Fisher, R. S., and Sahyoun, N. (1991). Nuclear and axonal localization of Ca^{2+}/calmodulin-dependent protein kinase type Gr in rat cerebellar cortex. *Proc. Natl. Acad. Sci. U.S.A.* **88**, 2850–2853.

Kapiloff, M. S. (1991). Calcium/calmodulin-dependent protein kinase mediates a pathway for transcriptional regulation. *Proc. Natl. Acad. Sci. U.S.A.* **88**, 3710–3714.

Knepel, W., Chafitz, J., and Habener, J. F. (1990). Transcriptional activation of the rat glucagon gene by the cyclic AMP-responsive element in pancreatic islet cells. *Mol. Cell. Biol.* **10**, 6799–6804.

Knepel, W., Vallejo, M., Chafitz, J. A., and Habener, J. F. (1991). The pancreatic islet-specific glucagon G3 transcription factors recognize control elements in the rat somatostatin and insulin-I genes. *Mol. Endocrinol.* **5**, 1457–1466.

Lee, J. C., and Edelman, A. M. (1994). A protein activator of Ca^{2+}-calmodulin-dependent protein kinase Ia. *J. Biol. Chem.* **269**, 2158–2164.

Lenzen, S., Freytag, S., and Panten, U. (1988). Inhibition of glucokinase by alloxan through interaction with SH groups in the sugar-binding site of the enzyme. *Mol. Pharmacol.* **34**, 395–400.

Madison, D. V., Malenka, R. C., and Nicoll, R. A. (1991). Mechanisms underlying long-term potentiation of synaptic transmission. *Annu. Rev. Neurosci.* **14**, 379–397.

Malinow, R., Schulman, H., and Tsien, R. W. (1989). Inhibition of postsynaptic PKC or CaMKII blocks induction but not expression of LTP. *Science* **245**, 862–866.

Mayer, P., Mohlig, M., Seidler, U., Rochlitz, H., Fahrmann, M., Schatz, H., Hidaka, H., and Pfeiffer, A. (1994). Characterization of γ- and δ-subunits of Ca^{2+}/calmodulin-dependent protein kinase II in rat gastric mucosal cell populations. *Biochem. J.* **297**, 157–162.

Meador, W. E., Means, A. R., and Quiocho, F. A. (1992). Target enzyme recognition by calmodulin: 2.4 A structure of a calmodulin-peptide complex. *Science* **257**, 1251–1255.

Meador, W. E., Means, A. R., and Quiocho, F. A. (1993). Modulation of calmodulin plasticity in molecular recognition on the basis of x-ray structures. *Science* **262**, 1718–1721.

Minami, H., Inoue, S., and Hidaka, H. (1994). The effect of KN-62, Ca^{2+}/calmodulin-dependent protein kinase II inhibitor on cell cycle. *Biochem. Biophys. Res. Commun.* **199**, 241–248.

Miyano, O., Kameshita, I., and Fujisawa, H. (1992). Purification and characterization of a brain-specific multifunctional calmodulin-dependent protein kinase from rat cerebellum. *J. Biol. Chem.* **267**, 1198–1203.

Mochizuki, H., Ito, T., and Hidaka, H. (1993b). Purification and characterization of Ca^{2+}/calmodulin-dependent protein kinase V from rat cerebrum. *J. Biol. Chem.* **268**, 9143–9147.

Mochizuki, H., Sugita, R., Ito, T., and Hidaka, H. (1993a). Phosphorylation of Ca^{2+}/calmodulin-dependent protein kinase V and regulation of its activity. *Biochem. Biophys. Res. Commun.* **197,** 1595–1600.

Naccache, P. H., Molski, T. F. P., Alobaidi, T., Becker, E. L., Showell, H. J., and Sha'afi, R. I. (1980). Calmodulin inhibitors block neutrophil degranulation at a step distal from mobilization of calcium. *Biochem. Biophys. Res. Commun.* **97,** 62–68.

Nairn, A. C., and Greengard, P. (1987). Purification and characterization of Ca^{2+}/calmodulin-dependent protein kinase I from bovine brain. *J. Biol. Chem.* **262,** 7273–7281.

Nairn, A. C., and Picciotto, M. R. (1994). Calcium/calmodulin-dependent protein kinases. *Semin. Cancer Biol.* **5,** 295–303.

Niki, I., Okazaki, K., Saito, M., Niki, A., Niki, H., Tamagawa, T., Iguchi, A., and Hidaka, H. (1993). Presence and possible involvement of Ca^{2+}/calmodulin-dependent protein kinase in insulin release from the rat pancreatic β cell. *Biochem. Biophys. Res. Commun.* **191,** 255–261.

Nishikawa, M., Tanaka, T., and Hidaka, H. (1980). Ca^{2+}-calmodulin dependent phosphorylation and platelet secretion. *Nature (London)* **287,** 863–865.

Obata, N. H., Okazaki, K., Kikkawa, F., Tomoda, Y., and Hidaka, H. (1995). Effect of KN-62, Ca^{2+}/calmodulin-dependent protein kinase II inhibitor, on adriamycin-resistance of human ovarian cancer cells. *Biochem. Biophys. Res. Commun.* **215,** 566–571.

Ohmstede, C.-A., Bland, M. M., Merrill, B. M., and Sahyoun, N. (1991). Relationship of genes encoding Ca^{2+}/calmodulin-dependent protein kinase Gr and calspermin: A gene within a gene. *Proc. Natl. Acad. Sci. U.S.A.* **88,** 5784–5788.

Ohmstede, C.-A., Jensen, K. F., and Sahyoun, N. E. (1989). Ca^{2+}/calmodulin-dependent protein kinase enriched in cerebellar granule cells. *J. Biol. Chem.* **264,** 5866–5875.

Ohno, S., Kawasaki, H., Imajoh, S., Suzuki, K., Inagaki, M., Yokokura, H., Sakoh, T., and Hidaka, H. (1987). Tissue-specific expression of three distinct types of rabbit protein kinase C. *Nature (London)* **325,** 161–166.

Okazaki, K., Ishikawa, T., Inui, M., Tada, M., Goshima, K., Okamoto, T., and Hidaka, H. (1994). KN-62, a specific C^{++}/calmodulin-dependent protein kinase inhibitor, reversibly depresses the rate of beating of cultured fetal mouse cardiac myocytes. *J. Pharmacol. Exp. Ther.* **270,** 1319–1324.

Okuno, S., and Fujisawa, H. (1993). Requirement of brain extract for the activity of brain calmodulin-dependent protein kinase IV expressed in *Escherichia coli. J. Biochem. (Tokyo)* **114,** 167–170.

Okuno, S., Kitani, T., and Fujisawa H. (1994). Purification and characterization of Ca^{2+}/CaM-dependent protein kinase IV kinase from rat brain. *J. Biochem. (Tokyo)* **116,** 923–930.

O'Neil, K. T., and DeGrado, W. F. (1990). How calmodulin binds its target: Sequence independent recognition of amphiphilic alpha-helices. *Trends Biochem. Sci.* **15,** 59–64.

Picciotto, M. R., Czernik, A. J., and Nairn, A. C. (1993). Calcium-calmodulin-dependent protein kinase I: cDNA cloning and identification of autophosphorylation site. *J. Biol. Chem.* **268,** 26512–26521.

Prpic, V., Basavappa, S., Liddle, R. A., and Mangel, A. W. (1994). Regulation of cholecystokinin secretion by calcium-dependent calmodulin kinase II: Differential effects of phenylalanine and cAMP. *Biochem. Biophys. Res. Commun.* **201,** 1483–1489.

Rall, T. W., and Sutherland, E. W. (1958). Formation of a cyclic adenine ribonucleotide by tissue particles. *J. Biol. Chem.* **232,** 1065–1076.

Ryden, T. A., and Beemon, K. (1989). Avian retroviral long terminal repeats bind CCAAT/ enhancer-binding protein. *Mol. Cell. Biol.* **9,** 1155–1164.

Saitoh, M., Ishikawa, T., Matsushima, S., Naka, M., and Hidaka, H. (1987). Selective inhibition of catalytic activity of smooth muscle myosin light chain kinase. *J. Biol. Chem.* **262,** 7796–7801.

Schulman, H. (1988). The multifunctional Ca^{2+}/calmodulin-dependent protein kinase. *In* "Advances in Second Messenger and Phosphoprotein Research" (P. Greengard and G. A. Robison, eds.), Vol. 22, pp. 39–112. Academic Press, New York.

Schulman, H. (1993). The multifunctional Ca^{2+}/calmodulin-dependent protein kinases. *Curr. Opin. Cell Biol.* **5**, 247–253.

Schwaninger, M., Lux, G., Blume, R., Oetjen, E., Hidaka, H., and Knepel, W. (1993). Membrane depolarization and calcium influx induce glucagon gene transcription in pancreatic islet cells through the cyclic AMP-responsive element. *J. Biol. Chem.* **268**, 5168–5177.

Sheng, M., Thompson, M. A., and Greenberg, M. E. (1991). CREB: A Ca^{2+}- regulated transcription factor phosphorylated by calmodulin-dependent kinases. *Science* **252**, 1427–1430.

Shimizu, H., Ito, M., Miyahara, M., Ichikawa, K., Okubo, S., Konishi, T., Naka, M., Tanaka, T., Hirano, K., Hartshorne, D. J., and Nakano, T. (1994). Characterization of the myosin-binding subunit of smooth muscle myosin phosphatase. *J. Biol. Chem.* **269**, 30407–30411.

Sikela, J. M., and Hahn, W. E. (1987). Screening an expression library with a ligand probe: Isolation and sequence of a cDNA corresponding to a brain calmodulin-binding protein. *Proc. Natl. Acad. Sci. U.S.A.* **84**, 3038–3042.

Sutherland, E. W., and Rall, T. W. (1958). Fractionation and characterization of a cyclic adenine ribonucleotide formed by tissue particles. *J. Biol. Chem.* **232**, 1077–1091.

Tan, S.-E., Wenthold, R. J., and Soderling, T. R. (1994). Phosphorylation of AMPA-type glutamate receptors by calcium/calmodulin-dependent protein kinase II and protein kinase C in cultured hippocampal neurons. *J. Neurosci.* **14**, 1123–1129.

Tanaka, T., and Hidaka, H. (1980). Hydrophobic regions function in calmodulin-enzyme(s) interactions. *J. Biol. Chem.* **255**, 11078–11080.

Tansey, M. G., Luby-Phelps, K., Kamm, K. E., and Stull, J. T. (1994). Ca^{2+}-dependent phosphorylation of myosin light chain kinase decreases the Ca^{2+} sensitivity of light chain phosphorylation within smooth muscle cells. *J. Biol. Chem.* **269**, 9912–9920.

Tansey, M. G., Word, R. A., Hidaka, H., Singer, H. A., Schworer, C. M., Kamm, K. E., and Stull, J. T. (1992). Phosphorylation of myosin light chain kinase by multifunctional calmodulin-dependent protein kinase II in smooth muscle cells. *J. Biol. Chem.* **267**, 12511–12516.

Tokumitsu, H., Chijiwa, T., Hagiwara, M., Mizutani, A., Terasawa, M., and Hidaka, H. (1990). KN-62, 1-[N,O-Bis(5-isoquinolinesulfonyl)-N-methyl-L-tyrosyl]-4-phenylpiperazine, a specific inhibitor of Ca^{2+}/calmodulin-dependent protein kinase II. *J. Biol. Chem.* **265**, 4315–4320.

Tokumitsu, H., Brickey, D. A., Glod, J., Hidaka, H., Sikela, J., and Soderling, T. R. (1994). Activation mechanism for Ca^{2+}/calmodulin-dependent protein kinase IV. *J. Biol. Chem.* **269**, 28640–28647.

Tsunoda, Y., Funasaka, M., Modlin, I. M., Hidaka, H., Fox, L. M., and Goldenring, J. R. (1992). An inhibitor of Ca^{2+}/calmodulin-dependent protein kinase II, KN-62, inhibits cholinergic-stimulated parietal cell secretion. *Am. J. Physiol.* **262**, G118–G122.

Walsh, D. A., Perkins, J. P., and Krebs, E. G. (1968). An adenosine 3′, 5′-monophosphate-dependent protein kinase from rabbit skeletal muscle. *J. Biol. Chem.* **243**, 3763–3765.

Walsh, M. P., and Hartshone, D. J. (1982). Actomyosin of smooth muscle. *In* "Calcium and Cell Function" (W. Y. Cheung, ed.), Vol. 3, pp. 223–269. Academic Press, Orlando, FL.

Wegner, M., Cao, Z., and Resenfeld, M. G. (1992). Calcium-regulated phosphorylation within the leucine zipper of C/EBPβ. *Science* **256**, 370–373.

Wenham, R. M., Landt, M., Walters, S. M., Hidaka, H., and Easom, R. A. (1992). Inhibition of insulin secretion by KN-62, a specific inhibitor of the multifunctional Ca^{2+}/calmodulin-dependent protein kinase II. *Biochem. Biophys. Res. Commun.* **189**, 128–133.

Wolff, D. J., and Brostrom, C. O. (1979). Properties and functions of the calcium-dependent regulator protein. *In* "Advances in Cyclic Nucleotide Research" (P. Greengard and G. A. Robison, eds.), Vol. 11, pp. 27–88. Raven Press, New York.

Yokokura, H., Picciotto, M. R., Nairn, A. C., and Hidaka, H. (1995). The regulatory region of calcium/calmodulin-dependent protein kinase I contains overlapping autoinhibitory and calmodulin-binding domains. *J. Biol. Chem.* **270**, 23851–23859.

B. E. Kemp*
J. A. Barden†
B. Kobe*
C. House*
M. W. Parker*

*St. Vincent's Institute of Medical Research
Fitzroy 3065, Victoria, Australia

†Department of Anatomy and Histology
University of Sydney
Sydney 2006, New South Wales, Australia

Intrasteric Regulation of Calmodulin-Dependent Protein Kinases

I. Introduction

It was apparent early in the study of protein kinases with the pioneering work of Krebs and Fischer and colleagues that phosphorylase kinase could be interconverted between a relatively inactive form and an activated form that differed in their kinetics of phosphorylase *b* phosphorylation (Fischer and Krebs, 1989). Phosphorylase kinase could be activated in a variety of ways; by phosphorylation by the cAMP-dependent protein kinase, by partial proteolysis with chymotrypsin, and by calcium. These experimental observations suggested that phosphorylase kinase existed in a latent inactive form that required either Ca^{2+} or cAMP as signals for activation.

Over the past 30 years, this theme repeatedly has been found for the regulation of many other protein kinases that require the binding of an allosteric regulator or phosphorylation by another protein kinase to stimulate their activity. In a number of cases, the protein kinase is inactive because part of its structure blocks access to the active site. The cAMP-dependent protein kinase regulatory subunits and the inhibitory protein PKI contain substrate-like mimics of the local phosphorylation site sequences that are responsible for inhibition; this led to the idea that the inhibitory structure was a pseudosubstrate. Studies on the regulation and substrate specificity of the myosin light-chain kinase, protein kinase C, phosphorylase kinase, and the calmodulin-dependent protein kinase II indicated that the pseudosubstrate model of regulation may be a widely used regulatory mechanism. Subsequent work has extended these concepts to a variety of other protein kinases including calmodulin-dependent protein kinases I, III, and IV, the p70 S6 kinase, and the MAP kinase-activating kinase; these have pseudosubstrate-like autoregulatory sequences outside their catalytic cores on either the carboxyl- or the amino-terminal ends. The first crystal structure of a protein kinase, that of the cAMP-dependent protein kinase with its bound inhibitor peptide, provided compelling evidence that inhibition could occur by a pseudosubstrate mechanism. The first crystal structure of an autoinhibitory protein kinase, that of twitchin kinase, showed that the autoinhibitory sequence makes a number of important contacts in the active site and does not just resemble a substrate.

We have used the term *intrasteric* regulation to describe the regulation in which the regulatory sequence acts directly at the active site taking advantage of its specific features (Kemp and Pearson, 1991). This is in contrast to allosteric regulation, which involves the binding of a regulator at an alternative site to the active site's subtrate binding site. Intrasteric regulation can be extended to include pseudosubstrate binding or phosphorylation events within the active site that are typically catalyzed by distinct protein kinases. The phosphorylation of bacterial isocitrate dehydrogenase at its active site is a good example of intrasteric regulation. In this case, a serine deep in the active site is phosphorylated and inhibits isocitrate binding without any major changes in the enzyme's structure. The cyclin-dependent kinase cdc2 is also regulated by intrasteric phosphorylation occurring at the ATP-binding loop and inhibits enzyme activity. Activation of a number of protein kinases including the cAMP-dependent protein kinase occurs by intrasteric phosphorylation on Thr197. Phosphorylation of Thr197 located in the substrate anchoring loop is essential for full enzyme activity (Taylor *et al.*, 1993). Other protein kinases requiring phosphorylation in the substrate anchoring loop (or activation loop) for maximal activity include the insulin receptor kinase and calmodulin-dependent protein kinases I and IV.

Extrasteric phosphorylation refers to phosphorylation that occurs distal to the active site and in this respect is the covalent modification equivalent of *allosteric* regulation. In this respect, classical allosteric regulation falls under the umbrella of extrasteric regulation and is the noncovalent counterpart of regulation of extrasteric phosphorylation. The smooth muscle myosin light-chain kinase and calmodulin-dependent protein kinase II are regulated by extrasteric phosphorylation that occurs in proximity to their respective calmodulin-binding sequences and modulates the calmodulin dependence of enzyme activation.

This chapter reviews the recently identified protein crystal structure of the autoinhibited twitchin kinase and discusses these findings in the context of the large body of the earlier biochemical studies that lead to the concepts of autoregulation of the Ca^{2+}/calmodulin-dependent protein kinase family.

II. Structure of Autoinhibited Twitchin Kinase

The giant muscle protein twitchin kinase encoded by the *Caenorhabditis elegans unc-22* gene contains 6839 residues (Benian *et al.*, 1993) with a catalytic core between residues 5950 and 6199. The catalytic core of twitchin kinase is 52% identical to smooth muscle myosin light-chain kinase sequence without insertions or deletions. Fragments of twitchin kinase containing the catalytic core have been successfully expressed in *Escherichia coli* (Lei *et al.*, 1994). When the C-terminal sequence outside the catalytic core is removed, the enzyme fragment is constitutively active using peptide substrates. The twitchin kinase fragment (residues 5890–6262) containing both the catalytic core and the C-terminal regulatory domain was crystallized (Hu *et al.*, 1994a), and its structure solved (Hu *et al.*, 1994b) using multiple isomorphic replacement.

A. Protein Kinase Fold

Twitchin kinase contains a canonical bilobate protein kinase structure (Fig. 1) similar to the cAMP-dependent protein kinase structure (Knighton *et al.*, 1993). After superimposition, the root mean square deviation between the $C\alpha$ atoms of the catalytic cores of twitchin kinase and the cAMP-dependent protein kinase is 1.9 Å. The twitchin kinase represents an open kinase structure with the small lobe rotated approximately 30° relative to the large lobe (Fig. 1). The B-helix of the cAMP-dependent protein kinase in the small lobe is absent as anticipated from molecular modeling of MLCK (Knighton *et al.*, 1992). The rotation of the small lobe in twitchin kinase is readily achieved by a hinge motion about adjacent Gly residues between the two lobes (6021 and 6022); hinging at the corresponding position has been proposed for the solution structure of the cAMP-dependent protein kinase (Olah *et al.*, 1993) with a bound inhibitor peptide. The rotation of the small lobe in twitchin kinase accommodates the path of the C-terminal tail that elicits a "pseudo-nucleotide-bound" conformation. This is most noticeable for the phosphate anchoring loop residues G^{5949}TGAFG^{5954} between β-strands S3 and S4 (Fig. 1) where the strand-turn-strand structure of twitchin kinase mimics the porcine heart cAMP-dependent protein kinase

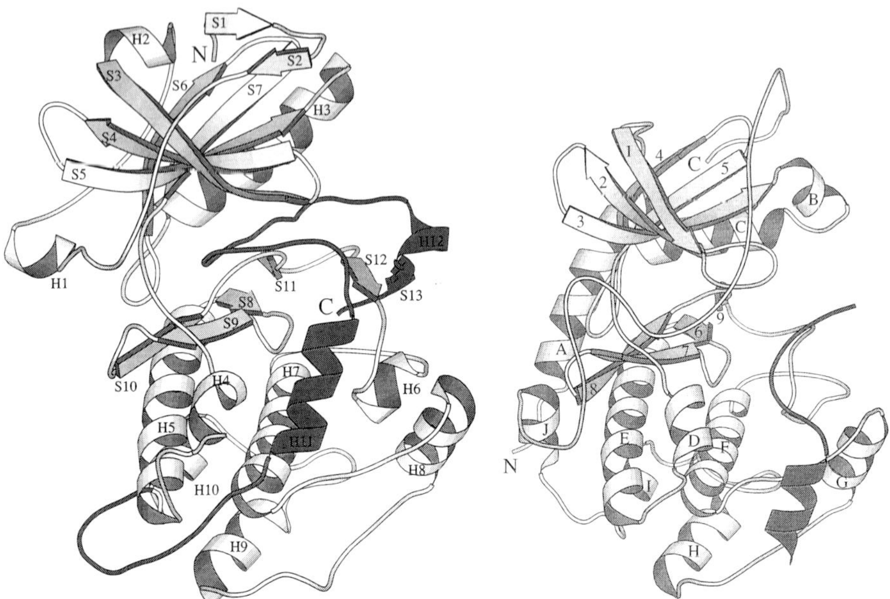

FIGURE I   Ribbon diagrams of the twitchin kinase fragment (residues 5894–6260, left) and the cAMP-dependent protein kianse catalytic subunit complexed with the PKI (5–24) peptide (right). The secondary structure elements in twitchin kinase are labeled H and S for α-helices and β-strands, respectively. PKI (5–24) in cAMP-dependent protein kinase and the autoinhibitory region in twitchin kinase are shaded dark. Diagrams were drawn using the program MOLSCRIPT (Kraulis, 1991).

complexed to MnAMP–PNP and the inhibitor peptide (Bossemeyer *et al.*, 1993). Another major difference between the twitchin kinase and cAMP-dependent protein kinase is the position of the helix H4 that is shifted 5 Å sideways relative to the position of the corresponding D-helix (Fig. 1). In the twitchin kinase, two uncharged residues, Val6098 and Leu6062, replace phosphorylated Thr197 and Arg165 of the cAMP-dependent protein kinase, respectively. On the N-terminal side of the catalytic core (residues 5894– 5940), there are a pair of antiparallel β-strands (S1 and S2) and a pair of short helices (H1 and H2).

B. Associated Noncatalytic *unc* Domains

In addition to the protein kinase domain, proteins from the smooth muscle myosin light-chain kinase superfamily contain class I and class II domains related to fibronectin type III and immunoglobulin C_2 domains (Fig. 2) (Olson *et al.*, 1990). In some smooth muscles, the C-terminal *unc* class II domain of the MLCK is independently transcribed off the MLCK gene and expressed as a protein called telokin. The crystal structure of telokin (Holden *et al.*, 1992) shows a β-barrel fold related to the C_2 domain of immunoglobulins. NMR studies also indicate that the *unc* class II domain of titin kinase situated amino terminal to the protein kinase core contains a β-structure (Politou *et al.*, 1994). It will be of interest to determine the juxtaposition of the class I and II domains with the catalytic core and autoregulatory structures (Fig. 3; see color insert).

C. Intrasteric Inhibition: Profile of Contacts

The 60-residue C-terminal tail of twitchin kinase follows the path of the substrate binding groove between the two lobes of the structure (Figs. 1 and 4). The major secondary structure features of the C-terminal tail are a prominent three-turn α-helix H11 and a single-turn α-helix H12. There are two short 3_{10} helices and a short β-strand between the α-helices (Fig. 4). The β-strand S13 folds under the substrate anchoring loop. The contacts between the C-terminal tail and the catalytic core can be divided into three classes: substrate-like, ATP binding site contacts, and contacts with the catalytic residues. A summary of some of the important contact residues between the autoinhibitory sequence and the catalytic core in twitchin and

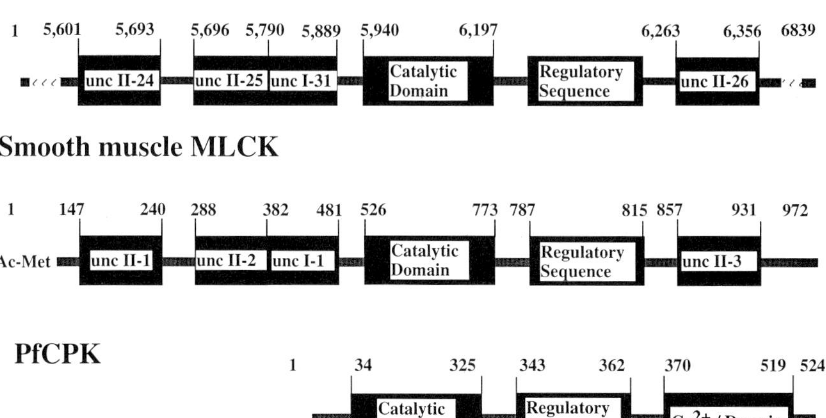

FIGURE 2   Comparison of the domain structure of the nematode twitchin kinase, smooth muscle myosin light-chain kinase (chicken gizzard), and the Ca²⁺-dependent *Plasmodium falciparum* kinase (PfCPK). All three kinases are autoregulated by regulatory sequences on the COOH-terminal ends of their catalytic domains. Both twitchin kinase and the smooth muscle myosin light-chain kinase have *unc* I (fibronectin type III) and *unc* II (immunoglobulin C_2) repeats. The calcium binding domain of PfCPK is 34% identical with chicken troponin C and 35% identical with calmodulin.

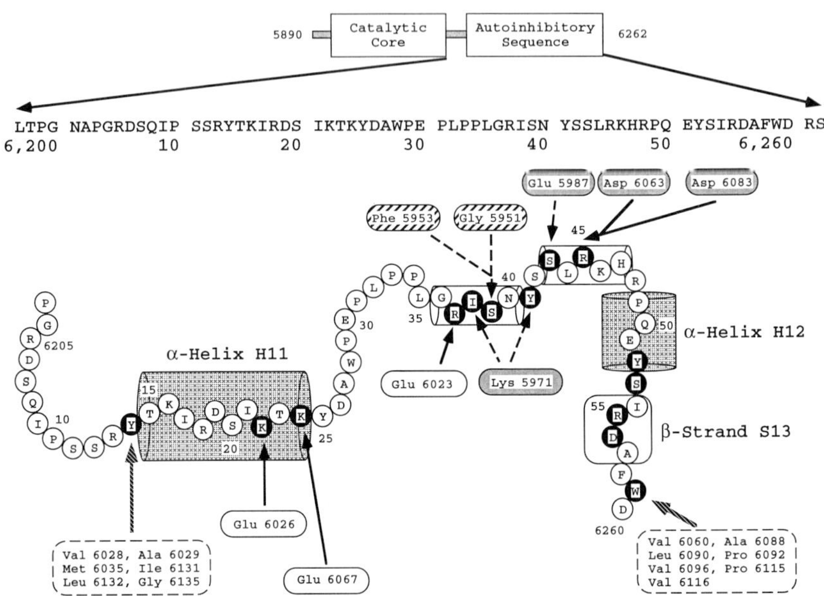

FIGURE 4   Summary of major contacts between the autoinhibitory sequence (6200–6262) and the catalytic core of nematode twitchin kinase. The secondary structure features (see Fig. 1) are shown as large cylinders [α-helix (H11 and H12)], small cylinders (3_{10} helix), and a square for the β-strand (S13). Basic residues in the autoinhibitory sequence making putative substrate-like electrostatic contacts to Glu6026, Glu6067, and Glu6023 in the catalytic core are shown. There is a steric effect with autoinhibitory sequence GRISNY[6241] occupying the position corresponding to ribose diphosphate of ATP in the cAMP-dependent protein kinase. Contacts are also made with prominent residues in the active site. These include Lys5971 and Glu5987, which make contacts with the backbone carbonyls of Ile6238 and Tyr6241, and both Gly5951 and Phe5593 of the phosphate anchor region are hydrogen bonded to the side chain of Ser6243 in the autoinhibitory sequence. Both Asp6063 and Asp6083, which are important in catalysis and magnesium binding in cAMP-dependent protein kinase, make an electrostatic contact with Arg6245. Tyr6214 in the autoinhibitory sequence is located in a hydrophobic pocket of the substrate binding groove equivalent to the P−11 site in cAMP-dependent protein kinase. On exiting the active site, the autoinhibitory sequence folds over the activation loop and Trp6259 is anchored in a hydrophobic pocket beneath it.

corresponding residues in the cAMP-dependent protein kinase catalytic subunit is given in Table I.

I. Substrate-like Contacts

In the case of the cAMP-dependent protein kinase, substrate-specificity determinants in the local phosphorylation site sequence are notated P+1, P+2 or P−1, P−2, etc. depending on their distance from the phosphate acceptor site P_0. Within the twitchin kinase helix H11, Lys6224 makes an electrostatic contact with Glu6067, the equivalent of Glu170 in the cAMP-dependent protein kinase that is the recognition site for the P−2 Arg in the

TABLE I Contacts between Twitchin Kinase Autoinhibitory Sequence and Catalytic Core

Autoinhibitory sequence	Twitchin core contact residues	cAMP-PK equivalent	Function of core residues in cAMP-PK
Y6214	V6028, A6029, M6035, I6131, L6132	Y235, F239	Hydrophobic pocket for P−11 aromatic residue
K6222	E6026	E130	Electrostatic contact with the P−3 Arg
K6224	E6067	E170	Electrostatic contact with the P−2 Arg
R6237	E6023	E127	Electrostatic contact with the P−3 Arg
I6238, Y6241	K5971	K72	Electrostatic contact with α- and β-PO$_4$
S6243	E5987	E91	Electrostatic contact with K72
S6239	G5951	G50	β-Phosphate anchor
R6245	D6063	D166	Catalytic base
	D6083	D184	Chelates activating Mg^{2+}

substrate (Fig. 4). Lys6222 in the twitchin autoregulatory sequence makes a contact with Glu6026 located in H4 helix, which is equivalent to the D-helix in the cAMP-dependent protein kinase structure. A Glu residue in this region is a conserved feature of the myosin light-chain kinase family but not in other protein kinases. The Arg6237 also forms an electrostatic interaction with Glu6023 in the H4 helix, which is equivalent to Glu127 in the cAMP-dependent protein kinase structure that is involved in recognition of the substrate's P−3 Arg. The inhibitor peptide in the cAMP-dependent protein kinase structure makes a contact with Glu203 through the substrate P−6 Arg. Although there is equivalent Glu6104 in twitchin kinase, it does not make contacts with the C-terminal tail residues in the structure. Thus, some but not all possible substrate-like contacts are made in the twitchin autoinhibited structure.

2. ATP Binding Site Contacts

The twitchin kinase structure does not contain a bound nucleotide nor can nucleotides be soaked into the structure. The backbone of the sequence GRISN^{6240} that forms the first 3$_{10}$ helix occupies the equivalent position of the ribose 5′-diphosphate of ATP (Fig. 4). Thus, the autoinhibitory sequence strictly blocks the ATP binding site. Some pseudosubstrate inhibitory peptides, including those of the myosin light-chain kinase and the calmodulin-dependent protein kinase II, act as both substrate and ATP antagonists. The twitchin kinase structure provides some clues as to how this may occur.

3. Catalytic Residue Contacts

The second 3$_{10}$ helix containing Arg6245 (residues SLRKH^{6247}) makes electrostatic contacts with both Asp6063 and Asp6083 in the catalytic core

(Fig. 4). These two Asp residues correspond to Asp166 and Asp184 in the cAMP-dependent protein kinase structure and are conserved for all protein kinases because of their crucial role in catalysis. In the cAMP-dependent protein kinase, Asp184 chelates the Mg^{2+} ion that bridges the β- and γ-phosphates and is thought to orient the γ-phosphate for transfer to the phosphate acceptor hydroxyl. Asp166 (equivalent of Asp6083 in twitchin) is the putative catalytic base involved in the transfer reaction. Thus, Arg6245 in the twitchin kinase autoregulatory sequence plays a key role in interacting with the conserved Asp residues that are crucial for catalysis. In the cAMP-dependent protein kinase, the conserved Lys72 anchors the α- and β-phosphates but in the absence of Mg^{2+}/ATP it forms an electrostatic contact with Glu91. With the nucleotide-free twitchin structure, these contacts have been rearranged so that Lys5971, equivalent to Lys72 of the cAMP-dependent protein kinase, forms hydrogen bonds with the backbone carbonyls of Ile6238 and Tyr6241 in the 3_{10} helix of the autoregulatory sequence that occupies the ribose 5′-diphosphate position. The Glu5987 (equivalent to Glu91 in cAMP-dependent protein kinase) forms a hydrogen bond with the side chain of Ser6243 in the second 3_{10} helix instead of making a contact with the equivalent of Lys5971 (Lys72). The autoinhibition of twitchin kinase thus involves many contacts with key residues responsible for catalysis (Fig. 4).

D. Implications for the Calmodulin-Dependent Protein Kinase Family

The twitchin kinase structure provides a useful framework for understanding the autoinhibition of members of the families of calmodulin-dependent and Ca^{2+}-dependent protein kinases. Inspection of the known autoregulatory sequences (Fig. 5) demonstrates some common features as well as some major differences. The twitchin kinase family is exceptional in having the Pro-rich segment inserted in the autoinhibitory sequence (PLPPL, for nematode twitchin). This sequence permits the sharp turn out of the substrate binding groove occupied by the H11 helix and penetrates deep into the active site to position the two short 3_{10} helices to contact the key catalytic residues. Many members of the calmodulin-dependent protein kinase family may not have this feature. In the skeletal muscle myosin light-chain kinase, for example, the autoinhibitory sequence does not appear to obstruct ATP binding as assessed with 5′-*p*-fluorosulphonylbenzoyl adenosine labeling (Kennelly *et al.*, 1991) and therefore is perhaps less likely to have these contacts. The inactive form of calmodulin-dependent protein kinase II does not bind ATP, so that in this instance there may be some contacts that mimic those of twitchin to block ATP binding (Schulman, 1993).

Enzyme		Autoregulatory sequence	
nematode TW	HPWLT	PGNAPGRDSQIPSSR**Y**TKI**R**DSIKTKYDAWPEPLPPLG**R**ISNYSSLRKHRPQEYSIRDAFWDRS	
Aplysia TW	HPWLK	GDHSNLTSRIPSSR**Y**NKI**R**QKIKEKYADWPAPQPAIG**R**IANFSSLRKHRPQEYQIYDSYFDRK	
projectin	HPWLT	GDHSAMKHEINRLG**Y**LSL**R**EKLRRKYEDFERFLLPIG**R**LSEYSSLRKLLMEKYKIHDAVQRAS	
smMLCK	HPWLQ	KDTKNMEAKKLSKDRMKK**Y**MAR**R**KWQKTGHAV	RAIG**R**LSSMAMISG
skMLCK	HPWLNN	LAEKAKRCNRRLKSQILLKK**Y**LMK**R**RWKKNFIAV.	SAAN**R**FKKISSSGALMALGV
CaMK I	HPWIA	GDTALDKNIHQSVSEQIKKN**F**AKS**K**WKQAFNRTA	VVRHM**R**KLQLGHQPGGTGTDS
CaMK II	HPW	QRERVASVVHRQETVDCLKK**F**NAR**R**KLKGAILTTMLATRNFSS**R**SMITKKGEGSQVKE	
CaMK IV	HPWVT	GKAANFVHMDTAQKKLQE**F**NAR**R**KLKAAVKAV	VASS**R**LGSASSSHTNIQESNKASS
Phos Kin γ	HPFF	QQYVVEEVRHFSPRGK**F**KVICLTVLASVRIY	YQYR**R**VKPVTREIVIRDPYALRPLRRLIDA
PfCPK	SKWIKKYANNINKSDQKTLCGALSNMRK**F**EGSQKLAQAAI		LFIGS**K**LTTL

FIGURE 5 Alignment of autoinhibitory sequences: nematode twitchin (nematode TW), *Aplysia* twitchin (*Aplysia* TW), *Drosophila* projectin, smooth muscle myosin light-chain kinase (smMLCK), skeletal muscle myosin light-chain kinase (skMLCK), calmodulin-dependent protein kinases I, II, and IV, phosphorylase kinase (Phos Kin), and *Plasmodium falciparum* calcium-dependent kinase (PfCPK). Putative conserved residues are shown in bold.

The alignment of the autoregulatory sequences reveals few highly conserved residues (Fig. 5). An aromatic residue present at the start of the H11 helix (Tyr6214) equivalent to the P−11 position in the cAMP-dependent protein kinase is highly conserved. Three residues downstream from the conserved aromatic residue is a highly conserved Arg; interestingly, in the twitchin kinase structure this residue serves no obvious purpose and is solvent exposed. However, in other members of the family it may be employed in contacts with the core. Also conserved is the twitchin kinase Arg6237 (first 3_{10} helix) that forms an electrostatic contact with Glu6203 (equivalent to Glu127 in the cAMP-dependent protein kinase). This substrate-like contact fits with the specificity requirements of many members of the calmodulin-dependent protein kinase family that require a basic residue in the P−3 position (Kemp and Pearson, 1990).

The rich diversity of contacts within the active site seen with the structure of twitchin kinase exemplifies well the features of intrasteric regulation. We can expect many of these features to also be found in other members of the calmodulin-dependent protein kinase family. It is clear, however, that there is considerable room for diversity in terms of the detailed interactions within the active site.

III. Autoregulated Protein Kinases

This section describes biochemical evidence for the intrasteric regulation of calmodulin-dependent protein kinase family members that has been gained from a diversity of approaches including mutagenesis, proteolysis, antibody, and synthetic peptide studies. In the majority of examples known, the pseudosubstrate autoinhibitory sequence is contiguous with the catalytic domain. In contrast, in cAMP-dependent protein kinase, the autoregulatory sequence is present on a separate protein (either regulatory subunit or PKI) and is encoded by a different gene. The calmodulin-dependent protein kinases have their pseudosubstrate autoregulatory sequences on the COOH-terminal side of the catalytic domains adjacent to or overlapping their calmodulin-binding sequences. Some members exist as multimeric complexes; calmodulin-dependent protein kinase II, for example, is a dodecamer complex. It appears that all autoinhibition is *cis* with the pseudosubstrate sequences folding back into the active site of their parent catalytic domain. However, we cannot entirely rule out the possibility of the pseudosubstrate sequence of one monomer inhibiting another in a *trans* interaction within a multimeric complex. The linking sequence between the catalytic domain and the pseudosubstrate sequence is often susceptible to proteolysis that leads to removal of the pseudosubstrate sequence and the generation of a constitutively active catalytic core.

A. Myosin Light-Chain Kinase/Twitchin Kinase Family

The myosin light-chain kinase family of protein kinases consists of structurally and functionally diverse proteins. It includes the largest known vertebrate protein, titin (3000 kDa), twitchin from the nematode (750 kDa) (Benian *et al.*, 1989), projectin (500 kDa) from *Drosophila,* as well as smooth muscle myosin light-chain kinase (130–155 kDa), skeletal muscle myosin light-chain kinase (87 kDa), and the *Dictyostelium* myosin light-chain kinase (34 kDa). In terms of sequence relationships, the *Dictyostelium* myosin light-chain kinase is the most remotely related member of the family and its sequence is more similar to DUN1 kinase from yeast, which is involved in DNA repair, than the myosin light-chain kinases.

The giant myosin-associated proteins of the titin/twitchin family are important cytoskeletal elements in muscle, where they may act as rulers for thick filament assembly (Trinick, 1994). These kinases contain catalytic domains similar to myosin light-chain kinases near their respective COOH-termini, and the rest is made up of super repeats of class I (fibronection type III) and class II (immunoglobulin C2) motifs (Benian *et al.*, 1989) (Fig. 2). Twitchin kinase was originally identified as the product of the *unc-22* gene in *C. elegans* (Moerman *et al.*, 1988). All homozygous *unc-22* mutants are characterized by impaired movement due to an uncoordinated twitch of their bodywall muscles, indicating a role of the gene product in muscle function. Smooth muscle myosin light-chain kinase has three class II and one class I motifs arranged around the catalytic domain in the same juxtaposition as that found in twitchin and titin (Fig. 2).

Smooth muscle myosin light-chain kinase has a well-defined role in regulating the initiation of smooth muscle contraction (Adelstein and Conti, 1975; Hoar *et al.*, 1979). Following an increase in cytosolic Ca^{2+}, the $Ca^{2+}/$

calmodulin complex binds to myosin light-chain kinase at a specific binding site between residues 796 and 813 (Lucas *et al.*, 1986) and activates it to cause phosphorylation of the 20-kDa regulatory myosin light chains at Ser19. This in turn leads to stimulation of the actin-activated myosin Mg–ATPase activity and muscle contraction.

Myosin light-chain kinase also occurs in nonmuscle cells, in which it is thought to participate in a number of processes including capping, cellular motility, organelle movement, and in mitotic cells, in which it is thought to participate in spindle function. In smooth muscle, the myosin light-chain kinase N-terminal 114 residues are bound to actin (Kanoh *et al.*, 1993), whereas the COOH terminus may be tethered to unphosphorylated myosin (heavy meromyosin fragment but not the myosin rod) (Shirinsky *et al.*, 1993). Indeed, it seems reasonable that the myosin light-chain kinase forms a mobile "latch" between the thin filament and the thick filament.

Skeletal muscle myosin light-chain kinase is also regulated by calmodulin (Blumenthal *et al.*, 1985). It is smaller (87 kDa) and does not have the class I and class II motifs characteristic of its smooth muscle counterpart. Initiation of contraction in striated (skeletal and cardiac) muscle is mediated via troponin on the thin filament; phosphorylation of the regulatory light chains serves to increase the rate of force development with a shift in the dependence of force development on the calcium concentration (Sweeney *et al.*, 1993).

All members of the myosin light-chain kinase family are autoinhibited but there is considerable variation in their mechanisms of activation. The *Dictyostelium* isoform is apparently activated by phosphorylation on its COOH-terminal tail near the pseudosubstrate sequence (Tan and Spudich, 1991) and is not regulated by calmodulin. *Aplysia* twitchin kinase binds calmodulin, as indicated by blotting experiments with labeled calmodulin (Heierhorst *et al.*, 1994) and calcium-dependent binding to a calmodulin affinity column; however, there is no evidence of kinase activation. Activation of the nematode twitchin has been reported (Lei *et al.*, 1994) but remains equivocal. The calmodulin binding sequence of *Aplysia* twitchin kinase has been suggested to be located in an equivalent position to that found in smooth muscle myosin light-chain kinase (Heierhorst *et al.*, 1994).

I. Autoregulatory Pseudosubstrate

Studies on the regulation of the myosin light-chain kinases family have played a major role in the development of the pseudosubstrate autoregulatory hypothesis (reviewed in Kemp and Pearson, 1991). Initially, it was recognized (Kemp *et al.*, 1987) that the basic residues present in the calmodulin binding domain (Lucas *et al.*, 1986) closely resembled the number and juxtaposition of the basic residues identified as specificity determinants in the myosin light-chain kinase (Kemp and Pearson, 1985).

| Substrate smMLC(1–22) | SSKRAKAKTTKKRPQRATSNV^{21}- | substrate |
| smMLCK(787–808) | -SKDRMKKYMARRKWQKTGHAV^{807}- | pseudosubstrate |

This led to the hypothesis that the sequence containing the calmodulin binding site may be positioned in the active site and render the myosin light-chain kinase inactive. Support for this concept was initially obtained using synthetic peptides corresponding to the pseudosubstrate sequence; these were found to inhibit smooth muscle myosin light-chain kinase (Kemp *et al.*, 1987). Similar results were also found for the skeletal muscle enzyme in which the peptide KRRWKKNFIAVSAANRFG, related to the skeletal muscle myosin light-chain kinase [skMLCK(577–594)], inhibited a calmodulin-independent chymotryptic fragment of the enzyme βC35 (Kennelly *et al.*, 1987). The peptides acted as both inhibitors and calmodulin antagonists so that the initial IC_{50} values underestimated the potency of the peptides as substrate inhibitors. More thorough kinetic studies (Ikebe, 1990) and the use of calmodulin-independent fragments of myosin light-chain kinase showed that the K_i values were in the nanomolar range (Pearson *et al.*, 1988).

Detailed studies using peptides and partial proteolysis have shown that the pseudosubstrate inhibitor and calmodulin-binding functions overlap but are not identical (Kemp and Pearson, 1991). The most potent calmodulin antagonist activity is contained within the sequence A^{796}RRKWQKTGHA-VRAIGRLSS^{815} as originally proposed by Lucas *et al.* (1986). COOH-terminal extension of peptides beyond Val807 did not enhance their substrate antagonist activities (Kemp *et al.*, 1987; Foster *et al.*, 1990; Pearson *et al.*, 1991a). However, N-terminal extension of the pseudosubstrate peptide to Leu774 greatly increased the potency as a substrate antagonist. Whereas the smMLCK(787–807) peptide has a K_i of 12 nM, the longer peptide smMLCK(774–807) has a K_i of 0.3 nM (Knighton *et al.*, 1992). The N-terminal extension is not itself an inhibitor but greatly boosts the potency of the pseudosubstrate sequence (Knighton *et al.*, 1992).

The pseudosubstrate peptide, smMLCK(787–807), also acts as a potent ATP antagonist with an IC_{50} of 149 nM (Pearson *et al.*, 1991a; Knighton *et al.*, 1992). Similar ATP antagonism is seen with the corresponding pseudosubstrate peptide from the calmodulin-dependent protein kinase II (Colbran *et al.*, 1988). We cannot exclude the possibility that the extended pseudosubstrate peptide naturally binds to the small lobe and modulates ATP binding. Nevertheless, this is surprising; active site labeling of the skeletal muscle myosin light-chain kinase by 5′-*p*-fluorosulphonylbenzoyl adenosine is relatively insensitive to the presence of calmodulin, indicating that access to the ATP binding site is not hindered by the pseudosubstrate inhibitor sequence in the intact myosin light-chain kinase (Kennelly *et al.*, 1991). This suggests that the pseudosubstrate peptide does not mask the ATP binding site in the small lobe or that the free pseudosubstrate peptide is able to bind to the active site in ways not available to its tethered counterpart.

The substitution of Tyr794 with Ala in the pseudosubstrate peptide sequence decreases potency 30-fold (Knighton *et al.*, 1992). This residue corresponds to Phe10 at the P-11 position in the inhibitor (PKI) of the cAMP-dependent protein kinase (Taylor *et al.*, 1993). The Tyr794 to Ala substitution also strongly reduces ATP antagonist activity of the pseudosubstrate peptide. Paradoxically, Trp800 appears to be important for both optimal pseudosubstrate peptide inhibition and for calmodulin binding and activation (Bagchi *et al.*, 1992a).

a. Proteolysis   A variety of biochemical approaches, each with its own strengths and shortcomings, has been used to test the pseudosubstrate hypothesis. Smooth muscle and skeletal myosin light-chain kinase (Tanaka *et al.*, 1980) are activated by partial proteolysis. For example, the tryptic digestion of smooth muscle myosin light-chain kinase results in an intermediate 64-kDa fragment that is inactive, and further digestion gives rise to a constitutively active 61-kDa fragment (Ikebe *et al.*, 1987). The 64-kDa fragment is generated by cleavage at Arg808 within the calmodulin binding sequence adjacent to the pseudosubstrate sequence (Pearson *et al.*, 1988); therefore, this fragment is no longer able to bind calmodulin. Inactive fragments can also be generated by either cleavage at Val807 by thermolysin or cleavage at Lys801 with Lys-C endoproteinase (Pearson *et al.*, 1991b).

These results suggest that a portion of the pseudosubstrate sequence (787–807) is sufficient for inhibition and that TGHAV^{807} is not essential for inhibition. The COOH-terminal residue of the constitutively active 61-kDa fragment is Lys779, positioned NH$_2$ terminal to the beginning of the pseudosubstrate sequence at Ser807 (Olson *et al.*, 1990). Although trypsin also cleaves myosin light-chain kinase on the N-terminal side of the catalytic domain at Thr283, the generation of the constitutively active 61-kDa fragment can be attributed entirely to the removal of residues on the COOH-terminal side of the catalytic core (Olson *et al.*, 1990). The peptide D^{777}TK^{779}NMEAKKLS^{787}KDRMKK^{793} was a 10,000-fold weaker inhibitor of the myosin light-chain kinase than the pseudosubstrate peptide 787–807, indicating that activation of the 64-kDa fragment of the myosin light-chain kinase was due to the removal of the pseudosubstrate sequence smMLCK(787–807) (Pearson *et al.*, 1991b). Ikebe *et al.*, (1989) suggested the inactive 64-kDa fragment was generated by the cleavage of the enzyme at Lys793 or Arg797; this result is inconsistent with peptide inhibition (Olson *et al.*, 1990), other mutational studies (Ito *et al.*, 1991), and the results of proteolysis (Olson *et al.*, 1990).

A 35-kDa chymotryptic fragment of the skeletal enzyme, skMLCK(256–584), does not bind calmodulin (Edelman *et al.*, 1985). The cleavage site is within the calmodulin binding sequence of this enzyme. Interestingly, it retains most of the pseudosubstrate sequence but becomes increasingly active during purification (Edelman *et al.*, 1985). This inherent instability made it difficult to recognize that only the COOH-terminal sequence of the 35-kDa fragment was responsible for autoinhibition of the intact skeletal muscle enzyme (Edelman *et al.*, 1985). Similarly, in nematode twitchin kinase the truncation of only a small portion of the autoinhibitory sequence results in constitutively active enzyme (albeit with low activity); by contrast, in calmodulin-dependent protein kinase II, a substantial section of the autoinhibitory sequence can be truncated without activating the enzyme (see below).

Skeletal muscle myosin light-chain kinase undergoes temperature- and time-dependent inactivation in the presence of calmodulin (Kennelly *et al.*, 1990). This instability could be prevented by the presence of either substrate or inhibitor peptides (Kennelly *et al.*, 1990). Similar inactivation is not observed for smooth muscle myosin light-chain kinase with calmodulin binding (Faux *et al.*, 1993), although the constitutively active 61-kDa fragment is particularly susceptible to thermal inactivation. Detailed structure/ function studies revealed that the pseudosubstrate peptide 787–807, SKDRMKKYMARRKWQKTGHAV, protects the enzyme fragment from inactivation. The shorter peptide, RRKWQK, was equipotent (73 nM PC$_{0.5}$; protective concentration) at protecting the enzyme; all four basic residues and the Trp were important for protection. However, the substitution of Trp800 in the full-length peptide did not alter its potency for protecting the enzyme from thermal inactivation suggesting that multiple interactions were involved (Faux *et al.*, 1993). It is now apparent that the overlapping pseudosubstrate/calmodulin binding sequence has a stabilizing structural role in addition to its regulatory role.

b. Mutation Analysis   The role of the pseudosubstrate sequence in regulating myosin light-chain kinase has been investigated using mutagenesis in several laboratories. The approaches have included truncation mutagenesis, point mutations, sequence inversions, exchanges to form chimeric molecules, and insertion of a phosphorylatable residue in the pseudosubstrate sequence.

Truncation mutagenesis from the COOH terminus of a fragment (447–800) of smooth muscle myosin light-chain kinase by Ito *et al.* (1991) showed that the pseudosubstrate sequence could be truncated up to Trp800 before there was substantial release from autoinhibition. Deletion of the residues Y[794]MARRKW[800] resulted in activation of the enzyme. Truncation of the nonmuscle myosin light-chain kinase up to Glu1067 (corresponding to Glu782 in smMLCK), deleting AKKL as well as the pseudosubstrate sequence, also resulted in constitutively active enzyme (Shoemaker *et al.*, 1990), which is consistent with proteolysis studies. Based on the knowledge of the substrate specificity of myosin light-chain kinase, one would have expected that mutation of the basic residues in the pseudosubstrate sequence would have rendered the enzyme constitutively active. This does not occur with either Ala substitutions or charge reversal of these basic residues (Shoemaker *et al.*, 1990). Thus, although the results of truncation mutagenesis and proteolysis experiments were consistent with the pseudosubstrate model, point mutation experiments were not consistent with it.

Yano *et al.* (1993) reported expression of truncated mutants of smooth muscle myosin light-chain kinase in *Escherichia coli*. Enzyme truncated to either Ala796 or Lys799 behaved as though it was autoinhibited and could be activated by treatment with trypsin. These results are inconsistent with the simple pseudosubstrate autoinhibitory model and other mutation studies (Ito *et al.*, 1991) but they are consistent with the group's earlier proteolysis studies (Ikebe *et al.*, 1989). In a subsequent study (Tanaka *et al.*, 1995), it was found that enzymes truncated to Lys793 or shorter fragments were constitutively active. Based on the truncation mutagenesis results, this suggested that the residues KYMA[796] were sufficient to inhibit the enzyme and it was concluded that the pseudosubstrate sequence in the myosin light-chain kinase was not critical for autoinhibition (Tanaka *et al.*, 1995). These results are similar to those reported earlier for calmodulin-dependent protein kinase II (Cruzalegui *et al.*, 1992) (see below).

If the myosin light-chain kinase is modeled on the twitchin kinase structure, this places the YMARRK[799] in the helix equivalent to H11 (see above) lying in the substrate binding groove. Based on the twitchin kinase structure, the RRK triplet is expected to make substrate-like side chain contacts if it corresponds precisely to IRD in the twitchin kinase structure. In the twitchin kinase structure, two Lys residues (**KTK**) and one Arg (**GRI**) make substrate-like contacts with the catalytic core (shown in bold). One of the lysines is replaced by Gly in myosin light-chain kinase. The Tyr (shown in bold) occupies the P−11 hydrophobic pocket (see above).

Twitchin	YTKIRDSIKTKYDAWPEPLPPLGRISNYSSLR
MLCK	YMARRKWQKTGHAV RAIGRLSS

The twitchin kinase structure shows that substrate-like contacts occur. The corresponding residues in the myosin light-chain kinase, however, do not seem to be essential from truncation and mutagenesis results.

The crucial point of the pseudosubstrate hypothesis is that the pseudosubstrate is located in the enzyme's active site substrate binding groove. This has been shown (Bagchi *et al.*, 1992) by incorporating the phosphorylation site sequence RATSNV within the pseudosubstrate/calmodulin binding sequence. In the absence of calmodulin, this mutant underwent autophosphorylation. Furthermore, the structure of twitchin kinase (Hu *et al.*, 1994b) is entirely consistent with the pseudosubstrate sequence being located in the active site. It is important to recognize from the twitchin kinase structure that numerous contacts are made between the catalytic core and the autoinhibitory sequence; accordingly, the effect of specific point mutations or truncations may be masked by other interactions.

B. Calmodulin-Dependent Protein Kinase II Family

The pioneering studies of neuronal protein phosphorylation by Greengard and colleagues led to the identification of a series of calmodulin-dependent protein kinases (reviewed in Hanson and Schulman, 1992). Many of these are also found in a variety of nonneuronal tissues and cell types. There are multiple gene products and in many instances multiple isoforms produced from each gene: calmodulin-dependent protein kinases Ia, Ib, II, III (eEF-2–calmodulin-dependent protein kinase), and IV.

These kinases are thought to play a central role in Ca^{2+}-dependent signal transduction in cells. Both calmodulin-dependent protein kinases I and II were identified as synapsin kinases. Calmodulin-dependent protein kinase III is a specific elongation factor-2 kinase (Mitsui et al., 1993) and does not readily phosphorylate other substrates. Calmodulin-dependent protein kinase IV also phosphorylates synapsin and has been purified from rat cerebellum (Miyano et al., 1992). There are multiple isoforms of the calmodulin-dependent protein kinases II (α, β, β', γ_B, γ_C, δ) and IV. Calmodulin-dependent protein kinase II is the most thoroughly studied member of the family (Schulman, 1993). This enzyme is widely distributed in most tissues and is especially abundant in brain, constituting approximately 0.3% of the total brain protein. Typically, the mammalian calmodulin-dependent protein kinase II isoforms have subunits of 50–62 kDa that occur in multimers (10–12 subunits) to give a native mass of 300–700 kDa. The enzyme is made up of an N-terminal catalytic domain, a central regulatory sequence, and a COOH-terminal domain responsible for association of the subunits into multimers. Upon activation with Ca^{2+}/calmodulin, the enzyme is autophosphorylated and this leads to a reduction in its calmodulin dependency (Meyer et al., 1992).

I. Autoinhibitory Pseudosubstrates

Calmodulin-dependent protein kinase II (Payne et al., 1988) contains a pseudosubstrate sequence overlapping its calmodulin binding sequence, analogous to myosin light-chain kinase. Residues 290–309 in the sequence M[281]HRQET[286]VDCL[290]KKFNARRKLKGAILTTMLA[309] adjacent to the major autophosphorylation site at Thr286 (Payne et al., 1988) were identified as the autoinhibitory pseudosubstrate sequence. The calmodulin binding sequence extends from Ala295 to Ala309 (Hanson and Schulman, 1992; Hanley et al., 1987).

There has been a large number of studies directed at understanding the function of autophosphorylation and locating the autoinhibitory sequence. This area has been extensively reviewed (Schulman, 1993); therefore, we will only highlight the problems and limitations of the biochemical approaches used to investigate the mechanism of autoinhibition. The peptide containing residues 281–309 was a competitive inhibitor of ATP. By contrast, L[290]KKFNARRKLKGAILTTMLA[309] was less potent and was competitive with respect to peptide substrate alone (Colbran et al., 1989; Colbran et al., 1988). The extended peptide 281–309 contained the autophosphorylation site at Thr286. This led Colbran and colleagues (Smith et al., 1992) to suggest a dual interaction between the ATP and protein substrate binding elements of the catalytic core with the extended autoinhibitory peptide. Extension of the myosin light-chain kinase pseudosubstrate peptide to include the linking residues to the catalytic core also enhanced ATP antagonism (Knighton *et al.*, 1992). We consider the dual-interaction hypothesis to be implausible because the N-terminal extension that enhances ATP antagonism is distal to the ATP binding site (Cruzalegui *et al.*, 1992). The possibility that the autoinhibitory sequence of calmodulin-dependent kinase II is oriented in the active site in the reverse direction seems unlikely.

The crystal structure of twitchin kinase (see above) provides an explanation for the enhanced ATP-antagonist activities of the extended autoinhibitory peptides. In the structure (Fig. 4), a segment of the autoinhibitory sequence (GRISNY[6241]) is positioned in the binding site of the ribose diphosphate of ATP. Thus, in the case of the calmodulin-dependent protein kinase II, it would be expected that the COOH-terminal end of the peptide 281–309 might be responsible for the ATP antagonism of this peptide. It seems likely that the segment 281–289 anchors the 281–309 peptide more firmly to the catalytic core than the peptide 290–309, permitting it to obstruct ATP binding. The structure of the twitchin kinase also emphasizes the importance of contacts between the autoinhibitory sequence and the key residues involved in enzyme catalysis such as Asp166 and Lys72 [cAMP-dependent protein kinase numbering (Knighton *et al.*, 1993)]. These contacts were not anticipated from any of the earlier mutation or synthetic peptide studies (see above).

Truncation mutagenesis studies have tended to be more illuminating (Cruzalegui *et al.*, 1992) than site-directed mutagenesis studies alone. The COOH-terminal truncation of the calmodulin-dependent protein kinase II to Ala294 resulted in an inactive form of the protein kinase. This was not expected because the truncated form of the enzyme contained only 5 of the 15 pseudosubstrate residues in the proposed pseudosubstrate sequence. Surprisingly, substitution of these residues (L[290]KKFN) with AAAAL lead to only partial (7%) Ca^{2+}-independent activity. Deletion of these residues resulted in somewhat higher (38%) Ca^{2+}-independent activity. Thus, one could conclude that the sequence L[290]KKFN was sufficient for autoinhibition of the calmodulin-dependent protein kinase II but paradoxically not essential when the remaining sequence was present. This paradox reflects the problem that we measure enzyme activity and not the binding energy of the autoinhibitory peptide for the catalytic core so that any marked changes in binding energy that did not cause a change in enzyme activity would be undetected. Deletion of the residues 283–289 (RQETVDC) resulted in 35.5% Ca^{2+}-independent activity, similar to the deletion of L[290]KKFN. We interpret this result to be a frameshift of the remaining autoinhibitory pseudosubstrate sequence rather than it being directly involved in contacts within the active site. The crystal structure of twitchin kinase shows that there are a large number of contacts between the autoinhibitory sequence and the catalytic core/active site; this explains why a segment of sequence can be both "sufficient but not essential" for inhibition of enzyme activity.

C. Calmodulin-Dependent Protein Kinases I and IV

Calmodulin-dependent protein kinase I was identified based on its ability to phosphorylate synapsin I (Nairn and Greengard, 1987). It has now been cloned from rat brain and has a mass of 43 kDa (Picciotto *et al.*, 1993). This enzyme phosphorylates the synapsin site 1 with the sequence RRRLSDS; the phosphorylation of the corresponding synthetic peptide indicates that all three Arg residues are important for recognition (Picciotto *et al.*, 1993). This sequence contains KKNF, similar to KKFN, that is important in autoregulation of calmodulin-dependent protein kinase II and it seems reasonable that it would have a similar function in calmodulin-dependent protein kinase I.

Calmodulin-dependent protein kinase IV has been cloned by several groups (Means *et al.*, 1991; Ohmstede *et al.*, 1991) and it is closely related to calmodulin-dependent protein kinase II but differs by being monomeric (Ohmstede *et al.*, 1989). Calmodulin-dependent protein kinase IV is regulated by a kinase kinase (Okuno and Fujisawa, 1993); it has been purified as a 68-kDa protein (Tokumitsu *et al.*, 1994). A distinct calmodulin-dependent protein kinase I activating kinase with an apparent mass of 52 kDa has also been described (Lee and Edelman, 1994). Mutagenesis studies have shown that the calmodulin-dependent protein kinase IV contains an autoregulatory pseudosubstrate sequence (Tokumitsu *et al.*, 1994) as was evident from sequence homology with calmodulin-dependent protein kinase II (Ohmstede *et al.*, 1991).

D. Calcium Binding Autoregulated Protein Kinase Family

Recently, several novel calcium-dependent protein kinases have been identified that contain a calcium-binding regulatory domain fused to the protein kinase catalytic core (Fig. 2). This configuration makes the enzyme directly dependent on Ca^{2+} concentration and independent of calmodulin. The soybean calcium-dependent protein kinase CDPK-α has been cloned (Harper *et al.*, 1991); the 57-kDa protein contains a catalytic domain (residues 41–328) fused to a calmodulin-like sequence with four calcium binding motifs (EF hands). Another isoform, AK1, has been cloned from *Arabidopsis thaliana* and encodes a protein of 72.6 kDa with an 116-residue N-terminal extension (Harper *et al.*, 1993). Other plant CDPK isoforms have been found in rice (SPK), carrot (CDPKc), and corn as well as in the parasite

Plasmodium falciparum (PfCPK) and are therefore likely to represent a substantial multigene family (Harper *et al.*, 1994).

I. Calcium Regulation

The calcium binding domains have been studied in several of the CDPKs. In the case of the parasite PfCPK, its calcium binding domain shares 35% homology with human calmodulin, 38% homology with falciparum calmodulin, and 32% homology with human troponin C (Zhao *et al.*, 1994a). Mutational studies with the PfCPK in the calcium binding sites indicate that sites I and II appear crucial for both calcium-induced conformational changes and activation of the enzyme. In contrast, mutations at sites III and IV in the calcium binding domain were found to have minor effects (Zhao *et al.*, 1994b).

2. Autoregulation

The CDPKs have been shown to contain an autoregulatory sequence in the junction between the calcium binding domain and the catalytic core that is relatively conserved between the isoforms. Mutations within this autoregulatory sequence or truncation render the APK1 constitutively active (Harper *et al.*, 1994). Furthermore, mutants with the calcium binding domain truncated but with the autoinhibitory sequence intact are inactive, analogous to the findings with myosin light-chain kinase (see above). The skeletal muscle MLCK pseudosubstrate peptide (577–587, KRRWKKN-FIAV) and the calmodulin-dependent kinase II pseudosubstrate peptide (209–309, LKKFNARRKLKGAILTTMLA) inhibited APK1 and the truncated mutant form of the enzyme (Harper *et al.*, 1994). Biochemical studies have also shown that activation of APK1 by calcium is influenced by phospholipids (Binder *et al.*, 1994), giving up to a 100-fold stimulation. Interestingly, phospholipid activation is a common feature of autoregulated proteins, both protein kinases, and others such as phenylalanine hydroxylase (Kemp and Pearson, 1991).

E. Phosphorylase Kinase

Phosphorylase kinase is a complex heterododecamer $(\alpha\beta\gamma\delta)_4$, where γ is the catalytic subunit and δ is a tightly bound calmodulin subunit. The enzyme is calcium dependent and can be further activated by a second molecule of calmodulin or troponin C. The γ-subunit is autoinhibited and synthetic peptide studies have suggested that its COOH-terminal sequence contains two calmodulin binding sequences (Dasgupta *et al.*, 1989).

I. Pseudosubstrates

Because the γ-subunit of phosphorylase kinase is autoregulated, it seems likely that it will contain an inhibitory sequence on its COOH-terminal tail similar to the other members of the calmodulin-dependent protein kinase subfamily. Constitutively active fragments of phosphorylase kinase-γ have been generated by proteolysis (Harris *et al.*, 1990) as well as by recombinant protein expression (Huang *et al.*, 1993). Cleavage with chymotrypsin at Phe298 resulted in a 34-kDa fragment with sixfold higher activity than the γ-subunit calmodulin complex that was calcium independent (Harris *et al.*, 1990). Evidence has been presented indicating that a sequence K[420]RNPGSQKRFPSNCGRD[436] derived from the β-subunit is responsible for autoinhibition (Sanchez and Carlson, 1993). It is not clear how autoregulation could be achieved through autoinhibitory sequences being contributed from both the C-terminal sequence of the γ-subunit as well as that from the β-subunit if they were both to bind in the substrate binding groove and active site. Despite phosphorylase kinase being the cradle of protein kinase discovery, it seems destined to reveal the complexities of its regulation painfully slowly.

IV. Choreography of Activation

Calmodulin serves as an intracellular Ca^{2+} receptor and transduces Ca^{2+} transients that frequently are generated in cells in response to hormones or growth factors. The highly conserved calmodulin molecule binds 4 mol of Ca^{2+} with high affinity and each Ca^{2+} is coordinated by a helix–loop–helix structure termed an EF hand (Kretsinger and Nockolds, 1973; Moncrief, *et al.*, 1990). The three-dimensional structure of Ca^{2+}/calmodulin revealed a dumbbell-shaped molecule in which each terminal globular domain containing two Ca^{2+} binding sites was separated by a solvent-exposed eight-turn α-helix (Babu *et al.*, 1985) (Fig. 6; see color insert). The Ca^{2+} binding sites in the C-terminal lobe of vertebrate calmodulin have a slightly higher affinity for Ca^{2+} than do the pairs of sites in the N-terminal lobe (Klee and Vanaman, 1982). Calcium binding to either member of a pair of sites greatly influences binding to the other, whereas one pair of sites is not markedly dependent on the other pair (Starovasnik *et al.*, 1992).

A. Mechanism of Activation by Calmodulin

The mechanism of activation of smooth muscle myosin light-chain kinase by calmodulin is the most extensively studied. The calmodulin binding region was identified by cyanogen bromide fragmentation of chicken smooth muscle myosin light-chain kinase (Lucas *et al.*, 1986) between amino acids Ser787 and Ser813 (Olson *et al.*, 1990) overlapping the autoregulatory domain. With the use of synthetic peptide analogs, it was shown that the autoregulatory domain could be roughly divided into three parts with the N-terminal two-thirds (787–807) containing the pseudosubstrate region

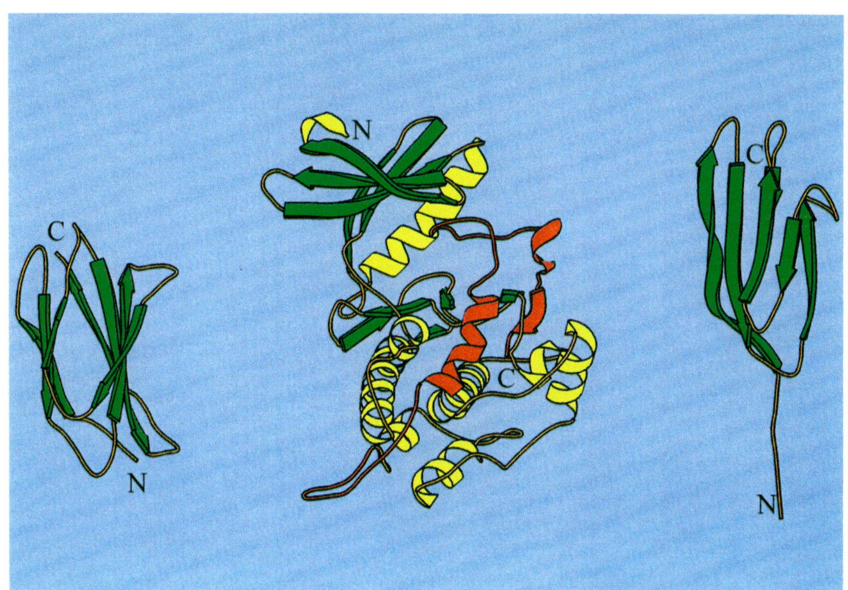

FIGURE 3 Twitchin kinase structural domains. The fibronectin type III (left) motif with β-strands is shown in green. Twitchin kinase fragment (residues 5894–6260, center) with β-strands (green), α-helices (yellow), and the autoregulatory COOH-terminal sequence (6200–6260) are shown in red. The structure of telokin (Holden *et al.*, 1992) (right) corresponding to *unc* II domain with β-strands is shown in green.

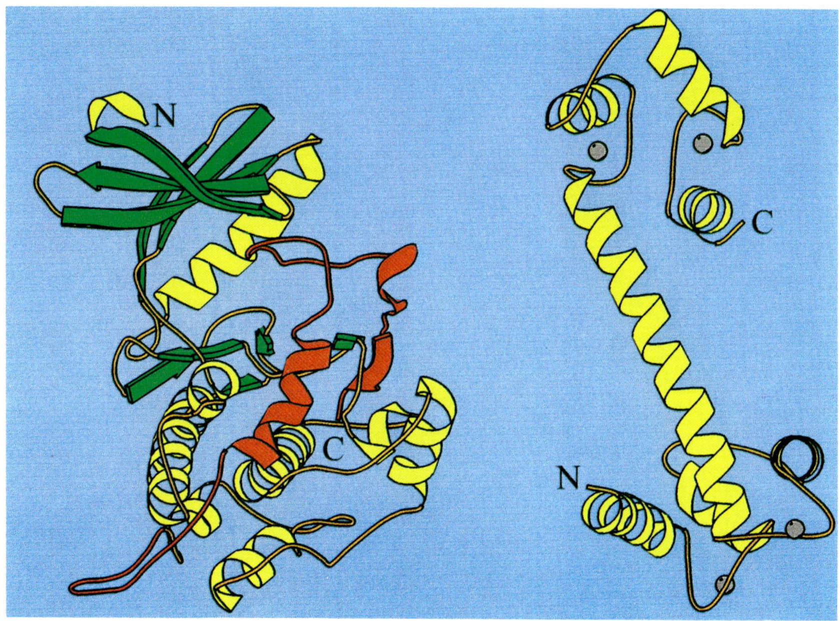

FIGURE 6 Ribbon diagrams of calmodulin and twitchin kinase. Calcium bound to the four binding sites in calmodulin is shown with gray spheres. Although twitchin kinase is not activated by calmodulin, the juxtaposed structures illustrate the major structural changes required to achieve activation.

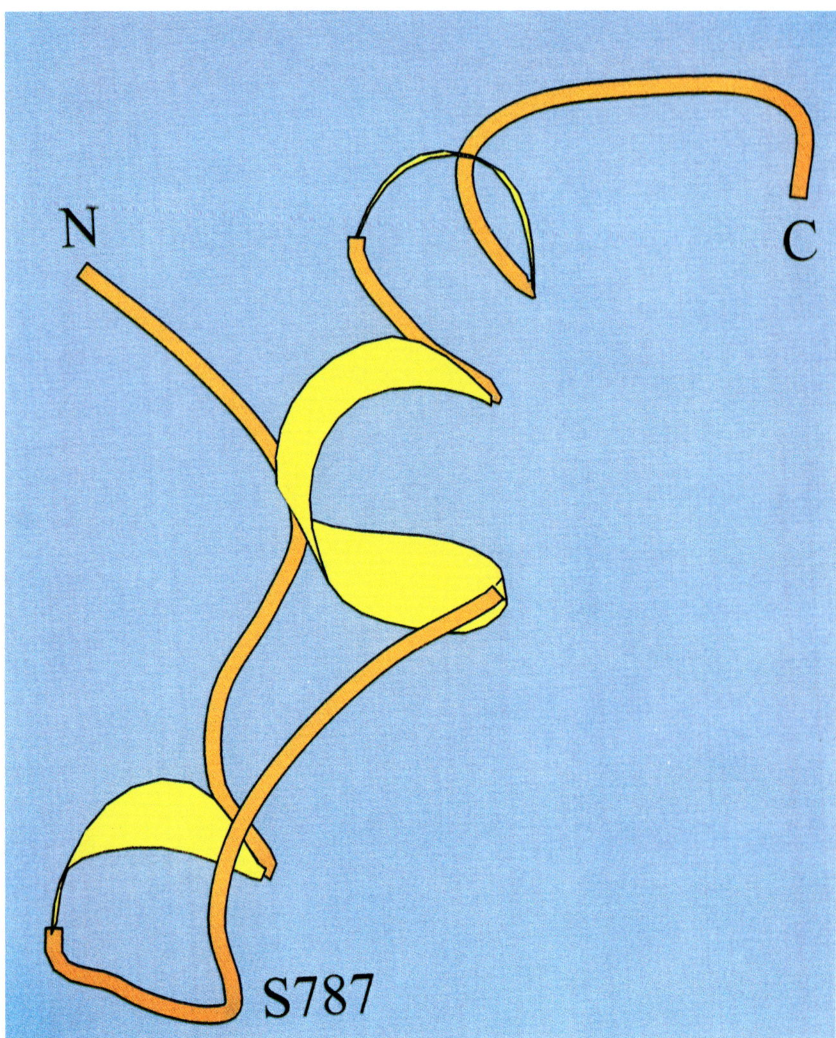

FIGURE 7 Ribbon structure of the COOH-terminal tail of smooth muscle myosin light-chain kinase. In this NMR structure of residues 774–813, some of the α-helix H11 is preserved; however, the NH$_2$-terminal section is collapsed onto the α-helix. The hinge in the structure occurs at Ser787, the beginning of the pseudosubstrate autoinhibitory sequence (Kemp *et al.*, 1987).

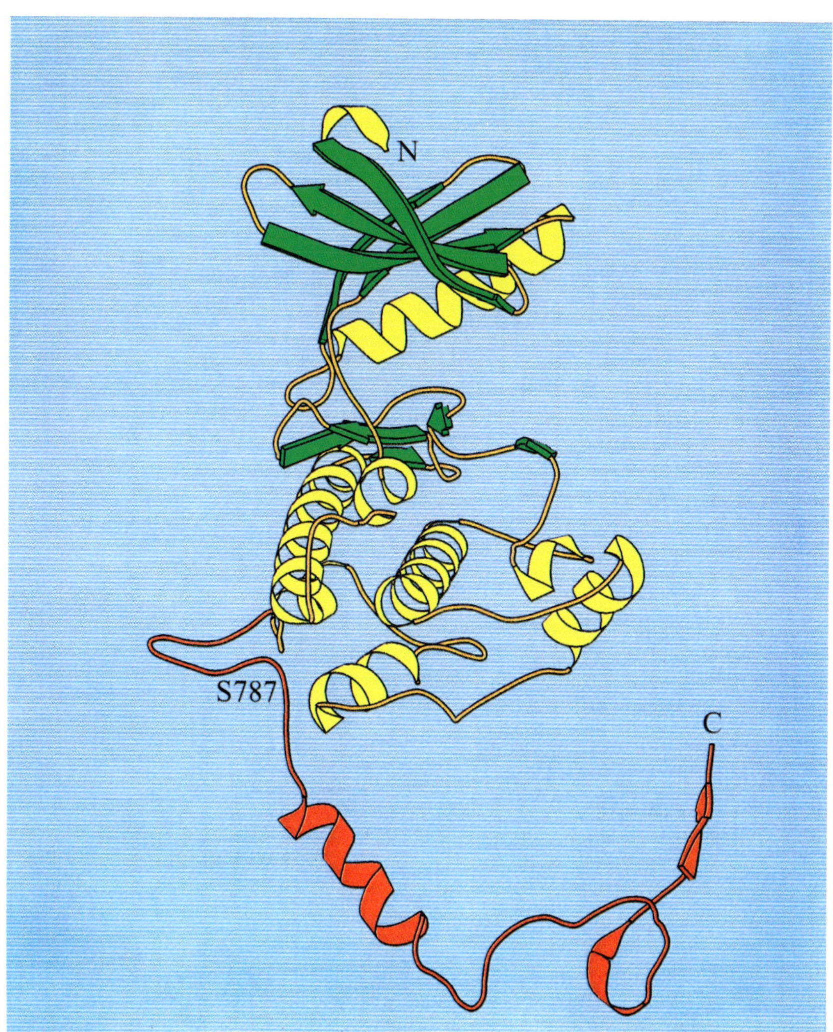

FIGURE 8 Ribbon structure of the smooth muscle myosin light-chain kinase model based on twitchin kinase. In this model the autoinhibitory sequence (shown in red) has been removed from the active site by rotation about the Ser-787 hypothetical hinge (see Figure 7) as may occur with activation by calmodulin. The α-helix H11 is shown, but may be longer with calmodulin bound.

the C-terminal two-thirds (796–813) being responsible for calmodulin binding (Lucas *et al.*, 1986). Bagchi *et al.* (1992a) expressed a 40-kDa fragment of smMLCK in bacteria that contained the entire catalytic and autoregulatory domain (Asp^{450}–Ser^{815}). Deletion of three residues from the C-terminal end of the calmodulin binding region, G^{811}RL^{813}, was sufficient to completely prevent calmodulin binding and consequently calmodulin-dependent activation of the enzyme. Subsequently, it was found that mutation of any one of these three amino acids would also eliminate calmodulin binding (Bagchi *et al.*, 1992a). Araki and Ikebe (1991) reported that a monoclonal antibody specific for amino acids between Arg808 and Ser815 activates the myosin light-chain kinase.

Extensive mutagenesis studies within the calmodulin binding sequence have examined the relationship between calmodulin binding and activation (Shoemaker *et al.*, 1990; Bagchi *et al.*, 1992a,b). Generally, it was found that residues in the autoregulatory sequence that encompasses the pseudo-substrate sequence are less important for calmodulin-dependent activation than are the residues from 808 to 813. The exception was Trp800; mutation of Trp800 to Ala completely eliminated calmodulin binding (Bagchi *et al.*, 1992a). Interestingly, chimeric enzymes in which the calmodulin-binding sequences of skeletal muscle myosin light-chain kinase or the multifunctional calmodulin-dependent protein kinase (CaM kinase) have been substituted into the smooth muscle myosin light-chain kinase reveal calmodulin-dependent activity (Leachman *et al.*, 1992; Shoemaker *et al.*, 1990). Although the three calmodulin domains vary considerably, all share a hydrophobic residue at the position normally occupied by Trp800. Shoemaker *et al.* (1990) reported that inversion of the calmodulin binding sequence in nonmuscle myosin light-chain kinase does not disrupt the calmodulin dependency. In the inverted sequence, an Ile occupies the position of Trp800 in the normal sequence and the overall basic residue-rich amphipathic nature is maintained. Thus, consistant with all reports to date, calmodulin-dependent activation of myosin light-chain kinase isoforms requires a hydrophobic residue in the fourth position (equivalent to Trp800).

A structural basis for calmodulin binding to myosin light-chain kinase has now been provided by the solution of the crystal structure of vertebrate Ca^{2+}/calmodulin complexed to a synthetic peptide corresponding to smMLCK (796–813) (Meador *et al.*, 1992). The calmodulin in the complex forms a tunnel diagonal to its long axis that completely engulfs the 18-amino-acid peptide, which is helical along its entire length. To accommodate the peptide, calmodulin has undergone a bend of 100°, due to the flexibility of residues 73–77 within the central helix, and a 120° twist between the two globular lobes of calmodulin. The close proximity of the two lobes of calmodulin creates a long arc of hydrophobic residues that is contiguous with the hydrophobic patch in each lobe. This arc interacts with the hydrophobic face of the helical peptide. The orientation has the NH$_2$-terminal lobe of calmodulin interacting with the COOH terminus of the peptide and vise versa, as was previously suggested from NMR experiments (Roth *et al.*, 1991). The peptide is anchored into position by Leu813 in the NH$_2$-terminal lobe with Trp800 in the COOH-terminal lobe. The only residue in the peptide that extensively interacts with the bend and the helices that occur immediately proximal and distal to the bend is Arg812. A considerable number of these contacts are hydrogen bonds and charge–coupling interactions. Thus, Arg812 serves as the pivotal residue for orchestrating the bend in calmodulin. Overall, there are about 185 contacts ($<$4 Å) between calmodulin and the peptide, with 80% of these being van Der Waals interactions.

A high-resolution structure of calmodulin bound to a calmodulin binding peptide based on that present in skeletal muscle myosin light-chain kinase has been solved by multidimensional NMR (Ikura *et al.*, 1992). Comparison of this structure and the crystal structure of the Ca^{2+}/calmodulin/smMLCK peptide complex reveals considerable similarities. Particularly relevant to the question of how calmodulin might activate its target enzymes is that, in both structures, the peptide binds in the same direction of the chain and the essential hydrophobic interactions are conserved. Thus, the critical importance of hydrophobic residues flanking the peptide has been confirmed in two high-resolution structures.

The crystal structure of calmodulin complexed with the calmodulin-dependent protein kinase II peptide has also been solved (Meador *et al.*, 1993). There are differences in the position of particular side chains and the entire orientation of the two domains has been shifted to accommodate the interactions between the hydrophobic and basic residues of the calmodulin-dependent protein kinase II peptide. The sequences of the calmodulin binding peptides whose structures have been solved that are bound to calmodulin are listed below with the helical sequence in italics.

smMLCK(796–815)	A*RRKWQKTGHAVRRAIGRL*SS	(crystal structure)
CamKII(290–314)	LKKFNA*RRKLKGAIL TTMLA TRNF*S	(crystal structure)
skMLCK(577–602)	K*RRWKKNFIAVSAANRFKKISS*GA	(NMR structure)

It has been pointed out (Finn and Forsen, 1995) that there is a preponderance of long-chain hydrophobic and basic residues involved in the peptide/calmodulin interaction and these provide the flexibility in precisely juxtaposing the different peptide sequences to calmodulin. Another level of adaptation is provided by the long tethering helix between the calmodulin domains that allows them to fine tune their orientation relative to one another (Finn and Forsen, 1995).

The pseudosubstrate of myosin light-chain kinase is most likely arranged in the active site so that the C-terminal residues that are critical for calmodulin binding (808–813) are available for calmodulin binding, whereas the N-terminal residues (787–807) are much more intimately associated with the catalytic core of the enzyme (Knighton *et al.*, 1992). Indeed, the molecular modeling predicts that amino acids 808–813 do not exist as an α-helix in the autoinhibited state of the enzyme. We can predict that the N-terminal portion of calmodulin initially binds to the C-terminal end of the autoregulatory region of smooth muscle myosin light-chain kinase with Arg812 and Leu813 playing prominent roles. This initial contact could result in a conformational change in the C-terminal portion of the calmodulin binding domain resulting in helix formation. Generation of the helix would disrupt the contacts with the active site and remove the pseudosubstrate such that the C-terminal lobe of calmodulin could twist and engulf Trp800. We expect that the surface of the N-terminal lobe of calmodulin would not be involved in direct binding to the calmodulin binding residues of the enzyme but may participate in removal of the pseudosubstrate. This is consistent with the observation that substitution of only three amino acids that exist in a cluster on the external surface of Ca^{2+}/calmodulin domain I converts the regulatory protein from agonist to potent antagonist (K_i 30 nM) (VanBerkem and Means, 1991). Validation of these putative interactions requires solution of a three-dimensional structure of Ca^{2+}/calmodulin complexed to smooth muscle myosin light-chain kinase. Interestingly, the details of the mechanism of calmodulin activation of the myosin light-chain kinase may be distinct from those of phosphorylase kinase (Farrar *et al.*, 1993).

B. Activation Hinge in the Autoinhibitory Sequence

At present, there is no detailed information on what happens to the myosin light-chain kinase autoinhibitory sequence during the activation process. Having bound to the target sequence in the enzyme via the calmodulin C-terminal domain, it is thought that the calmodulin N-terminal domain then participates to effectively chelate the peptide sequence (Finn and Forsen, 1995). A number of questions are raised. Does the form of calmodulin bound to the enzyme mimic the structure of the camodulin myosin light-chain kinase peptide complex? Does the entire C-terminal tail of the enzyme get removed from the surface of the myosin light-chain kinase or just the autoregulatory pseudosubstrate sequence? If just the autoregulatory pseudosubstrate sequence is removed, how does this happen and what residues are required to hinge it out?

Because we have not yet been able to crystallize myosin light-chain kinase, we have undertaken NMR structural studies with synthetic peptides corresponding to the autoinhibitory sequence of the myosin light-chain kinase, L^{774}QKDTKN^{780}MEAKKLSKDR^{790}MKKYMARRKW^{800}QKTGHA-VRAI^{810}GRL (autoinhibitory sequence underlined). The NMR structure preserves some of the H11 helix of the twitchin kinase structure in this region (Barden *et al.*, 1995). In the NMR structure, the N-terminal segment of the structure, L^{774}QKDTKN^{780}MEAKKL, is collapsed on the major helix and clearly does not reflect the likely corresponding structure in the intact enzyme

(Fig. 7; see color insert). The sequence, LS^{787}KD, forms a turn at the beginning of the autoinhibitory helical sequence in the collapsed structure. Significantly, this may well provide a clue to the mechanism because it may act as the hinge involved in the movement of helix H11 out of substrate binding groove. This hypothetical model of activation is illustrated in Figure 8 (see color insert) in which the autoinhibitory sequence is hinged out of the active site at Ser^{787}. It is important, however, to bear in mind that the myosin light-chain kinase contains class I and class II domains outside the catalytic core and that each end of the enzyme is likely to be tethered to the actin and myosin filaments, respectively. The activation mechanism may therefore more closely resemble the unraveling of a loose knot rather than the entire C terminus of the enzyme, including the class II domain, being shifted.

V. Conclusion

The calmodulin-dependent protein kinase family represents a diverse array of important regulatory enzymes that impinge on the control of almost all known physiological processes. Protein crystallography studies have provided important insights into the mechanism of autoinhibition of this family of protein kinases and the mechanism of how calmodulin binds peptide sequences. The novel CDPKs may be the most revealing for understanding the activation mechanism because they have their calcium binding domains fused to the catalytic domain and may therefore be more amenable to structural analysis. The interplay between calmodulin activation and phosphorylation of the substrate anchoring loop by activating protein kinases as is seen for calmodulin-dependent protein kinases I and IV seems likely to be exploited in other members of the calmodulin-dependent protein kinase family, although not in the myosin light-chain kinase family because there is no conserved phosphate acceptor. Thr or Ser in the substrate anchoring loop.

The most striking feature of the twitchin autoinhibitory structure is just how comprehensively the enzyme has been turned off. Why does twitchin kinase have to be so completely inhibited, and is this physiologically important? Twitchin demonstrates the great diversity of contacts that can be used in autoinhibition, and it will be of great interest to observe the similarities and differences in the mechanism of autoinhibition that are revealed as more structures of these important regulatory enzymes are determined.

References

Adelstein, R. S., and Conti, M. A. (1975). Phosphorylation of platelet myosin increases actin-activated myosin ATPase activity. *Nature* **256**, 597–598.

Araki, Y., and Ikebe, M. (1991). Activation of smooth muscle myosin light chain kinase activity by a monoclonal antibody which recognizes the calmodulin-binding region. *Biochem. J.* 275, 679–684.

Babu, Y. S., Sack, J. S., Greenbough, T. J., Bugg, C. E., Means, A. R., and Cook, W. J. (1985). Three-dimensional structure of calmodulin. *Nature* 315, 37–40.

Bagchi, I. C., Huang, Q., and Means, A. R. (1992a). Identification of amino acids essential for calmodulin binding and activation of smooth muscle myosin light chain kinase. *J. Biol. Chem.* 267, 3024–3029.

Bagchi, I. C., Kemp, B. E., and Means, A. R. (1992b). Intrasteric regulation of myosin light chain kinase: The pseudosubstrate prototope binds to the active site. *Mol. Endocrinol.* 6, 621–626.

Barden, J. A., Sehgal, P., and Kemp, B. E. (1996). Structure of the pseudosubstrate recognition site of chicken smooth muscle myosin light chain kinase. *Biochem. Biophys. Acta,* 1292, 106–112.

Benian, G. M., Kiff, J. E., N., Neckleman, Moerman, D. G., and Waterson, R. H. (1989). Sequence of an unusually large protein implicated in regulation of myosin activity in *C. elegans. Nature (London)* 342, 45–50.

Benian, G. M., L'Hernault, S. W., and Morris, M. E. (1993). Addition sequence complexity in the muscle gene, *unc-22,* and its encoded protein, twitchin, of *Caenorhabditis elegans. Genetics* 134, 1097–1104.

Binder, B. M., Harper, J. F., and Sussman, M. R. (1994). Characterization of an Arabidopsis calmodulin-like domain protein kinase purified from *Escherichia coli* using an affinity sandwich technique. *Biochemistry* 33, 2033–2041.

Blumenthal, D. K., Takio, K., Edelman, A. M., Charbonneau, H., Titani, K., Walsh, K. A., and Krebs, E. G. (1985). Identification of the calmodulin-binding domain of skeletal muscle myosin light chain kinase. *Proc. Natl. Acad. Sci. U.S.A.* 82, 3187–3191.

Bossemeyer, D., Engh, R. A., Kinzel, V., Ponstingl, H., and Huber, R. (1993). Phosphotransferase and substrate binding mechanism of the cAMP-dependent protein kinase catalytic subunit from porcine heart as deduced from the 2.0Å structure of the complex with Mn^{2+} adenylyl imidodiphosphate and inhibitor peptide PKI(5-24). *EMBO J.* 12, 849–859.

Colbran, R. J., Fong, Y. L., Schworer, C. M., and Solderling, T. R. (1988). Regulatory interactions of the calmodulin-binding, inhibitory, and autophosphorylation domains of Ca^{2+}/calmodulin-dependent protein kinase II. *J. Biol. Chem.* 263, 18145–18151.

Colbran, R. J., Smith, M. K., Schworer, C. M., Fong, Y. L., and Soderling, T. R. (1989). Regulatory domain of calcium/calmodulin-dependent protein kinase II. *J. Biol. Chem.* 264, 4800–4804.

Cruzalegui, F. H., Kapiloff, M. S., Morfin, J.-P., Kemp, B. E., Rosenfeld, M. G., and Means, A. R. (1992). Regulation of intrasteric inhibition of the multifunctional calcium/calmodulin-dependent protein kinase. *Proc. Natl. Acad. Sci. U.S.A.* 89, 12127–12131.

Dasgupta, M., Honeycutt, T., and Blumenthal, D. K. (1989). The γ-subunit of skeletal muscle phosphorylase kinase contains two noncontiguous domains that act in concert to bind calmodulin. *J. Biol. Chem.* 264, 17156–17163.

Edelman, A. M., Takio, K., Blumenthal, D. K., Hanson, R. S., Walsh, K. A., Titani, K., and Krebs, E. G. (1985). Characterization of the calmodulin-binding and catalytic domains in skeletal muscle myosin light chain kinase. *J. Biol. Chem.* 260, 11275–11285.

Farrar, Y. J., Lukas, T. J., Craig, T. A., Watterson, D. M., and Carlson, G. M. (1993). Features of calmodulin that are important in the activation of the catalytic subunit of phosphorylase kinase. *J. Biol. Chem.* 268, 4120–4125.

Faux, M. C., Mitchelhill, K. I., Katsis, F., Wettenhall, R. E. H., and Kemp, B. E. (1993). Chicken smooth muscle myosin light chain kinase is acetylated on its NH_2-terminal methionine. *Mol. Cell. Biochem.* 127/128, 81–91.

Finn, B. E., and Forsen, S. (1995). The evolving model of calmodulin structure, function and activation. *Structure* 3, 7–11.

Fischer, E. H., and Krebs, E. G. (1989). Commentary. *Biochim. Biophys. Acta* **1000**, 297–301.

Foster, C. J., Johnston, S. A., Sunday, B., and Gaeta, F. C. A. (1990). Potent peptide inhibitors of smooth muscle myosin light chain kinase: Mapping of the pseudosubstrate and calmodulin binding domains. *Arch. Biochem. Biophys.* **280**, 397–404.

Hanley, R. M., Means, A. R., Ono, T., Kemp, B. E., Burgin, K. E., Waxman, N., and Kelley, P. T. (1987). Functional analysis of a complementary DNA for the 50-kilodalton subunit of calmodulin kinase II. *Science* **237**, 293–297.

Hanson, P. I., and Schulman, H. (1992). Neuronal Ca^{2+}/calmodulin-dependent protein kinases. *Annu. Rev. Biochem.* **61**, 559–601.

Harper, J. F., Huang, J.-F., and Lloyd, S. J. (1994). Genetic identification of and autoinhibitor in CDPK, a protein kinase with a calmodulin-like domain. *Biochemistry* **33**, 7267–7277.

Harper, J. F., Binder, B. M., and Sussman, M. R. (1993). Calcium and lipid regulation of an Arabidopsis protein kinase expressed in *E. coli*. *Biochemistry* **32**, 3282–3290.

Harper, J. F., Sussman, M. R., Schaller, G. E., Putnam-Evans, C., Charbonneau, H., and Harmon, A. C. (1991). A calcium-dependent protein kinase with a regulatory domain similar to calmodulin. *Science* **252**, 951–953.

Harris, W. R., Malencik, D. A., Johnson, C. M., Carr, S. A., Roberts, G. D., Byles, C. A., Anderson, S. R., Heilmeyer, L. M., Fischer, E. H., and Crabb, J. W. (1990). Purification and characterization of catalytic fragments of phosphorylase kinase gamma subunit missing a calmodulin-binding domain. *J. Biol. Chem.* **265**, 11740–11745.

Heierhorst, J., Probst, W., Vilim, F. S., Buku, A., and Weiss, K. R. (1994). Autophosphorylation of molluscan twitchin and interaction of its kinase domain with calcium/calmodulin. *J. Biol. Chem.* **269**, 21086–21093.

Hoar, P. E., Kerrick, W. G., and Cassidy, R. L. (1979). Chicken gizzard: Relation between calcium activated phosphorylation and contraction. *Science* **204**, 503–506.

Holden, H. M., Ito, M., Hartshorne, D. J., and Rayment, I. (1992). X-ray structure determination of telokin, the C-terminal domain of myosin light chain kinase at 2.8Å resolution. *J. Mol. Biol.* **227**, 840–851.

Hu, S.-H., Lei, J. Y., Wilce, C. J., Benian, G. M., Valenzuela, M. R. H., Parker, M. W., and Kemp, B. E. (1994a). Crystallization and preliminary X-ray analysis of the auto-inhibited twitchin kinase. *J. Mol. Biol.* **236**, 1259–1261.

Hu, S.-H., Parker, M. W., Lei, J. Y., Wilce, M. C. J., Benian, G. M., and Kemp, B. E. (1994b). Insights into autoregulation from the crystal structure of twitchin kinase. *Nature* **369**, 581–584.

Huang, C. Y. F., Yuan, C.-J., Livanova, N. B., and Graves, D. J. (1993). Expression, purification, characterization and deletion muations of phosphorylase kinase gamma subunit: Identification of an inhibitory domain in the gamma subunit. *Mol. Cell. Biochem.* **127/128**, 7–18.

Ikebe, M. (1990). Mode of inhibition of smooth muscle myosin light chain kinase by synthetic peptide analogs of the regulatory site. *Biochem. Biophys. Res. Commun.* **168**, 714–720.

Ikebe, M., Maruta, S., and Reardon, S. (1989). Location of the inhibitory region of smooth muscle myosin light chain kinase. *J. Biol. Chem.* **264**, 6967–6971.

Ikebe, M., Stepinska, M., Kemp, B. E., Means, A. R., and Hartshorne, D. J. (1987). Proteolysis of smooth muscle myosin light chain kinase. *J. Biol. Chem.* **260**, 13828–13834.

Ikura, M., Clore, G. M., Gronenborn, A. M., Zhu, G., Klee, C. B., and Bax, A. (1992). solution structure of a calmodulin-target peptide complex by multidimensional NMR. *Science* **256**, 632–638.

Ito, M., Guerriero, V. J., Chen, X., and Hartshorne, D. J. (1991). Definition of the inhibitory domain of smooth muscle myosin light chain kinase by site-directed mutagenesis. *Biochemistry* **30**, 3498–3503.

Kanoh, S., Ito, M., Kawano, Y., and Hartshorne, D. J. (1993). Actin binding peptide from smooth muscle myosin light chain kinase. *Biochemistry* **37**, 8902–8907.

Kemp, B. E., and Pearson, R. B. (1985). Spatial requirements for location of basic residues in peptide substrates for smooth muscle myosin light chain kinase. *J. Biol. Chem.* **260**, 3355–3359.

Kemp, B. E., and Pearson, R. B. (1990). Protein kinase recognition sequence motifs. *Trends Biochem. Sci.* **15**, 342–346.

Kemp, B. E., and Pearson, R. B. (1991). Intrasteric regulation of protein kinases and phosphatases. *Biochim. Biophys. Acta* **1094**, 67–76.

Kemp, B. E., Pearson, R. B., Guerriero, V., Bagchi, I. C., and Means, A. R. (1987). The calmodulin binding domain of chicken smooth muscle myosin light chain kinase contains a pseudosubstrate sequence. *J. Biol. Chem.* **262**, 2542–2548.

Kennelly, P. J., Colburn, J. C., Lorenzen, J., Edelman, A. M., Stull, J. T., and Krebs, E. G. (1991). Activation mechanism of rabbit skeletal muscle myosin light chain kinase. *FEBS Lett.* **286**, 217–220.

Kennelly, P. J., Starovasnik, M. A., Edelman, A. M., and Krebs, E. G. (1990). Modulation of the stability of rabbit skeletal muscle myosin light chain kinase through the calmodulin-binding domain. *J. Biol. Chem.* **265**, 1742–1749.

Kennelly, P. J., Edelman, A. M., Blumenthal, D. K., and Krebs, E. G. (1987). Rabbit skeletal muscle myosin light chain kinase. *J. Biol. Chem.* **262**, 11958–11963.

Klee, C. B., and Vanaman, T. C. (1982). Calmodulin. *Adv. Prot. Chem.* **35**, 213–321.

Knighton, D. R., Bell, S. M., Zheng, J., Ten Eyck, L. F., Xuong, N.-H., Taylor, S. S., and Sowadski, J. M. (1993). 2.0 Å refined crystal structure of the catalytic subunit of cAMP-dependent protein kinase complexed with a peptide inhibitor and detergent. *Acta Crystallogr.* **D49**, 357–361.

Knighton, D. R., Pearson, R. B., Sowadski, J. M., Means, A. R., Ten Eyck, L. F., Taylor, S. S., and Kemp, B. E. (1992). Structural basis of the intrasteric regulation of myosin light chain kinases. *Science* **258**, 130–135.

Kraulis, P. J. (1991). MOLSCRIPT: A program to produce both detailed and schematic plots of protein structures. *J. Appl. Crystallogr.* **24**, 946–950.

Kretsinger, R. H., and Nockolds, C. E. (1973). Carp muscle calcium-binding protein. II. Structure determination and general description. *J. Biol. Chem.* **248**, 3313–3326.

Leachman, S. A., Gallagher, P. J., and Herring, B. P., McPhaul, M. J., and Stull, J. T. (1992). Biochemical properties of chimeric skeletal and smooth muscle myosin light chain kinases. *J. Biol. Chem.* **267**, 4930–4938.

Lee, J. C., and Edelman, A. M. (1994). A protein activator of Ca^{2+}-calmodulin-dependent protein kinase Ia. *J. Biol. Chem.* **269**, 2158–2164.

Lei, J., Tang, X., Pohl, J., and Benian, G. M. (1994). The protein kinase domain of twitchin has protein kinase activity and an autoinhibitory region. *J. Biol. Chem.* **269**, 21078–21085.

Lucas, T. J., Burgess, W. H., Prendergast, F. G., Lau, W., and Watterson, D. M. (1986). Calmodulin binding domains: Characterisation of a phosphorylation and calmodulin binding site from myosin light chain kinase. *Biochemistry* **25**, 1450–1464.

Meador, W. E., Means, A. R., and Quiocho, F. A. (1992). Target enzyme recognition by calmodulin: 2.4 Å structure of a calmodulin–peptide complex. *Science* **257**, 1251–1255.

Meador, W. E., Means, A. R., and Quiocho, F. A. (1993). Modulation of calmodulin plasticity in molecular recognition on the basis of x-ray structures. *Science* **262**, 1718–1721.

Means, A. R., Cruzalegui, F., LeMagueresse, B., Needleman, D. S., Slaughter, G. R., and Ono, T. (1991). A novel Ca^{2+}/calmodulin-dependent protein kinase and a male germ cell-specific calmodulin-binding protein are derived from the same gene. *Mol. Cell. Biol.* **11**, 3960–3971.

Meyer, T., Hanson, P. I., Stryer, L., and Schulman, H. (1992). Calmodulin trapping by calcium-calmodulin dependent protein kinase. *Science* **256**, 1199–1202.

Mitsui, K., Brady, M., Palfrey, H. C., and Nairn, A. C. (1993). Purification and characterization of calmodulin-dependent protein kinase III from rabbit reticulocytes and rat pancreas. *J. Biol. Chem.* **268**, 13422–13433.

Miyano, O., Kameshita, I., and Fujisawa, H. (1992). Purification and characterization of a brain-specific multifunctional calmodulin-dependent protein kinase from rat cerebellum. *J. Biol. Chem.* **267**, 1198–1203.

Moerman, D. G., Benian, G. M., Barstead, R. J., Schriefer, L. A., and Waterston, R. H. (1988). Identification and intracellular localization of the unc-22 gene product of *Caenorhabditis elegans*. *Genes Dev.* **2**, 93–105.

Moncrief, N. D., Kretsinger, R. H., and Goodman, M. (1990). Evolution of EF-hand calcium-modulated proteins. I. Relationships based on amino acid sequences. *J. Mol. Evol.* **30**, 522–562.

Nairn, A. C., and Greengard, P. (1987). Purification and characterization of Ca^{2+}/calmodulin-dependent protein kinase I from bovine brain. *J. Biol. Chem.* **262**, 7273–7281.

Ohmstede, C.-A., Bland, M. M., Merrill, B. M., and Sahyoun, N. (1991). Relationship of genes encoding Ca^{2+}/calmodulin-dependent protein kinase Gr and calspermin: A gene within a gene. *Proc. Natl. Acad. Sci. U.S.A.* **88**, 5784–5788.

Ohmstede, C. A., Jensen, K. F., and Sahyoun, N. (1989). Ca^{2+}/calmodulin-dependent protein kinase enriched in cerebellar granule cells. Identification of a novel neuronal calmodulin-dependent protein kinase. *J. Biol. Chem.* **264**, 5866–5875.

Okuno, S., and Fujisawa, H. (1993). Requirement of brain extract for the activity of brain calmodulin-dependent protein kinase IV expressed in *Escherichia coli. J. Biochem. (Tokyo)* **114**, 167–170.

Olah, G. A., Mitchell, R. D., Sosnich, T. R., Walsh, D. A., and Trewhella, J. (1993). Solution structure of the cAMP-dependent protein kinase catalytic subunit and its contraction upon binding the protein kinase inhibitor peptide. *Biochemistry* **32**, 3649–3657.

Olson, N. J., Pearson, R. B., Needleman, D. S., Hurwitz, M. Y., Kemp, B. E., and Means, A. R. (1990). Regulatory and structural motifs of chicken gizzard myosin light chain kinase. *Proc. Natl. Acad. Sci. U.S.A.* **87**, 2284–2288.

Payne, M. E., Fong, Y.-L., Ono, T., Colbran, R. J., Kemp, B. E., Soderling, T. R., and Means, A. R. (1988). Calcium/calmodulin-dependent protein kinase II: Characterization of distinct calmodulin binding and inhibitory domains. *J. Biol. Chem.* **263**, 7190–7195.

Pearson, R. B., Hunt, J. T., Mitchelhill, K. I., and Kemp, B. E. (1991a). Myosin light chain kinase autoregulatory pseudosubstrate prototope. *Peptide Res.* **4**, 147–157.

Pearson, R. B., Ito, M., Morrice, N. A., Smith, A. J., Condron, R., Wettenhall, R. E. H., Kemp, B. E., and Hartshorne, D. J. (1991b). Proteolytic cleavage sites in smooth muscle myosin light chain kinase and their relation to structural and regulatory domains. *Eur. J. Biochem.* **200**, 723–730.

Pearson, R. B., Wettenhall, R. E. H., Means, A. R., Hartshorne, D. J., and Kemp, B. E. (1988). Autoregulation by enzyme pseudosubstrate prototopes: Myosin light chain kinase. *Science* **241**, 970–973.

Picciotto, M. R., Czernik, A. J., and Nairn, A. C. (1993). Calcium/calmodulin-dependent protein kinase I. cDNA cloning and identification of autophosphorylation site. *J. Biol. Chem.* **268**, 26512–26521.

Politou, A. S., Gautel, M., Pfuhl, M., Labeit, S., and Pastore, A. (1994). Immunoglobulin-type domains of titin: Same fold, different stability? *Biochemistry* **33**, 4730–4737.

Roth, S. M., Schneider, D. M., Strobes, L. A., VanBerkum, M. F. A., Means, A. R., and Wand, A. J. (1991). Structure of smooth muscle myosin light chain kinase calmodulin-binding domain peptide bound to calmodulin. *Biochemistry* **30**, 10078–10084.

Sanchez, V. E., and Carlson, G. M. (1993). Isolation of an autoinhibitory region from the regulatory beta-subunt of phosphorylase kinase. *J. Biol. Chem.* **268**, 17889–17895.

Schulman, H. (1993). The multifunctional Ca^{2+}/calmodulin-dependent protein kinases. *Curr. Opin. Cell. Biol.* **5**, 247–253.

Shirinsky, V. P., Vorotnikov, A. V., Birukov, K. G., Nanaev, A. K., Collinge, M., Lukas, T. J., Sellers, J. R., and Watterson, D. M. (1993). A kinase-related protein stabilizes unphosphorylated smooth muscle myosin minifilaments in the presence of ATP. *J. Biol. Chem.* **268**, 16578–16583.

Shoemaker, M. O., Lau, W., Shattuck, R. L., Kwiatkowski, A. P., Matrisian, P. E., Guerra-Santos, L., Wilson, E., Lucas, T. J., Eldik, L. J. V., and Watterson, D. M. (1990). Use of DNA sequence and mutant analysis and antisense oligonucleotides to examine the molecular basis of nonmuscle myosin light chain kinase autoinhibition, calmodulin recognition, and activity. *J. Cell Biol.* **111**, 1107–1125.

Smith, M. K., Colbran, R. J., Bickey, D. A., and Soderling, T. R. (1992). Functional determinants in the autoinhibitory domain of calcium/calmodulin-dependent protein kinase II. *J. Biol. Chem.* **267**, 1761–1768.

Starovasnik, M. A., Su, D.-R., Beckingham, M. A., and Klevit, R. E. (1992). A series of point mutations reveal interactions between the calcium-binding sites of calmodulin. *Prot. Sci.* **1**, 245–253.

Sweeney, H. L., Bowman, B. F., and Stull, J. T. (1993). Myosin light chain phosphorylation in vertebrate striated muscle: Regulation and function. *Am. J. Physiol.* **253**, C1085–C1095.

Tan, J. L., and Spudich, J. A. (1991). Characterization and bacterial expression of the *Dictyostelium* myosin light chain kinase. *J. Biol. Chem.* **266**, 16044–16049.

Tanaka, M., Ikebe, R., Matsuura, M., and Ikebe, M. (1995). Pseudosubstrate sequence may not be critical for autoinhibition of smooth muscle myosin light chain kinase. *EMBO J.* **14**, 2839–2846.

Tanaka, T., Naka, M., and Hidaka, H. (1980). Activation of myosin light chain kinase by trypsin. *Biochem. Biophys. Res. Commun.* **92**, 313–318.

Taylor, S. S., Knighton, D. R., Zheng, J., Sowadski, J. M., Gibbs, C. S., and Zoller, M. J. (1993). A template for the protein kinase family. *Trends Biochem. Sci.* **18**, 84–89.

Tokumitsu, H., Brickey, D. A., Glod, J., Hidaka, H., Sikela, J., and Soderling, T. R. (1994). Activation mechanism for Ca^{2+}/calmodulin-dependent protein kinase IV. *J. Biol. Chem.* **269**, 28640–28647.

Trinick, J. (1994). Titin and nebulin: Protein rulers in muscle. *Trends Biochem. Sci.* **226**, 405–409.

VanBerkem, M. F. A., and Means, A. R. (1991). Three amino acid substitution in domain I of calmodulin prevent the activation of chicken smooth muscle myosin light chain kinase. *J. Biol. Chem.* **266**, 21488–21495.

Yano, K., Araki, Y., Hales, S. J., Tanaka, M., and Ikebe, M. (1993). Boundary of the autoinhibitory region of smooth muscle myosin light chain kinase. *Biochemistry* **32**, 12054–12061.

Zhao, Y., Kappes, B., and Franklin, R. M. (1994a). Gene structure and expression of an unusual protein kinase from *Plasmodium falciparum* homologous at its carboxyl terminus with the EF hand calcium-binding proteins. *J. Biol. Chem.* **268**, 4347–4354.

Zhao, Y., Pokutta, S., Maurer, P., Lindt, M., Franklin, R. M., and Kappes, B. (1994b). Calcium-binding properties of a calcium-dependent protein kinase from *Plasmodium falciparum* and the significance of individual calcium-binding sites for kinase activation. *Biochemistry* **33**, 3714–3721.

Marina R. Picciotto*
Kent L. Nastiuk†
Angus C. Nairn†

*Department of Psychiatry
Yale University School of Medicine
New Haven, Connecticut 06508

†Laboratory of Molecular and Cellular Neuroscience
The Rockefeller University
New York, New York 10021

Structure, Regulation, and Function of Calcium/Calmodulin-Dependent Protein Kinase I

I. Introduction

Ca^{2+} is an essential intracellular second messenger in eukaryotic systems regulating processes such as muscle contraction, neurotransmitter release, gene expression, and cell proliferation (Davis, 1992). Many of the effects of Ca^{2+}, in turn, result from its interaction with calmodulin (CaM) to form a complex that is able to regulate a variety of intracellular processes. In particular, a large number of the effects of Ca^{2+}/CaM are mediated through the regulation of protein phosphorylation and dephosphorylation. A family of Ca^{2+}/CaM-dependent protein kinases (CaM kinases) has been identified (for reviews, see Schulman and Lou, 1989; Kemp and Stull, 1990; Heilmeyer, 1991; Hanson and Schulman, 1992; Lu and Means, 1993; Nairn and Picciotto, 1994; Braun and Schulman, 1995; Nairn and Palfrey, 1996). The first two members identified, myosin light-chain kinase (MLCK) and phosphorylase kinase, were found to be highly specific for their respective substrates, myosin P light chain and phosphorylase. This led to the initial belief that the CaM kinases were monofunctional kinases that were distinct from the multifunctional protein kinases such as cAMP-dependent protein kinase (PKA) or protein kinase C (PKC). CaM kinase II was subsequently purified, however, and was found to phosphorylate many substrates with diverse functions. Since then, CaM kinases I and IV have also been shown to be widely distributed and are likely to be multifunctional protein kinases as well. New members of the family also continue to be described including EF-2 kinase (originally termed CaM kinase III), an enzyme that specifically phosphorylates elongation factor 2. Furthermore, enzymes that phosphorylate and activate CaM kinases I and IV as part of a kinase cascade are themselves CaM kinases. Thus, it is now clear that the CaM kinase family comprises both mono- and multifunctional enzymes that are essential for transducing calcium signals in all eukaryotic cell types.

CaM kinase I appears to be the simplest member of the CaM kinase family yet described. As such, it can be considered as a model for the other enzymes in this family. In this chapter, we focus on the structure, regulation, and function of CaM kinase I and highlight certain features that may be generalized to the other members of the CaM kinase family.

II. Structure of CaM Kinases

The CaM kinases share many common features. They each have a highly conserved catalytic domain and a COOH-terminal regulatory domain that is not as highly conserved among family members. In addition, several of the enzymes have NH_2- and/or COOH-terminal extensions that are involved in interaction with other subunits or in targeting to specific subcellular locations. Extensive comparisons of the catalytic domains of both serine/ threonine and tyrosine protein kinases (Hanks *et al.*, 1988; Hanks and Quinn, 1991; Hanks and Hunter, 1995a,b) reveal strongly conserved amino acid residues that correspond to structural and functional domains present in all kinases. To characterize structural features shared by the CaM kinase family, we undertook a more detailed comparison of the catalytic domains of CaM kinase I and 18 protein kinases representing either species variants of defined CaM kinases or enzymes that are related to the CaM kinase family based on amino acid identity. We also compared the CaM kinases to the cyclic nucleotide-dependent protein kinases and to other representative serine/threonine protein kinases. Table I lists the enzymes used in this comparison and their overall amino acid identity to CaM kinase I, determined using the ALIGN program (Version 1.02, Scientific and Educational Software, 1989). Even at this gross level of analysis, it is clear that the multifunctional enzymes, CaM kinases I, II, and IV, are more closely related to each other than to the specific CaM kinases, phosphorylase kinase and MLCK.

TABLE I Comparison of Amino Acid Sequences of CaM Kinases

Name	Description	Identical aa	% Identity	GenBank No.	Species	GXG-8	KHPN	HPWX
Camk I	Calcium/calmodulin-dependent kinase I	257	100	L24907	*R. norvegicus*	20	74	276
Camk IV	Calcium/calmodulin-dependent kinase IV	132	51	M64757	*R. norvegicus*	43	93	296
psk H1	Protein serine kinase H1	129	50	X71875	*H. sapiens*	141	194	398
aspCmk	Aspergillus camk2 homolog	125	49	M74120	*E. nidulans*	23	75	278
kIG5	Rat kinase ig5	123	48	L22557	*R. norvegicus*	23	55	286
rCamk II	Calcium/calmodulin-dependent kinase II	119	46	J02942	*R. norvegicus*	13	68	271
yCmk1	Yeast camk like	119	46	X57782	*S. cerevisiae*	37	94	298
yCmk2	Yeast camk like	117	46	D90376	*S. cerevisiae*	47	104	309
dCamk II	Drosophila camk2 homolog	115	45	D13330	*D. melanogaster*	14	69	272
Phosk	Phosphorylase kinase	108	42	M73808	*R. norvegicus*	24	86	291
cornCmk	Corn calcium-dependent kinase with cam domain	107	42	L15390	*Z. mays*	16	72	273
RCK2	Yeast camk like	107	42	X71065	*S. cerevisiae*	163	255	483
MAPKAP	MAP kinase activated protein kinase	102	40	X82220	*R. norvegicus*	1	53	254
carCmk	Carrot camk like	101	39	X56599	*D. carota*	1	31	232
AMPk	AMP-dependent kinase	100	39	Z29486	*R. norvegicus*	16	71	273
aplCmk	Apple camk 2 homolog	96	37	Z17313	*M. domestica*	12	106	308
rsk	rsk, p90	95	37	M99169	*R. norvegicus*	418	468	675
racA	rac, α	93	36	D30040	*R. norvegicus*	150	206	408
MLCK	Myosin light-chain kinase	92	36	J03886	*R. norvegicus*	299	352	554
PKA	Protein kinase A, β-catalytic	91	35	D10770	*R. norvegicus*	44	100	298
PKC	Protein kinase C, β1	91	35	M19007	*R. norvegicus*	342	399	600
Twitchin	Twitchin	90	35	Z30161	*A. californica*	47	100	302
PKG	Protein kinase G	89	35	Z36276	*R. norvegicus*	453	509	711
CDK4	Cyclin-dependent kinase 4	80	31	L11007	*R. norvegicus*	6	64	278
CASk2	Casein kinase 2	75	29	L15618	*R. norvegicus*	39	89	324

Note. The protein kinases used to produce the dendogram and for the amino acid sequence alignment examined are described. Identical aa is the number of identical amino acids and percentage identity of each sequence to CaM kinase I over the 257 amino acids from domains IV–XI. The GenBank accession number and the species from which the sequences were derived are indicated. The final three columns indicate the amino acid number in each sequence that defines domain I (GXG-8, the eighth amino acid toward the N-terminal end of the GXGXXG ATP binding motif), domain IV (the conserved KHPN motif), and the end of the aligned sequences in domain XI (the conserved HPWX motif).

Surprisingly, the serine/threonine protein kinases identified as being CaM kinase-like vary in identity to CaM kinase I from 50% for psk H1 to 36% for MLCK, the latter being equivalent to that of the cyclic nucleotide-dependent kinases. A calculated phylogenetic tree based on the conserved regions of the CaM kinase family members identifies several subgroups (Fig. 1). The tree was constructed with Geneworks (Version 2.45, Intelligenetics, 1995) using the UPGMA method, which is robust for such presumably highly conserved molecules that have diverged long ago (Nei, 1987). Notably, the multifunctional CaM kinases group together as do the sequences of CaM kinase-like enzymes from plants, whereas the specific CaM kinases appear to be at least as distant evolutionarily from the multifunctional

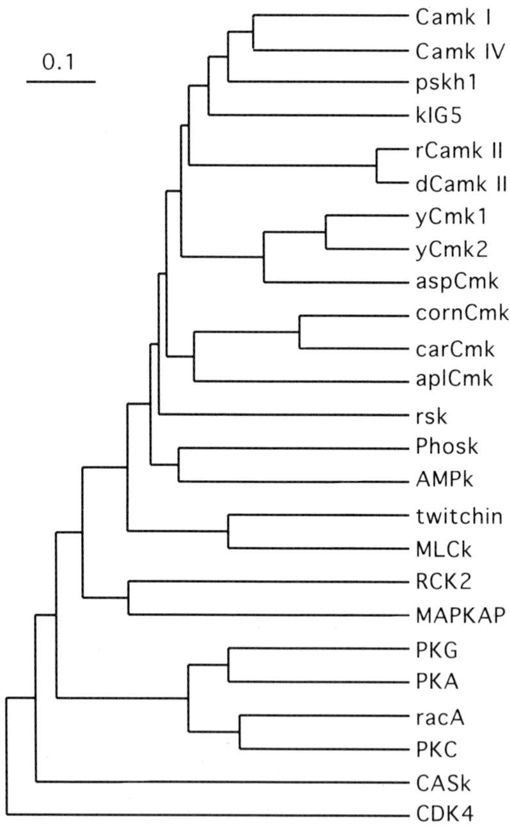

FIGURE I   Dendrogram of CaM kinases and CaM kinase-like enzymes, with representatives of other classes of serine/threonine kinases. The alignment was restricted to the amino acid sequences between domain IV and domain XI because large insertions in domains I–III in several kinases complicated the analysis. Branch lengths indicate standard genetic distance between members, representing the proportion of amino acids substituted over the length of the molecule to reach a theoretical progenitor kinase.

enzymes such as rsk or AMP-dependent protein kinase. Similar conclusions have been made in a previous overlapping study (Hanks and Hunter, 1995a).

Alignment of the kinase catalytic domains using Geneworks reveals the conserved protein kinase domains originally defined by Hanks *et al.* (1988) (Fig. 2). The presumed structure and function of each domain and its associated highly conserved amino acid sequence has been reviewed in detail and is not discussed here (for reviews, see Hanks *et al.*, 1988; Hanks and Quinn, 1991; Hanks and Hunter, 1995a). There are a number of regions that are highly conserved among the CaM kinases. In Figure 1, these are shown by the darker shading to distinguish them from regions conserved in all protein kinases (shown in lighter shading). Notable are the conserved residues HPNI(V/I) in domain IV, the GGELF and DRI sequences in domain V,

A

```
                  I       YG            S                          R                IM               I
Consensus    Y.....LG.G .FS.V..... ..T...---- -----.A.K. ...KK...-- --........ ..E..VL... -----.HPNIV
Camk I       YDFRDVLGTG AFSEVILAED KRTQKL---- -----VAIKC -IAKKALE-- ----GKEGSM ENEIAVLHKI -----KHPNIV
Camk IV      FEVESELGRG ATSIVYRCKQ KGTQKP---- -----YALKV -LKKTVD--- ------KKIV RTEIGVLLRL -----SHPNII
psk H1       YDIKALIGRG SFSRVVRVEH RATRQP---- -----YAIKM -IETKYRE-- -----GREVC ESELRVLRRV -----RHANII
kIG5         YDLGQVVKTE EFCEIFRAKD KTTGKL---- -----HTCKK F-QKRDGR-- ----KVRKAA KNEIGILKMV -----KHPNIL
rCamk II     YQLFEELGKG AFSVVRRCVK VLAGQE---- -----YAAKI INTKKLS--- ---ARDHQKL EREARICRLL -----KHPNIV
dCamk II     YDIKEELGKG AFSIVKRCVQ KSTGFE---- -----FAAKI INTKKLT--- ---ARDFQKL EREARICRKL -----HHPNIV
yCmk1        YVFGKTLGAG TFGVVRVRKN TETGED---- -----VAVKI V-IKKALK-- -GNKVQLEAL YDELDILQRL -----HHPNIV
yCmk2        YIFGRTLGAG SFGVVRQARK LSTNED---- -----VAIKI L-LKKALQ-- -GNNVQLQML YEELSILQKL -----SHPNIV
aspCmk       YRFGRTLGAG TYGIVREA-D -CSSGK---- -----VAVKI I-LKRNVR-- -GNERMV--- YDELDLLQKL -----NHPHIV
cornCmk      YALGRKLGQG QFGTTYLCTD -ATGG----- -----LRCKS I-AKRNVI-- --TKEDVEDV RREIQIMHHL A----GHRNVV
carCmk       ---------- ---------- ---------- -----ACKS I-LKRKLV-- --SKNDKEDI KREIQILQHL S----GQPNIV
aplCmk       YEISEILGRG GFSVVRKGIS RKSSSSSDKT D----VAIKT L--KRPFAPS -{}-ISNVLL TNEILVMRKI VENVSPHPNVI
rsk          YIVKETIGVG SYSVCKRCVH KATNME---- -----YAVKV I-DKS----- ----KRDPSE EIEILLRYG- -----QHPNII
Phosk        YDPKDIIGRG VSSVVRRCVH RATGDE---- -----FAVKI MEVSAERLSL EQLEEVRDAT RREMHILRQV -----GHPHII
AMPk         YVLGDTLGVG TFGKVKIGEH QLTGHK---- -----VAVKI L-NRQKIRSL ----DVVGKI KREIQNLKLF -----RHPHII
twitchin     YDILEELGSG AFGVVHRCVE KATGRV---- -----FEAKF I-NTPYPL-- -----DKYTV KNEISIMNQL -----HHPKLI
MLCK         MNSKEALGGG KFGAVCTCTE RSTGLK---- -----LAAKV I--KKQTPK- -----DKEMV LLEIEVMNQL -----NHRNLI
RCK2         YKLINKIGEG AFSKVFRAIP AKNSSHEFLT KNYKAVAIKV I--KKADLSS -{}-SSRDQA LKEVALHKTV SA---GCSQIV
MAPKAP       --------LG INGKVLRIFD KRTQ------ -------QKF A-LKMLQ--- ---DCPKARR EVELHWRAS- -----QCPHIV

PKG          LEIIATLGVG GFGRVELVKV KNENIA---- -----FAMKC I-RKKHIV-- --DTKQQEHV YSEKRILEEL -----CSPFIV
PKA          FERKKTLGTG SFGRVMLVKH KATEQY---- -----YAMKI L-DKQKVV-- --KLKQIEHT LNEKRILQA- ----VEFPFLV
racA         FEYLKLLGKG TFGKVILVKE KATGRY---- ----YAMKI L--KKEVI-- -VAKDEVAHT LTENRVLQN- ----SRHPFLT
PKC          FNFLMVLGKG SFGKVMLSER KGTDEL---- -----YAVKI L--KKDVV-- -IQDDDVECT MVFKRVLALP ----GKPPFLT

CDK4         YEPVAEIGVG AYGTVYKARD PHSGHF---- -----VALKS V--RVPNGG-A AGGGLPVST VREVALLRRL EAF--EHPNVV
CASk2        YQLVRKLGRG KYSEVFEAIN ITNNEK---- -----VVVKI L--KPVKK-- ---- --KKI KREIKILENL R----GGPNII
                  (I)                        (II)                      (III)              (IV)
```

FIGURE 2   Multiple alignment of amino acid sequence of protein kinase catalytic domains. The domain numbers defined by Hanks *et al.* (1988) are indicated at the bottom of each set of sequences. Regions of high sequence conservation within the CaM kinase family are indicated by dark shadowing (>13 of 19 amino acids identical). Regions of high sequence conservation among all kinases are indicated by the lighter stippling [>21 of 25 amino acids, including a majority of each class (CaM kinases, cyclic nucleotide-dependent kinases, and other serine/threonine protein kinases)]. A consensus is indicated on the top line when a majority of amino acids at a given position are identical. Two similar amino acids on top of one another and separated by a line represent a consensus if they are jointly present at >13 of 19 positions in the CaM kinases. (A) Domains I–IV were aligned independently from the remainder of the catalytic domains because large insertions complicated the alignment procedure. Additional insertions were also removed from aplCmk and RCK2 (indicated in the sequence by braces). (B) Overlapping alignment beginning with domain IV and continuing with domains V–XI.

B

```
                    I                                                    E
Consensus   .HPNIV.L.D ..E.------ ---......V .EL..GGELF DRI...G.-- .SE.DA.... .QI..AV.YL H.-..

Camk I      KHPNIVALDD IYES------ ---GGHLYLI MQLVSGGELF DRIVEKGF-- YTERDGSRLI FQVLDAVKYI HD-LG
Camk IV     SHPNIIKLKE IFET------ ---PTEISLV LELVTGGELF DRIVFKGY- YSERDAADAV KQILEAVAYL HE-NG
psk H1      RHANIIQLVE VFET------ ---QERVYMV MELATGGELF DRIIAKGS-- FTERDATRVL QMVLDGVRYL HA-LG
kIG5        KHPNILQLVD VFVT------ ---RKEYFIF LELATGREVF DWILDQGY-- YSERDTSNVV RQVLEAVAYL HS-LK
rCamk II    KHPNIVRLHD SISE------ ---EGHHYLI FDLVTGGELF EDIVAREY-- YSEADASHCI QQILEAVLHC HQ-MG
dCamk II    HHPNIVRLHD SIQE------ ---ENYHYLV FDLVTGGELF EDIVAREF-- YSEADASHCI QQILESVNHC HQ-NG
yCmk1       HHPNIVAFKD WFES------ ---KDKFYII TQLAKGGELF DRILKKGK-- FTEEDAVRIL VEILSAVKYM HS-QN
yCmk2       SHPNIVSFKD WFES------ ---KDKFYIV TQLATGGELF DRILSRGK-- FTEVDAVEII VQILGAVEYM HS-KN
aspCmk      NHPHIVHFVD WFES------ ---KDKFYIV TQLATGGELF DRICEYGK-- FTEKDASQTI RQVLDAVNYL HQ-RN
cornCmk     GHRNVVAIKG AYED------ ---QLYVHIV MEFCAGGELF DRIIQRG-H- YSERKAAELT RIIVGVVEAC HS-LG
carCmk      GQPNIVEFKG VFED------ ---RQSVHLV MELCAGGELF DRIIAQG-H- YSERAAATIC RQIVNVVHVC HF-MG
aplCmk      PHPNVIDLYD VYED------ ---ENGVHLV LELCSGGELF DRIVKQE-R- YSEVGAAAVV RQIAQGLAAL HR-SN
S6k         QHPNIITLKD VYDD------ ---SKHVYLV TELMRGGELL DKILRQKF-- FSEREASFVL YTISKTVEYL HS-QG
Phosk       GHPHIITLID SYES------ ---SSFMFLV FDLMRKGELF DYLTEKVA-- LSEKETRSIM RSLLEAVNFL HV-NN
AMPk        RHPHIIKLYQ VIST------ ---PTDFFMV MEYVSGGELF DYICKHG-R- VEEVEARRLF QQILSAVDYC HR-HM
twitchin    HHPKLINLHD AFED------ ---KYDMVLI LEFLSGGELF DRIAAEDYK- MSEAEVINYM RQACEGLKHM HE-HS
MLCK        NHRNLIQLYS AIET------ ---SHEIILF MEYIEGGELF ERIVDEDYQ- LTEVDTMVFV RQICDGILFM HK-MR
RCK2        GCSQIVAFID FQET------ ---DSYYYII QELLTGGEIF GEIVRLTY-- FSEDLSRHVI KQLALAVKHM HSLGV
MAPKAP      QCPHIVDIVD VYENLYAG-- ---RKCLLIV MECLDGGELF SRIQDRGDQA FTEREASEIM KSIGEAIQYL HS-IN

PKG         CSPFIVKLYR TFKD------ ---NKYVYML LEACLGGELW SILRDRGS-- FDEPTSKFCV ACVTEAFDYL HR-LG
PKA         EFPFLVRLEY SFKD------ ---NSNLYMV MEYVPGGEMF SHLRRIGR-- FSEPHARFYA AQIVLTFEYL HS-LD
racA        RHPFLTALKY SFQT------ ---HDRLCFV MEYANGGELF FHLSRERV-- FSEDRARFYG AEIVSALDYL HSEKN
PKC         KPPFLTQLHS CFQT------ ---MDRLYFV MEYVNGGDLM YHIQQVGR-- FKEPHAVFYA AEIAIGLFFL QS-KG

CDK4        EHPNVVRLMD V---CATSRT DRDIKVTLVF EHIDQDLRTY LDKAPPPG-- LPVETIKDLM RQFLSGLDFL HA-NC
CASk2       KDPVSRT--- ---------- ---PALVFEH VNNTDFKQLY QTLTDYDIRF Y--------M YEILKALDYC HS-MG
               (IV)                        (V)                        (VIa)


                    V                   I                       S   F S  V
Consensus   IVHRDLKPEN LL......... ...K..DFGL AK..-..... ......GTP. Y.APEVL--- ------....Y G..VD

Camk I      IVHRDLKPEN LLYYSLDEDS K-IMISDFGL SK-M-EDPGS VLSTACGTPG YVAPEVL--- -----AQKPY SKAVD
Camk IV     IVHRDLKPEN LLYATPAPDA P-LKIADFGL SK-I-VEHQV LMKTVCGTPG YCAPEIL--- -----RGCAY GPEVD
psk H1      ITHRDLKPEN LLYYHPGTDS K-IIITDFGL ASARKKGDDC LMKTTCGTPE YIAPEVL--- -----VRKPY TNSVD
kIG5        IVHRNLKLEN LVYYNRLKNS K-IVISDFHL AK----LENG LIKEPCGTPE YLAPEVV--- -----GRQRY GRPVD
rCamk II    VVHRDLKPEN LLLASKLKGA A-VKLADFGL AIEVEGEQQA WFGFA-GTPG YLSPEVL--- -----RKDPY GKPVD
dCamk II    VVHRDLKPEN LLLASKAKGA A-VKLADFGL AIEVQGDHQA WFGFA-GTPG YLSPEVL--- -----KKEPY GKSVD
yCmk1       IVHRDLKPEN LLYIDKSDES PLVV-ADFGI AKTLKSDEEL LYKAG-TALG YVAPEVL--- -----TQDGH GKPCD
yCmk2       IVHRDLKPEN VLYVDKSENS PLVI-ADFGI AKQLKGEEDL IYKAA-GSLG YVAPEVL--- -----TQDGH GKPCD
aspCmk      IVHRDLKPEN LLYLTRDLDS QLVL-ADFGI AKMLDNPAEV LTSMA-GSFG YAAPEVM--- -----LKQGH GKAVD
cornCmk     MHRDLKPEN FLLSN-KDDD MSLKAIDFGL SVFF--KPGQ IFTDVVGSPY YVAPEVLRK- --------RY GPEAD
carCmk      MHRDLKPEN FLLSS-KDKD AMLKATDFGL SVFI--EEGK VYRNIVGSAY YVAPEVLRR- --------SY GKEID
aplCmk      IVHRDLKPEN CLFLD-NTVD SPLKIMDFGL SSVE--EFTD PVVGLFGSID YVSPEALSQ- --------GQV TSKSD
S6k         VVHRDLKPSN ILYVDESGNP ECLRICDFGF AKQL-RAENG LLMTPCYTAN FVAPEVL--- -----KRQGY DEGCD
Phosk       IVHRDLKPEN ILLDDNMQ-- --IRLSDFGF SCHLEPGEKL R--ELCGTPG YLAPEI--K CSMDETHPGY GKEVD
AMPk        VVHRDLKPEN VLLDAQMN-- --AKIADFGL SNMMSDGEFL R--TSCGSPN YAAPEVI--- ----SGRLYA GPEVD
twitchin    IVHLDIKPEN IMCET-KK-A SSVKIIDFGL ATKL--NPDE IVKVTTATAE FAAPEIVDR- -------EPV GFYTD
MLCK        VLHLDLKPEN ILCVN-TT-G HLVKIIDFGL ARRY--NPNE KLKVNFGTPE FLSPEVVNY- -------DQI SDKTD
RCK2        VVHRDIKPEN LLFGPI-{}- GIVKLADFGL SKQ---IFSK NTKTPCGTVG YTAPEVVKD- -------EHY SMKVD
MAPKAP      IAHRDVKPEN LLYTS-KRPN AILKLTDFGF AKET--TSHN SLTTPCYTPY YVAPEVLGP- -------EKY DKSCD

PKG         IIYRDLKPEN LILDADGY-- --LKLVDFGF AK-K-IGSGQ KTWTFCGTPE YVAPEVI--- -----LNKGH DFSVD
PKA         LIYRDLKPEN LLIDHQGY-- --IQVTDFGF AK-R-VKG-- RTWTLCGTPE YLAPEII--- -----LSKGY NKAVD
racA        VVYRDLKLEN LMLDKDGH-- --IKITDFGL CK-EGIKDGA TMKTFCGTPE YLAPEVL--- -----EDNDY GRAVD
PKC         IIYRDLKLDN VMLDSEGH-- --IKIADFGM CK-ENIWDGV TTKTFCGTPD YIAPEII--- -----AYQPY GKSVD

CDK4        IVHRDLKPEN ILV----TSN GTVKLADFGL ARIY--SYQM ALTPVVVTLW YRAPEVLLQ- -------STY ATPVD
CASk2       IMHRDVKPHN VMI---DHEH RKLRLIDWGL AEFY--HPGQ EYNVRVASRY FKGPELLVDY -------QMY DYSLD
               (VIb)                 (VII)                   (VIII)                (IX)
```

FIGURE 2—*Continued*

B (*cont.*)

```
                            T    C                                                            NFV
Consensus    .W..GVI.YI LLSG.PPF.. .........--- -.-....I.. G--------- -.........-- .W...S..AK DLI..

Camk I       CWSIGVIAYI LLCGYPPFYD ENDA------ -K-LFEQILK A--------- -EYEFDSP-- YWDDISDSAK DFIRH
Camk IV      MWSVGIITYI LLCGFEPFYD ERGD------ -QFMFRRILN C--------- -EYYFISP-- WWDEVSLNAK DLVKK
psk H1       MWALGVIAYI LLSGTMPFED DNRTRLY--- -----RQILR G--------- -KYSYSGE-- PWPSVSNLAK DFIDR
kIG5         CWAIGVIMYI LLSGNPPFYE EVEEDDYENH DKNLFRKILA G--------- -DYEFDSP-- YWDDISQAAK DLVTR
rCamk II     LWACGVILYI LLVGYPPFWD EDQHRLY--- -----QQIKA G--------- -AYDFPSP-- EWDTVTPEAK DLINK
dCamk II     IWACGVILYI LLVGYPPFWD EDQHRLY--- -----SQIKA G--------- -AYDYPSP-- EWDTVTPEAK NLINQ
yCmk1        IWSIGVITYT LLCGYATI-D RKVQDF---- ----LDECTT G--------- -EYPVKFHRP YWDSVSNQAK QFILK
yCmk2        IWSIGVITYT LLCGYSPFIA ESVEGF---- ----MEECTA S--------- -RYPVTFHMP YWDNISIDVK RFILK
aspCmk       IWSLGVITYT LLCGYSPFRS ENLTDL---- ----IEECRS G--------- -RVVFHER-- YWKDVSKDAK DFILS
cornCmk      VWTAGVILYI LLCGVPPFWA ETQ------- -QGLFDAVLK GVIDFDLDP- ---------- -WPVISESAK DLIRR
carCmk       IWSAGVILYI LLSGVPPFWA ENE------- -KGIFDAILE GVIDFESEP- ---------- -WPSVSNSAK DLVRK
aplCmk       MWALGVILYI LLSGYPPFIA QSN------- -RQKQQMIMA GEFSFYEKT- ---------- -WKGFLCQPK QLISS
S6k          IWSLGVLLYT MLAGYTPFAN GPSDT----- PEEILTRISS G--------- -KFTLSGG-- NWNTVSETAK DLVSK
Phosk        LWACGVILFT LLAGSPPFWH RRQILM---- ----LRMIME GQYQFSSP-- ---------- EWDDRSNTVK DLIAK
AMPK         IWSCGVILYA LLCGTLPFDD EHVPTL---- ----FKKIRG G--VFYIP-- ---------- EYLNRS--IA TLLMH
twitchin     MWAIGVLGYV LLSGLSPFAG EDD------- -LETLQNVKR CDWEFDEDA- ---------- -FSSVSPEAK DFIKN
MLCK         MWSLGVITYM LLSGLSPFLG DDD------- -TETLNNVLS ANWYFDEET- ---------- -FEAVSDEAK DFVSN
RCK2         MWGIGCVLYT MLCGFPPFYD EKI------- -DTLTEKISR GEYTFLKPW- ---------- -WDEISAGAK NAVSK
MAPKAP       MWSLGVIMYI LLCGYPPFYS NHG----LAI SPGMKTRIRM GQYEFPNPE- ---------- -WSEVSEEVK MLIRN

PKG          FWSLGILVYE LLTGNPPFSG IDQMMT---- ----YNLILK G--------I EKMDFP---- --RKITRRPE DLIRR
PKA          WWALGVLIYE MAAGYPPFFA DQPIQI---- ----YEKIVS G--------- -KVRFP---- --SHFSSDLK DLLRN
racA         WWGLGVVMYE MMCGRLPFYN QDHEKL---- ----FELILM E--------- -EIRFP---- --RTLGPEAK SLLSG
PKC          WWAFGVLLYE MLAGQAPFEG EDEDEL---- ----FQSIME H--------- -NVAYP---- --KSMSKEAV AICKG

CDK4         MWSVGCIFAE MFRRKPLFCG NSE------- -ADQLG-{}- GAFSPRGPRP V-------QS VVPEMEESGA QLLLE
CASk2        MWSLGCMLAS MIFRKEPFFH GHD------- ---NYD-{}- ERFVHSENQ- ---------- --HLVSPEAL DFLDK
                  (IX)                                  (X)                           (XI)
```

```
                       Q
Consensus    LL..DP..R. -----TA.E. L.HPW.

Camk I       LMEKDPEKRF -----TCEQA LQHPWI
Camk IV      LIVLDPKKRL -----TTFQA LQHPWV
psk H1       LLTVDPGARM -----TALQA LRHPWV
kIG5         LMEVEQDQRI -----TAEEA ISHEWI
rCamk II     MLTINPSKRI -----TAAEA LKHPWI
dCamk II     MLTVNPNKRI -----TAAEA LKHPWI
yCmk1        ALNLDPSKRP -----TAAEL LEDPWI
yCmk2        ALRLNPADRP -----TATEL LDDPWI
aspCmk       LLQVDPAQRP -----TSEEA LKHPWL
cornCmk      MLNPIPSRRL -----TAHEV LCHPWI
carCmk       MLTQDPRRRI -----TSAQV LDHPWM
aplCmk       LLKVDPDKRP -----SAQEL LDHPWV
S6k          MLHVDPHQRL -----TAKQV LQHPWI
Phosk        LLQVDPNARL -----TAEQA LQHPFF
AMPk         MLQVDPLKRA -----TIKDI REHEPF
twitchin     LLQKEPRKRL -----TVHDA LEHPWL
MLCK         LITKDQSARM -----SAEQC LAHPWL
RCK2         LLELEPSKRY -----DIEQF LDDPWL
MAPKAP       LLKTEPTQRM -----TITEF MNHPWI

PKG          LCRQNPTERL GNLKNGINDI KKHRWL
PKA          LLQVDLTKRF GNLKNGVSDI KTHKWF
racA         LLKKDPTQRL GGGSEDAKEI MQHRFF
PKC          LMTKHPGKRL GCGPEGERDI KEHAFF

CDK4         MLTFNPLKRI -----SAFRA LQHSYL
CASk2        LLRYDHQSRL -----TAREA MEHPYF
                  (XI)
```

FIGURE 2—*Continued*

the Y(I/T)LL(S/C)GXXPF motif in domain IX, and the R(X)TA(X)E/Q(X)L (X)HPW motif in domain XI. The structural conservation of these amino acids in the CaM kinase family is likely to be related to the functional properties of these enzymes, namely the fact that they are regulated by an autoinhibitory mechanism and that they bind to and are activated by Ca^{2+}/ CaM (see below). Alternatively, the conservation may be related to the recognition and binding of the kinases to their respective peptide substrates, which in the CaM kinases examined in detail to date involve basic amino acids within their respective substrate recognition motifs (see below).

III. Characterization of CaM Kinase I

CaM kinase I was first identified in brain extracts as an activity that was able to phosphorylate site 1 of the neuronal protein, synapsin I (Huttner *et al.*, 1981; Kennedy and Greengard, 1981). The enzyme was subsequently purified from bovine brain and its basic biochemical properties were examined (Nairn and Greengard, 1987). CaM kinase I is active as a monomer, with various molecular weights having been assigned to the enzyme. The preparation from bovine brain contained two major polypeptides of 37 and 39 kDa and a minor species of 42 kDa (Nairn and Greengard, 1987). Two other enzyme preparations, termed CaM kinase Ia and CaM kinase Ib, have also been purified from rat brain based on their ability to phosphorylate site 1 of synapsin I (DeRemer *et al.*, 1992a,b). CaM kinase Ia has a molecular weight of 42 kDa and is phosphorylated on threonine by a regulatory protein kinase resulting in a 15-fold increase in kinase activity (see below). CaM kinase Ib, however, is a 39-kDa protein, is active in the absence of any regulatory protein other than CaM, is autophosphorylated on both serine and threonine residues, but is not activated by autophosphorylation. However, treatment of CaM kinase Ib with protein phosphatase results in an inactive species, suggesting that CaM kinase Ib may also require phosphorylation by an activating protein kinase for full activity (see below).

Based on amino acid sequencing of the 37- and 39-kDa polypeptides of bovine CaM kinase I, a full length cDNA for rat CaM kinase I was cloned and a molecular weight of 41,643 deduced from the predicted amino acid sequence (Picciotto *et al.*, 1993; Cho *et al.*, 1994). The amino acid sequences of several peptides isolated from the purified bovine brain enzyme were found to be completely conserved in rat CaM kinase I. In addition, several peptide sequences obtained from an enzyme purified from rat brain and designated CaM kinase V were also found in rat CaM kinase I (Mochizuki *et al.*, 1993a). Based on the high level of identity of the amino acid sequences, CaM kinase V appears to be either identical or closely related to CaM kinase I. A human form of the enzyme was also cloned that has a molecular weight of 44,337 and is 97% identical at the amino acid level to rat CaM kinase

I (Haribabu *et al.*, 1995). Although it appears that CaM kinase I/V, CaM kinase Ia, and CaM kinase Ib are highly related, and the enzyme is likely to be highly conserved in various species, it is not clear if the various forms of the enzyme isolated are in fact the same entity. CaM kinase I is encoded by a single gene in the rat (Picciotto *et al.*, 1993) and human (Haribabu *et al.*, 1995) genome, but alternative splicing to produce multiple forms of the enzyme is possible.

IV. Regulation of CaM Kinase I

Irrespective of the exact relationship of CaM kinase I/V and CaM kinases Ia and Ib, CaM kinase I is the smallest member of the CaM kinase family. The enzyme consists largely of a catalytic domain (residues 1–275) and a regulatory domain (residues 280–320) (Fig. 3). Based on extensive biochemical and structural analyses of the CaM kinases, as well as other second messenger-regulated protein kinases, it has been proposed that this class of enzyme is regulated by an intrasteric autoinhibitory mechanism (Kemp and Pearson, 1991) (see also Chapter 12). The central feature of this mechanism is that amino acid residues within the regulatory domain bind to the active site of the kinase and sterically block the access of peptide substrate and/ or MgATP. In certain cases, e.g., PKA, the inhibitory domain appears to act as a pseudosubstrate in that the amino acid sequence of the inhibitory domain is very similar to that of the peptide substrates for this kinase. However, in the case of the CaM kinases it is not yet clear if regulation of enzyme activity occurs strictly via a pseudosubstrate mechanism or via a more general autoinhibitory mechanism (see below).

Recent biochemical studies of CaM kinase I have investigated in detail the structural basis for the interaction of CaM with the regulatory domain and for autoinhibition (Haribabu *et al.*, 1995; Yokokura *et al.*, 1995). Following comparison of the amino acid sequence of the regulatory domain of CaM kinase I with that of other CaM kinases and CaM-binding proteins, a series of COOH-terminal truncated mutants of the clone encoding rat CaM kinase I were prepared (Yokokura *et al.*, 1995) (Fig. 3). CaM binding was successively abolished by removing amino acids 321 to 305. The truncation mutants 1–305 or 1–299 were found to be completely unable to bind Ca^{2+}/CaM and to be inactive. Truncation to residues 293 or 277 resulted in a constitutively active enzyme that could no longer bind Ca^{2+}/CaM. These results defined two separate but overlapping domains within the regulatory domain of CaM kinase I. Residues 305–325 are necessary for binding of Ca^{2+}/CaM but are not necessary for autoinhibition of activity. In contrast, residues between 293 and 305 are not essential for Ca^{2+}/CaM binding, but are able to autoinhibit kinase activity in the absence of the Ca^{2+}/CaM binding

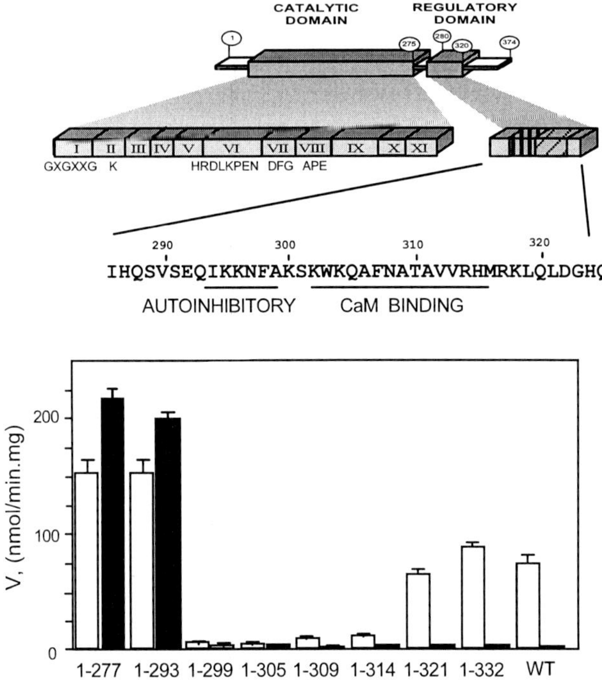

FIGURE 3 Domain structure and regulation of CaM kinase I. (Top) The CaM kinases have a highly conserved catalytic domain and a less highly conserved regulatory region that contains an autoinhibitory domain and an overlapping CaM binding domain. The catalytic domain can be divided into subdomains I–XI based on homology with other protein kinases (Hanks *et al.*, 1988; Hanks and Quinn, 1991). Oligonucleotides synthesized corresponding to the HRD motif and the APE motif were used to isolate CaM kinase I by polymerase chain reaction (Picciotto *et al.*, 1993). (Middle) Amino acid sequence of the regulatory domain of CaM kinase I highlighting the minimal regions involved in autoinhibition and binding to CaM. (Bottom) Ca^{2+}/CaM-independent and -dependent protein kinase activities of various truncated mutants of CaM kinase I. Protein kinase activities were measured in the presence of 1 mM EGTA (solid bars) or 0.5 mM Ca^{2+} plus CaM (open bars). See text and Yokokura *et al.* (1995) for further discussion.

domain. Similar conclusions have also been obtained from studies of human CaM kinase I (Haribabu *et al.*, 1995).

In general, CaM binding domains are not highly conserved at the level of primary structure; however, there is conservation at the level of secondary structure in that they are all predicted to exist as amphipathic α-helices (O'Neil and DeGrado, 1990; Crivici and Ikura, 1995). Analysis of residues 303–316 of CaM kinase I predicts an α-helical conformation (Yokokura *et al.*, 1995). Studies of the structure of a number of complexes of CaM together with peptides representing CaM binding domains of MLCK and CaM kinase II have confirmed the existence of an α-helical structure (Ikura *et al.*, 1992; Meador *et al.*, 1992, 1993; for review, see Crivici and Ikura,

1995). Upon binding of Ca^{2+} to its four binding sites, the two lobes of CaM fold around the helical peptide. Two hydrophobic amino acids 8–12 residues apart are necessary to bind to hydrophobic pockets in CaM, although there are extensive contacts between many residues in the two molecules. The variability in the exact location of the principal hydrophobic amino acids in the different structures is tolerated because the central helix of CaM is highly flexible and extends to accommodate the particular binding peptide. In CaM kinase I, W303 aligns with conserved tryptophan residues in several other CaM-binding proteins (Fig. 4). Thus, the amino acid residues, K^{300}SK-WKQ^{305}, in CaM kinase I are likely to represent the NH$_2$-terminal end of the CaM binding domain. Correspondingly, amino acid residues, T^{310}AVV^{313}, are likely to represent the COOH-terminal end of the CaM-

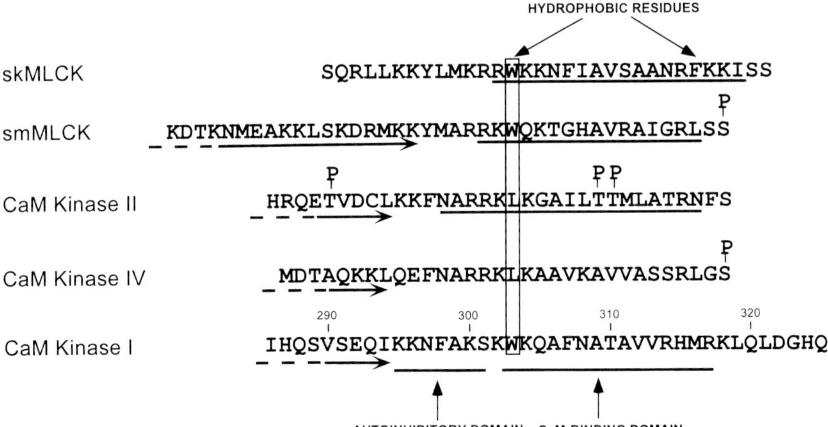

FIGURE 4 Comparison of autoinhibitory and CaM binding domains of CaM kinases. The amino acid sequences surrounding the CaM binding domains of the CaM kinases are compared (the numbers shown are for CaM kinase I only). Direct experimental evidence has been obtained that shows the position of the CaM binding domain for skeletal and smooth muscle MLCK and CaM kinase II (underlined) (Ikura *et al.*, 1992; Meador *et al.*, 1992, 1993). Truncation mutagenesis and peptide analysis have been used to identify the CaM-binding domains of CaM kinases I, II, IV, and MLCK (Cruzalegui *et al.*, 1992; Yano *et al.*, 1993; Cruzalegui and Means, 1993; Yokokura *et al.*, 1995). Each of these sequences defines an amphipathic α-helix in which basic and hydrophobic residues are important for binding to CaM. The sequences are aligned based on the position of the hydrophobic residue (boxed) believed to be at the NH$_2$-terminal end of the CaM binding domain in each enzyme. A second hydrophobic amino acid 8–12 residues COOH terminal to the first hydrophobic amino acid represents the end of the CaM-binding domain. The underlining with the arrows indicates the COOH-terminal limit of truncation mutants that show constitutive activity (for CaM kinase I, see Figure 3). The minimal autoinhibitory domain, which lies between residues at the COOH terminal of these truncation sites and the NH$_2$-terminal end of the CaM binding domains, is indicated. P indicates phosphorylation sites that are important in regulating enzyme activity or CaM binding. Note that the lysine residue doublet found close to the autoinhibitory domain in all the sequences is apparently not necessary for autoinhibition of CaM kinase IV.

binding domain. However, the amino acid residues, $H^{315}MRKLQL^{321}$, contribute to binding of CaM in some way, perhaps by stabilizing the structure of the CaM binding domain (Yokokura *et al.*, 1995).

V. Substrate Specificity of CaM Kinase I and Autoinhibition by a Pseudosubstrate Mechanism

The results from the studies of truncation mutants of CaM kinase I clearly indicate that the autoinhibitory domain is close to or overlaps the NH_2-terminal end of the CaM binding domain. An evaluation of the precise nature of the mechanism of autoinhibition requires consideration of the substrate specificity of CaM kinase I. Although several substrates have now been found for CaM kinase I, the best characterized remains site 1 of synapsin I. Analysis of the phosphorylation of synthetic peptides based on the sequence of site 1 of synapsin I has determined that CaM kinase I has a preference for (Hyd)(X)RRR(X)S(XX)(Hyd), with the arginine residue in the P−3 position being especially important (Lee *et al.*, 1994; Picciotto *et al.*, unpublished results). In addition, hydrophobic residues at the P−5 and P+4 positions are also required (Lee *et al.*, 1994). This consensus phosphorylation site sequence overlaps that of PKA, which prefers basic amino acids at the P−2 and P−3 positions as well as a hydrophobic residue at the P+1 position (Pearson and Kemp, 1991). The overlapping substrate specificity of CaM kinase I and PKA raises the possibility that these two enzymes may be a point of convergence for second messenger systems transducing Ca^{2+} and cAMP signals. In support of this possibility, it is notable that the residue phosphorylated by CaM kinase I in each of its substrates thus far identified (see below) can also be phosphorylated by PKA. It is also of interest that the substrate recognition motif for CaM kinase I is essentially identical to that of the AMP-activated protein kinase and the related enzymes, plant HMG-CoA reductase and yeast SNF1 (Weekes *et al.*, 1993; Dale *et al.*, 1995). Furthermore, the requirement for a hydrophobic residue at the P−5 position and a basic residue at the P−3 position is also found for MAPKAP kinase-2 (Stokoe *et al.*, 1993). It is probable that each of these enzymes has appropriate residues close to their active sites that interact with the basic and hydrophobic amino acids found in their substrates. In PKA, the glutamate residues, E^{127} and E^{331}, interact with the basic residue at the P−3 position, and E^{170} and E^{230} interact with the basic residue at the P−2 position (Knighton *et al.*, 1991a,b). In CaM kinase I, E^{102} and E^{145} correspond respectively to E^{127} and E^{170} in PKA, with no residues obviously corresponding to E^{230} and E^{331}. E^{102} is present within the motif GGELF in domain IV, and E^{145} is present in the motif HRDLKPEN in domain VIb (Fig. 2, see above). Notably, based on comparison of the amino acid sequences of their catalytic domains, AMP-activated protein kinase, MAPKAP kinase-2 (Fig. 2), HMG-CoA reductase, and SNF1 (Hanks and Hunter, 1995a; Hardie and Hanks, 1995) fall into the family of CaM kinase-like enzymes, although none of these proteins is regulated by CaM. In each enzyme, acidic residues are found at positions equivalent to E^{127} and E^{170} (of PKA) within conserved GGELF and HRD(X)K-PEN motifs. Each of these enzymes is also likely to have one or more hydrophobic pockets that bind the $P-5$ and $P+4$ hydrophobic side chains found in their substrates. The conserved hydrophobic motif found in the CaM kinase-like enzymes within domain IX (Fig. 2) may be important in this respect.

Comparison of amino acids 294–305 of CaM kinase I with the consensus phosphorylation site sequence for the enzyme revealed several common features that support the possibility that the autoinhibitory domain of CaM kinase I interacts with the catalytic domain through a pseudosubstrate mechanism (see Yokokura *et al.*, 1995, for more discussion). Three alignments of the autoinhibitory domain CaM kinase I were possible for maintaining the necessary features of a pseudosubstrate, namely hydrophobic residues at the $P-5$ and $P+4$ positions and a basic residue at the $P-3$ position. These potential alignments, together with the observations that the substrate specificity of CaM kinase I depends on hydrophobic amino acids (Lee *et al.*, 1994), suggest that at least in the case of CaM kinase I, the mechanism of autoinhibition may depend on hydrophobic as well as basic amino acids.

Comparison of the results obtained with CaM kinase I with those of other CaM kinases reveals that there is general agreement concerning the position of the Ca^{2+}/CaM binding and autoinhibitory domains of the various enzymes (Fig. 4). Mutagenesis studies of CaM kinase II (Cruzalegui *et al.*, 1992) and CaM kinase IV (Cruzalegui and Means, 1993) have suggested regulation by an autoinhibitory mechanism. In the case of CaM kinase II, the presence of residues $K^{291}KFN^{294}$ (Fig. 4) was sufficient to maintain the enzyme in an inactive state, as has been seen with the equivalent residues of CaM kinase I. Replacement of KKFN by AAAL in CaM kinase II had little effect on Ca^{2+}/CaM-dependent activation of the enzyme, although deletion of the amino acids generated a partially Ca^{2+}/CaM-independent kinase. Based on these results, it was concluded that autoinhibition of CaM kinase II required determinants in addition to residues 291–294. In the case of CaM kinase IV, truncation at L^{313} ($K^{311}KL^{313}$, Fig. 4) generated a fully active kinase that did not require Ca^{2+}/CaM. Notably, truncation of CaM kinases I, II, and IV at an analogous hydrophobic residue (Fig. 4) resulted in an active enzyme; however, the position of the truncation placed a pair of lysine residues as part of the autoinhibitory domain (CaM kinases I and II) or not (CaM kinase IV). The relative importance of hydrophobic and basic residues will hopefully be resolved by further detailed structural studies of each of these enzymes.

VI. Regulation of CaM Kinase I by a
Phosphorylation Cascade

There is now significant evidence to support the idea that the physiological regulation of CaM kinase I requires phosphorylation by a distinct CaM kinase I kinase that is itself a CaM kinase, and that this is part of a novel CaM kinase cascade (DeRemer *et al.*, 1992b; Mochizuki *et al.*, 1993b; Lee and Edelman, 1994, 1995; Sugita *et al.*, 1994; Haribabu *et al.*, 1995). As discussed previously, CaM kinase Ia was originally shown to be activated in the presence of an activator protein (DeRemer *et al.*, 1992b). The activator protein (M_r ~52 kDa) has been purified to near homogeneity and shown to be a novel CaM kinase (CaM kinase I kinase) that specifically phosphorylates CaM kinase Ia (Lee and Edelman, 1994, 1995). In parallel studies, the activity defined as CaM kinase V was also found to be subject to regulation by a CaM-dependent protein kinase of M_r ~64 kDa (Mochizuki *et al.*, 1993b; Sugita *et al.*, 1994).

The site on CaM kinase I phosphorylated by CaM kinase I kinase has been identified as T^{177} in both direct (Picciotto *et al.*, 1993; Sugita *et al.*, 1994) and indirect studies (Mochizuki *et al.*, 1993b; Haribabu *et al.*, 1995). The original description of T^{177} as a site of autophosphorylation (Picciotto *et al.*, 1993) is incorrect and resulted from phosphorylation of CaM kinase I by low levels of contaminating CaM kinase I kinase, although the recombinant enzyme does phosphorylate T^{177} at a very slow rate (M. Picciotto and A. Nairn, unpublished results). Notably, T^{177}, located between domains VII and VIII, is in an equivalent position for activating phosphorylation sites in PKA, PKC, PKG, the MAP kinases such as ERK1, and the cyclin-dependent kinases, such as cdc2 kinase (Fig. 5). X-ray crystallographic studies of PKA indicate that the phosphorylated threonine (T^{197} in PKA) is present within the T- or activation loop, and that phosphorylation is critical for the correct alignment of the active site of the enzyme (Knighton *et al.*, 1991a,b). Phosphorylation of CaM kinase I, or other kinases, within the activation loop is likely to lead to corresponding changes in active site conformation. A mechanism of activation similar to CaM kinase I has also been found for CaM kinase IV, which exhibits low basal Ca^{2+}/CaM-dependent activity and is phosphorylated by a CaM kinase IV kinase at a corresponding threonine in the activation loop (Fig. 5) (see below for further discussion). In contrast, CaM kinase II, MLCK, and phosphorylase kinase do not have phosphorylatable residues in the activation loop (Fig. 5) and are fully active in the presence of only Ca^{2+}/CaM.

It is important to note that phosphorylated CaM kinase I is still dependent for activity on Ca^{2+}/CaM. CaM kinase I kinase also appears to be a CaM-dependent protein kinase based on its binding to CaM–Sepharose (Mochizuki *et al.*, 1993b; Sugita *et al.*, 1994; Lee and Edelman, 1994, 1995) and the fact that the enzyme is activated by Ca^{2+}/CaM when the

```
                                P
CAMK  I      DFGLSKMEDPG----SVLSTACGTPGYVAPE
CAMK  IV     DFGLSKIVEHQ----VLMKTVCGTPGYCAPE

PKA          DFGFAKRVK------GRTWTLCGTPEYLAPE
PKC          DFGMCKEHMMDG---VTTRTFCGTPDYIAPE
PKG          DFGFAKKIGFG----KKTWTFCGTPEYVAPE
AMP-K        DFGLSNMMSDGE----FLRTSCGSPNYAAPE

cdc2         DFGLARSFGVP----LRNYTHEIVTLWRAPE
ERK1         DFGLARIADPEHDT-GFLTEYVATRWYRAPE

CaMKIV K     DFGVSNQFEGND---AQLSSTAGTPAFMAPE

CAMK  II     DFGLAIEVE-GE--QQAWFGFAGTPGYLSPE
Phosk        DFGFSCQLDPGE----KLREVCGTPSYLAPE
MLCKsk       DFGLARRYNPNE----KLKVNFGTPEFLSPE
MLCKsm       DFGLARRLESAG----SLKVLFGTPEFVAPE
TWITCH       DFGLTAHLDPKQ----SVKVTTGTAEFAAPE
CMK1         DFGIAKTLKSDE---ELLYKAGTALGYVAPE
CMK2         DFGIAKQLKGEED---LIYKAAGSLGYVAPE
```

FIGURE 5 Comparison of amino acid sequences between domains VII and VIII of CaM kinase I and other serine/threonine protein kinases. The amino acid sequences surrounding the threonine phosphorylated in CaM kinases I and IV (CAMK I and CAMK IV), PKA, PKC, PKG, AMP-activated protein kinase (AMP-K), and cdc2 kinase (cdc2), and the threonine and tyrosine phosphorylated in MAP kinase (ERK1) are shown (phosphorylation site indicated by P and outlined in box) [see Table I, Figure 2, and Hardie and Hanks (1995) for information on the amino acid sequences]. There is no corresponding phosphorylatable residue in the other CaM kinases (see Table I for abbreviations and other details). CaM kinase IV kinase (CaM KIV K) has been cloned (Tokumitsu *et al.*, 1995) and it is not yet known if it is activated by phosphorylation within the activation loop. The glutamate residue in phosphorylase kinase may be functionally equivalent to the phosphorylated threonine (Owen *et al.*, 1995).

constitutively active truncated form of CaM kinase I (1–294) is used as a substrate (Haribabu *et al.*, 1995). Furthermore, studies have revealed an intriguing parallel relationship between the activation of CaM kinase I by CaM kinase I kinase and the activation of the structurally related AMP-activated protein kinase (AMPK) by its activator (Hawley *et al.*, 1995). AMPK is activated by an upstream kinase (termed AMP kinase kinase or AMPKK) probably at a threonine residue within the activation loop (Fig. 5), although this has yet to be directly demonstrated. Although CaM kinase I is more efficiently activated by CaM kinase I kinase, it can also be phosphorylated and activated by AMPKK. AMPK can also be phosphorylated and activated by CaM kinase I kinase to some extent. Using AMPK, which does not bind CaM, as a substrate, CaM kinase I kinase was clearly shown to be dependent on CaM for activity. CaM kinase I kinase does display some basal activity in the absence of Ca^{2+}/CaM, however, which raises questions of the exact mechanism of regulation of the enzyme (Haribabu *et al.*, 1995; Hawley *et al.*, 1995). The study of the interrelationship of the activation of AMPK and CaM kinase I by their activating kinases has also shown that CaM must be bound to CaM kinase I for phosphorylation of T^{177} to occur. Thus, at a minimum, two independent mechanisms are required for CaM kinase I to be fully activated, and CaM is required at three different levels. First, CaM must bind to CaM kinase I kinase for it to be active. Next, its substrate, CaM kinase I, must bind CaM to become phosphorylated at T^{177}, which results in the orientation of the activation loop of the enzyme. Finally, CaM kinase I still requires CaM to phosphorylate its downstream target substrates. This suggests that the activation of CaM kinase I is extremely sensitive to regulation by the second messenger, Ca^{2+}.

Studies have indicated that CaM kinase IV is also regulated by an activating kinase (Okuno and Fujisawa, 1993; Tokumitsu *et al.*, 1994, 1995; Okuno *et al.*, 1995; Kameshita and Fujisawa, 1995; Selbert *et al.*, 1995). A novel CaM kinase IV kinase of M_r 68 kDa has been isolated and cloned (Tokumitsu *et al.*, 1995). The enzyme shows homology to the CaM kinase family and binds CaM. Notably, CaM kinase IV kinase is able to phosphorylate and activate CaM kinase I (Tokumitsu *et al.*, 1995). In addition, CaM kinase I kinase is able to phosphorylate the activation loop threonine of CaM kinase IV (T^{196}) and stimulate enzyme activity (Selbert *et al.*, 1995). Clearly, detailed analysis of the two enzymes will be necessary, including amino acid sequencing of CaM kinase I kinase and kinetic analysis of phosphorylation of each substrate by each kinase preparation, to address the question of whether these two activities represent different proteins. The fact that the best characterized CaM kinase I kinase preparation has a M_r of ~ 52 kDa and that AMPKK is able to phosphorylate CaM kinase I suggests the possibility that a family of structurally related activating enzymes exists. In this respect, it is also notable that the mechanism of activation of CaM kinases I and IV resembles that of the well-established MAP kinase cascade in which several isozymes of MAP kinase are activated by multiple upstream kinases within parallel and distinct pathways (Blumer and Johnson, 1994).

A question related to the relationship of the CaM kinases I and IV kinases concerns the specificity of the enzymes and whether they phosphorylate other substrates besides CaM kinases I and IV. As discussed previously, the CaM kinases characterized to date all recognize basic amino acids within their substrate recognition motifs. It is likely that the glutamate residue within the GGELF motif in domain V, which is conserved in all the CaM kinases, is involved in binding substrates. Notably, this motif is not conserved in CaM kinase IV kinase (Tokumitsu *et al.*, 1995), suggesting that this enzyme will display a distinct substrate specificity from other CaM kinase family members. It is also of interest that CaM kinase I kinase fails to phosphorylate the inactive truncated form of CaM kinase I, but does phosphorylate the constitutively active shorter form of the enzyme (Haribabu *et al.*, 1995). This suggests that the recognition of T^{177} by CaM kinase I kinase requires the correct orientation of the activation loop and that this is dependent on the removal of the autoinhibitory domain by binding of CaM.

It is not yet clear why an enzyme whose activity is tightly regulated by Ca^{2+}/CaM must also be subject to regulation by a cascade of CaM kinases.

By analogy with the MAP kinase pathway, it appears possible that such a cascade will allow input by signaling pathways in addition to Ca^{2+}. An important question that requires study is whether CaM kinase I kinase (or CaM kinase IV kinase) is itself regulated by other phosphorylation events. Notably, CaM kinase IV kinase contains the amino acid sequence SST within its activation loop at an equivalent position to T^{177} of CaM kinase I (Fig. 5), suggesting that it may require phosphorylation for full activity. Further characterization of the upstream kinase(s) and its effects on the enzymatic properties of CaM kinases I and IV should shed light on this issue.

VII. Cellular and Subcellular Distribution of CaM Kinase I

When CaM kinase I was first purified from bovine brain based on its ability to phosphorylate synapsin I, it was thought to be a strictly neuronal protein with a single substrate. The enzyme was subsequently cloned, antibodies were raised against the protein, and RNAse protection, and immunocytochemistry and immunoblotting have demonstrated that CaM kinase I is present in many nonneuronal tissues and cell types (Picciotto *et al.*, 1993, 1995; Ito *et al.*, 1994). In addition to being widely distributed, substrates other than synapsins I and II have now been identified, at least *in vitro* (see below). Given the wide cellular distribution of the enzyme and the fact that nonneuronal tissues do not express the synapsins, it is likely that CaM kinase I is a multifunctional kinase with as yet unidentified substrates.

CaM kinase I activity is highest in brain tissue with lower levels being found in nonneuronal tissues (Nairn and Greengard, 1987). CaM kinase I RNA is present at high levels in brain, adrenal gland, liver, and lung and is expressed in all tissues examined to date (Picciotto *et al.*, 1993). Immunoblotting analysis detected high levels of two major immunoreactive species of M_r ~38 and 42 kDa, and minor immunoreactive species, in brain, liver, and intestine, with significant levels being found in all other tissues examined (Picciotto *et al.*, 1995). Relatively high levels of CaM kinase I protein were also detected in a variety of cultured cell lines. As discussed previously, it appears possible that although CaM kinase I is the product of a single gene, multiple isoforms may exist. To date, immunoblotting studies in rat have not indicated any differential distribution of the different immunoreactive species. However, various immunoreactive bands of M_r 35–43 kDa were detected in brain tissue from different species.

Immunocytochemistry has been used to analyze the localization of CaM kinase I within the brain (Ito *et al.*, 1994; Picciotto *et al.*, 1995). CaM kinase I was found to be widely expressed, was not homogeneously distributed (that is, the staining did not follow cell density in the brain), and the immunoreactivity appeared to be found primarily in neurons rather than in glia (Picciotto *et al.*, 1995) (see Figure 6). Densely stained regions in the telencephalon included layers II–VI of the isocortex, CA1–CA3 in the hippocampus, several amygdaloid nuclei (with the exception of the basolateral nucleus, which was practically devoid of staining), and the medial septum and diagonal band nuclei. In the diencephalon, CaM kinase I immunoreactivity was high in the medial and lateral geniculate thalamic nuclei but very low in the anterior and ventral posterior nuclei, and the hypothalamus was very highly labeled. In general, more intense CaM kinase I staining was found in the caudal regions of the brain. In the brain stem, many nuclei were highly labeled including the substantia nigra, interpeduncular nucleus, optic tectum, central gray, hypoglossal nucleus, and area postrema. The molecular layer was more intensely stained that the granular cell layer in the cerebellum, and both cell bodies and dendrites of Purkinje cells were highly labeled. Finally, many neurons throughout the spinal cord were labeled including somatic motor neurons of the ventral horn.

Immunocytochemistry has also been used to study the subcellular distribution of the enzyme (Piciotto *et al.*, 1995). In neurons, CaM kinase I appears to be predominantly cytosolic and to be excluded from the nucleus. Translocation of CaM kinase I to the nucleus after activation cannot be excluded. In general, staining was most intense in neuropil. However, in

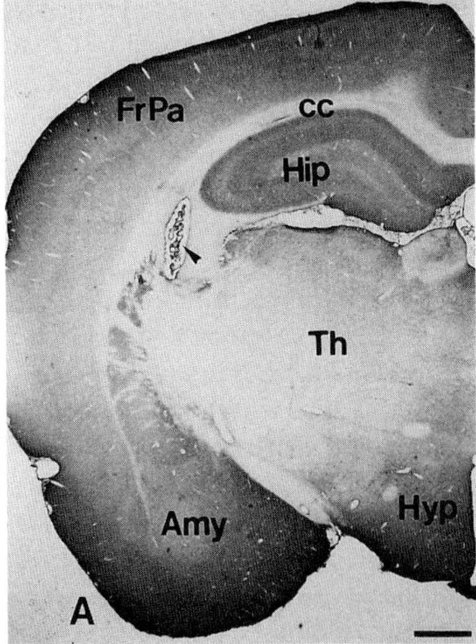

FIGURE 6   Photomicrograph showing CaM kinase I immunoreactivity in brain. A coronal section of rat brain is shown (Bregma level 3.0 mm). Staining is shown without (A) or with (B) preadsorption of the antibody with CaM kinase I. Amy, amygdala; cc, corpus callosum; FrPa, frontoparietal cortex; Hip, hippocampus; Hyp, hypothalamus; Th, thalamus. The arrowhead indicates the choroid plexus. Scale bar, 1 mm.

several brain regions, particularly pyramidal cells in the hippocampus, Purkinje cells in the cerebellum, and motor neurons in the spinal cord, cell bodies were highly labeled. Interestingly, although both cell bodies and dendrites appear to contain CaM kinase I immunoreactivity, the protein appears to be highly concentrated in axons, particularly in motor neurons, the lateral hypothalamus, the facial nerve, and Purkinje cell axons. Experiments in lamprey suggest that phosphorylation of site 1 of synapsin I (the site phosphorylated by CaM kinase I or PKA) appears to regulate axonal transport of that protein (Pieribone *et al.*, 1993), providing a possible explanation for the high levels of CaM kinase I in axons. This link remains to be explored.

An important issue that remains to be investigated is the relationship between the distribution of CaM kinase I protein and its enzyme activity. The levels of CaM kinase I mRNA or protein are not absolutely correlated with enzyme activity (Nairn and Greengard, 1987; Picciotto *et al.*, 1993, 1995). The discrepancy may be due to differential expression of CaM kinase I and CaM kinase I kinase. Another possibility is that the high activity in brain tissue (assayed as phosphorylation of site 1 of synapsin I) includes the activity of CaM kinase IV, which has also been shown to phosphorylate site 1 of synapsin I and is highly localized to brain tissue (Cruzalegui and Means, 1993). Examination of the localization of CaM kinase I kinase will help to resolve this issue.

VIII. Substrates for CaM Kinase I

The physiological function of CaM kinase I is not yet well described. Synapsins I and II, the best characterized substrates for the kinase, are synaptic vesicle-associated proteins believed to be involved in the regulation of neurotransmitter release through their ability to crosslink the vesicles to actin filaments at the nerve terminal (for review, see Greengard *et al.*, 1993). The synapsins are phosphorylated on site 1 by CaM kinase I and sites 2 and 3 (only present in synapsin I) by CaM kinase II. Although phosphorylation by CaM kinase II has been found to regulate several properties of synapsin I, phosphorylation by CaM kinase I has not been found to affect the interaction of either synapsins I or II with synaptic vesicles or actin *in vitro*. Synapsins I and II have also been shown to play a role in synaptogenesis (Han *et al.*, 1991; Greengard *et al.*, 1993); however, the role of CaM kinase I in these processes is yet to be studied. In general, the pattern of CaM kinase I expression, particularly in the hindbrain and spinal cord, is very similar to that of synapsin I (De Camilli *et al.*, 1983; Südhof *et al.*, 1989), supporting a possible role for phosphorylation of the synapsins by CaM kinase I at the synapse. The high expression of CaM kinase I in axons (Picciotto *et al.*, 1995) also suggests a role of the kinase outside the synapse. Studies of axonal transport of synapsin I in lamprey neurons have shown that phosphorylation of synapsin I at site 1 inhibits its axonal transport (Pieribone *et al.*, 1993). Although there are several possible explanations of this result, an exciting possibility is that phosphorylation of site 1 of synapsins I or II by CaM kinase I (or PKA) may act to regulate axonal transport of synaptic vesicle proteins in response to elevated Ca^{2+} or to some aspect of synapsin I or II localization at the synapse. Finally, the localization of CaM kinase I in areas other than axons or the synapse argues that additional neuronal substrates for CaM kinase I are likely to be found.

Potential physiological substrates for CaM kinase I include two proteins that the enzyme has been found to phosphorylate only *in vitro*. CaM kinase I has been shown to phosphorylate S^{133} of CREB, the cyclic AMP response element-binding protein (Sheng *et al.*, 1991) that binds to a DNA motif called CRE (the cAMP response element) and that increases transcription from several genes in response to elevation of cAMP (Armstrong and Montminy, 1993). Phosphorylation of S^{133} in CREB by PKA greatly increases its ability to stimulate transcription but does not change its ability to bind to CRE. Increased transcription of CRE-containing genes also occurs in response to elevation of Ca^{2+}, and S^{133} is phosphorylated under these conditions indicating that phosphorylation of CREB by a CaM kinase may be important. CaM kinase I, CaM kinase II, and CaM kinase IV all phosphorylate S^{133} in CREB *in vitro*, raising the question of which CaM kinase acts *in vivo* (Sheng *et al.*, 1991; Cruzalegui and Means, 1993). The amino acid sequence surrounding S^{133} of CREB conforms well to the consensus amino acid sequence for CaM kinase I. In addition, the phosphorylation sites for CaM kinase I overlap that of PKA (see above), suggesting that CaM kinase I may be a candidate for the enzyme that phosphorylates CREB *in vivo*. However, CREB is restricted to the nucleus and the subcellular distribution of CaM kinase I appears to be predominantly cytosolic. Of the CaM kinases that can phosphorylate CREB *in vitro*, CaM kinase IV is best characterized as being localized to the nucleus (Jensen *et al.*, 1991), although certain isozymes of CaM kinase II contain nuclear targeting sequences and appear to be involved in regulation of transcription (Srinivasan *et al.*, 1994; Nghiem *et al.*, 1994). Furthermore, overexpression of recombinant CaM kinase IV has been found to regulate CREB-dependent gene expression (Enslen *et al.*, 1994, 1995), and this is further stimulated by coexpression of CaM kinase IV with CaM kinase IV kinase (Tokumitsu *et al.*, 1995). It is possible, however, that translocation of CaM kinase I to the nucleus occurs in response to phosphorylation by CaM kinase I kinase by analogy to MAP kinase (Lenormand *et al.*, 1993).

Finally, CaM kinase I has been found to phosphorylate CF-2, a portion of the R domain of CFTR, the cystic fibrosis transmembrane regulator chloride channel (Picciotto *et al.*, 1992). Ca^{2+} is known to regulate chloride flux in epithelial cells and CaM kinase I is found in several tissues that are defective in electrolyte regulation in CF, raising the possibility that CaM kinase I might regulate CFTR chloride channel activity.

IX. Concluding Remarks

CaM kinases are key components of signal transduction pathways involving Ca^{2+} in all eukaryotic cells. To date, six distinct CaM kinases have been well characterized and found to contain conserved catalytic and regulatory domains that likely reflect their common mechanism of autoinhibition and activation by Ca^{2+}/CaM. It is likely, however, that additional CaM kinases will be identified in the future. For example, enzymes involved in the activation of CaM kinases I and IV are themselves CaM kinases. Several of these enzymes are highly specific, showing limited cellular expression and phosphorylating only one substrate, whereas others are widely expressed and multifunctional. Recent studies of CaM kinase I indicate that it is widely expressed in a variety of tissues and cell types and appears to be a multifunctional enzyme. Furthermore, the enzyme is regulated by a complex mechanism involving a direct interaction with CaM as well as by phosphorylation by a novel CaM-dependent protein kinase. Future studies will identify the physiological significance of this regulatory mechanism as well as identify downstream substrates for the enzyme.

The crystal structure of CaM kinase I was recently determined (Goldberg *et al.*, 1996). The results obtained provide a detailed understanding of the mechanism of autoinhibition and activation of CaM kinase I as well as of the related CaM kinases. Recent results also indicate that in intact cells both CaM kinases I and IV can activate CREB, consistent with a role for these kinases in mediating transcriptional responses to Ca^{2+} signaling (Sun *et al.*, 1996).

References

Armstrong, R. C., and Montminy, M. R. (1993). Transsynaptic control of gene expression. *Annu. Rev. Neurosci.* **16,** 17–29.

Blumer, K. J., and Johnson, G. L. (1994). Diversity in function and regulation of MAP kinase pathways. *Trends Biochem. Sci.* **19,** 236–240.

Braun, A. P., and Schulman, H. (1995). The multifunctional calcium/calmodulin-dependent protein kinase: From form to function. *Annu. Rev. Physiol.* **57,** 417–445.

Cho, F. S., Phillips, K. S., Bogucki, B., and Weaver, T. W. (1994). Characterization of a rat cDNA clone encoding calcium/calmodulin-dependent protein kinase I. *Biochim. Biophys. Acta Mol. Cell. Res.* **1224,** 156–160.

Crivici, A., and Ikura, M. (1995). Molecular and structural basis of target recognition by calmodulin. *Annu. Rev. Biophys. Biomol. Struct.* **24,** 85–116.

Cruzalegui, F. H., and Means, A. R. (1993). Biochemical characterization of the multifunctional Ca/calmodulin-dependent protein kinase type IV expressed in insect cells. *J. Biol. Chem.* **268,** 26161–26179.

Cruzalegui, F. H., Kapiloff, M. S., Morfin, J.-P., Kemp, B. E., Rosenfeld, M. G., and Means, A. R. (1992). Regulation of intrasteric inhibition of the multifunctional calcium/calmodulin-dependent protein kinase. *Proc. Natl. Acad. Sci. U.S.A.* **89**, 12127–12131.

Dale, S., Wilson, W. A., Edelman, A. M., and Hardie, D. G. (1995). Similar substrate recognition motifs for mammalian AMP-activated protein kinase, higher plant HMG-CoA reductase kinase-A, yeast SNF1, and mammalian calmodulin-dependent protein kinase I. *FEBS Lett.* **361**, 191–195.

Davis, T. N. (1992). What's new with calcium. *Cell* **71**, 557–564.

De Camilli, P., Cameron, R., and Greengard, P. (1983). Synapsin I (protein I), a nerve terminal specific phosphoprotein 1. Its general distribution in synapses of the central and peripheral nervous system demonstrated by immuno-fluorescence in frozen and plastic sections. *J. Cell Biol.* **96**, 1337–1354.

DeRemer, M. F., Saeli, R. J., Brautigan, D. L., and Edelman, A. M. (1992a). Ca^{2+}–calmodulin-dependent protein kinases Ia and Ib from rat brain. II. Enzymatic characteristics and regulation of activities by phosphorylation and dephosphorylation. *J. Biol. Chem.* **267**, 13466–13471.

DeRemer, M. F., Saeli, R. J., and Edelman, A. M. (1992b). Ca^{2+}–calmodulin-dependent protein kinase Ia and Ib from rat brain. I. Identification, purification, and structural comparisons. *J. Biol. Chem.* **267**, 13460–13465.

Enslen, H., Sun, P., Brickey, D. A., Soderling, S. H., Klamo, E., and Soderling, T. R. (1994). Characterization of Ca^{2+}/calmodulin-dependent protein kinase IV. Role in transcriptional regulation. *J. Biol. Chem.* **269**, 15520–15527.

Enslen, H., Tokumitsu, H., and Soderling, T. R. (1995). Phosphorylation of CREB by CaM-kinase IV activated by CaM-kinase IV kinase. *Biochem. Biophys. Res. Commun.* **207**, 1038–1043.

Goldberg, J., Nairn, A. C., and Kuriyan, J. (1996). Structural basis for the autoinhibition of calcium calmodulin-dependent protein kinase I. *Cell* **84**, 875–887.

Greengard, P., Valtorta, F., Czernik, A. J., and Benfenati, F. (1993). Synaptic vesicle phosphoproteins and regulation of synaptic function. *Science* **259**, 780–785.

Han, H.-Q., Nichols, R. A., Rubin, M. R., Bähler, M., and Greengard, P. (1991). Induction of formation of presynaptic terminals in neuroblastoma cells by synapsin IIb. *Nature* **349**, 697–700.

Hanks, S. K., and Hunter, T. (1995a). The eukaryotic protein kinase superfamily. In "The Protein Kinase FactsBook: Protein–Serine Kinases" (D. G. Hardie and S. K. Hanks, eds.), pp. 7–47. Academic Press, London.

Hanks, S. K., and Hunter, T. (1995b). Protein kinases 6: The eukaryotic protein kinase superfamily: Kinase (catalytic) domain structure and classification. *FASEB J.* **9**, 576–596.

Hanks, S. K., and Quinn, A. M. (1991). Protein kinase catalytic domain sequence database: Identification of conserved features of primary structure and classification of family members. *Methods Enzymol.* **200**, 38–62.

Hanks, S. K., Quinn, A. M., and Hunter, T. (1988). The protein kinase family: Conserved features and deduced phylogeny of the catalytic domains. *Science* **241**, 42–52.

Hanson, P. I., and Schulman, H. (1992). Neuronal Ca^{2+}/calmodulin-dependent protein kinases. *Annu. Rev. Biochem.* **61**, 559–601.

Hardie, D. G., and Hanks, S. K. (eds.) (1995). "The Protein Kinase FactsBook: Protein–Serine Kinases." Academic Press, London.

Haribabu, B., Hook, S. S., Selbert, M. A., Goldstein, E. G., Tomhave, E. D., Edelman, A. M., Snyderman, R., and Means, A. R. (1995). Human calcium-calmodulin dependent protein kinase I: cDNA cloning, domain structure and activation by phosphorylation at threonine-177 by calcium–calmodulin dependent protein kinase I kinase. *EMBO J.* **14**, 3679–3686.

Hawley, S. A., Selbert, M. A., Goldstein, E. G., Edelman, A. M., Carling, D., and Hardie, D. G. (1995). 5'-AMP activates the AMP-activated protein kinase cascade, and Ca^{2+}/

calmodulin activates the calmodulin-dependent protein kinase I cascade, via three independent mechanism. *J. Biol. Chem.* **270**, 27186–27191.

Heilmeyer, L. M. G., Jr. (1991). Molecular basis of signal integration in phosphorylase kinase. *Biochim. Biophys. Acta Mol. Cell. Res.* **1094**, 168–174.

Huttner, W. B., DeGennaro, L. J., and Greengard, P. (1981). Differential phosphorylation of multiple sites in purified protein I by cyclic AMP-dependent and calcium-dependent protein kinases. *J. Biol. Chem.* **256**, 1482–1488.

Ikura, M., Clore, G. M., Gronenborn, A. M., Zhu, G., Klee, C. B., and Bax, A. (1992). Solution structure of a calmodulin–target peptide complex by multidimensional NMR. *Science* **256**, 632–638.

Ito, T., Mochizuki, H., Kato, M., Nimura, Y., Hanai, T., Usuda, N., and Hidaka, H. (1994). Ca^{2+}/calmodulin-dependent protein kinase V: Tissue distribution and immunohistochemical localization in rat brain. *Arch. Biochem. Biophys.* **312**, 278–284.

Jensen, K. F., Ohmstede, C.-A., Fisher, R. S., and Sahyoun, N. (1991). Nuclear and axonal localization of Ca^{2+}/calmodulin-dependent protein kinase type Gr in rat cerebellar cortex. *Proc. Natl. Acad. Sci. U.S.A.* **88**, 2850–2853.

Kameshita, I., and Fujisawa, H. (1995). Preparation and characterization of calmodulin-dependent protein kinase IV (CaM-kinase IV) free of CaM-kinase IV kinase from rat cerebral cortex. *J. Biochem. (Tokyo)* **117**, 85–90.

Kemp, B. E., and Pearson, R. B. (1991). Intrasteric regulation of protein kinases and phosphatases. *Biochim. Biophys. Acta Mol. Cell. Res.* **1094**, 67–76.

Kemp, B. E., and Stull, J. T. (1990). Myosin light chain kinases. In "Peptides and Protein Phosphorylation," pp. 115–134. CRC Press, Boca Raton, FL.

Kennedy, M. B., and Greengard, P. (1981). Two calcium/calmodulin-dependent protein kinases which are highly concentrated in brain phosphorylate protein I at distinct sites. *Proc. Natl. Acad. Sci. U.S.A.* **78**, 1293–1297.

Knighton, D. R., Zheng, J., Ten Eyck, L. F., Ashford, V. A., Xuong, N.-H., Taylor, S. S., and Sowadski, J. M. (1991a). Crystal structure of the catalytic subunit of cyclic adenosine monophosphate-dependent protein kinase. *Science* **253**, 407–414.

Knighton, D. R., Zheng, J., Ten Eyck, L. F., Xuong, N.-H., Taylor, S. S., and Sowadski, J. M. (1991b). Structure of a peptide inhibitor bound to the catalytic subunit of cyclic adenosine monophosphate-dependent protein kinase. *Science* **253**, 414–420.

Lee, J. C., and Edelman, A. M. (1994). A protein activator of Ca^{2+}/calmodulin-dependent protein kinase Ia. *J. Biol. Chem.* **269**, 2158–2164.

Lee, J. C., and Edelman, A. M. (1995). Activation of Ca^{2+}–calmodulin-dependent protein kinase Ia is due to direct phosphorylation by its activator. *Biochem. Biophys. Res. Commun.* **210**, 631–637.

Lee, J. C., Kwon, Y.-G., Lawrence, D. S., and Edelman, A. M. (1994). A requirement of hydrophobic and basic amino acid residues for substrate recognition by Ca^{2+}/calmodulin-dependent protein kinase Ia. *Proc. Natl. Acad. Sci. U.S.A.* **91**, 6413–6417.

Lenormand, P., Sardet, C., Pagès, G., L'Allemain, G., Brunet, A., and Pouysségur, J. (1993). Growth factors induce nuclear translocation of MAP kinases ($p42^{mapk}$ and $p44^{mapk}$) but not of their activator MAP kinase kinase ($p45^{mapkk}$) in fibroblasts. *J. Cell. Biol.* **122**, 1079–1088.

Lu, K. P., and Means, A. R. (1993). Regulation of the cell cycle by calcium and calmodulin. *Endocrinol. Rev.* **14**, 40–58.

Meador, W. E., Means, A. R., and Quiocho, F. A. (1992). Target enzyme recognition by calmodulin: 2.4 Å Structure of a calmodulin–peptide complex. *Science* **257**, 1251–1255.

Meador, W. E., Means, A. R., and Quiocho, F. A. (1993). Modulation of calmodulin plasticity in molecular recognition on the basis of X-ray structures. *Science* **262**, 1718–1721.

Mochizuki, H., Ito, T., and Hidaka, H. (1993a). Purification and characterization of Ca^{2+}/calmodulin-dependent protein kinase V from rat cerebrum. *J. Biol. Chem.* **268**, 9143–9147.

Mochizuki, H., Sugita, R., Ito, T., and Hidaka, H. (1993b). Phosphorylation of Ca^{2+}/calmodulin-dependent protein kinase V and regulation of its activity. *Biochem. Biophys. Res. Commun.* **197**, 1595–1600.

Nairn, A. C., and Greengard, P. (1987). Purification and characterization of Ca^{2+}/calmodulin-dependent protein kinase I from bovine brain. *J. Biol. Chem.* **262**, 7273–7281.

Nairn, A. C., and Palfrey, H. C. (1996). Regulation of protein synthesis by calcium. In "Translational Control" (J. W. B. Hershey, M. B. Mathews, and N. Sonenberg, eds.), pp. 295–318. Cold Spring Harbor Laboratory Press, Cold Spring Harbor, NY.

Nairn, A. C., and Picciotto, M. R. (1994). Calcium/calmodulin-dependent protein kinases. *Semin. Cancer Biol.* **5**, 407.

Nei, M. (1987). "Phylogenetic Trees. Molecular Evolutionary Genetics," pp. 287–326. Columbia University Press, New York.

Nghiem, P., Ollick, T., Gardner, P., and Schulman, H. (1994). Interleukin-2 transcriptional block by multifunctional Ca^{2+}/calmodulin kinase. *Nature* **371**, 347–350.

O'Neil, K. T., and DeGrado, W. F. (1990). How calmodulin binds its targets: Sequence independent recognition of amphiphilic α-helices. *TIBS* **15**, 59–64.

Okuno, S., and Fujisawa, H. (1993). Requirement of brain extract for the activity of brain calmodulin-dependent protein kinase IV expressed in *Escherichia coli. J. Biochem. (Tokyo)* **114**, 167–170.

Okuno, S., Kitani, T., and Fujisawa, H. (1995). Full activation of brain calmodulin-dependent protein kinase IV requires phosphorylation of the amino-terminal serine-rich region by calmodulin-dependent protein kinase IV kinase. *J. Biochem. (Tokyo)* **117**, 686–690.

Owen, D. J., Noble, M. E., Garmen, E. F., Papagerorgiou, A. C., and Johnson, L. N. (1995). Two structures of the catalytic domain of phosphorylase kinase: An active protein kinase complexed with substrate analogue and product. *Structure* **3**, 467–482.

Pearson, R. B., and Kemp, B. E. (1991). Protein kinase phosphorylation site sequences and consensus specificity motifs: Tabulations. *Methods Enzymol.* **200**, 62–81.

Picciotto, M. R., Cohn, J. A., Bertuzzi, G., Greengard, P., and Nairn, A. C. (1992). Phosphorylation of the cystic fibrosis transmembrane conductance regulator. *J. Biol. Chem.* **267**, 12742–12752.

Picciotto, M. R., Czernik, A. J., and Nairn, A. C. (1993). Calcium/calmodulin-dependent protein kinase I: cDNA cloning and identification of autophosphorylation site. *J. Biol. Chem.* **268**, 26512–26521.

Picciotto, M. R., Zoli, M., Bertuzzi, G., and Nairn, A. C. (1995). Immunochemical localization of calcium/calmodulin-dependent protein kinase I. *Synapse* **20**, 75–84.

Pieribone, V. A., Muller, T. H., Czernik, A. J., Brodin, L., Grillner, S., and Greengard, P. (1993). Transport of synapsin I in axons of the lamprey spinal cord. *Soc. Neurosci. Abstr.* **19**, 902.

Schulman, H., and Lou, L. L. (1989). Multifunctional Ca^{2+}/calmodulin-dependent protein kinase: Domain structure and regulation. *TIBS* **14**, 62–66.

Selbert, M. A., Anderson, K. A., Huang, Q.-H., Goldstein, E. G., Means, A. R., and Edelman, A. M. (1995). Phosphorylation and activation of Ca^{2+}–calmodulin-dependent protein kinase IV by Ca^{2+}–calmodulin-dependent protein kinase Ia kinase. Phosphorylation of threonine 196 is essential for activation. *J. Biol. Chem.* **270**, 17616–17621.

Sheng, M., Thompson, M. A., and Greenberg, M. E. (1991). CREB: A Ca^{2+}-regulated transcription factor phosphorylated by calmodulin-dependent kinases. *Science* **252**, 1427–1430.

Srinivasan, M., Edman, C. F., and Schulman, H. (1994). Alternative splicing introduces a nuclear localization signal that targets multifunctional CaM kinase to the nucleus. *J. Cell. Biol.* **126**, 839–852.

Stokoe, D., Caudwell, B., Cohen, P. T. W., and Cohen, P. (1993). The substrate specificity and structure of mitogen-activated protein (MAP) kinase-activated protein kinase-2. *Biochem. J.* **296**, 843–849.

Sudhof, T. C., Czernik, A. J., Kao, H.-T., Takei, K., Johnston, P. A., Horiuchi, A., Kanazir, S. D., Wagner, M. A., Perin, M. S., De Camilli, P., and Greengard, P. (1989). Synapsins: Mosaics of shared and individual domains in a family of synaptic vesicle phosphoproteins. *Science* **245**, 1474–1480.

Sugita, R., Mochizuki, H., Ito, T., Yokokura, H., Kobayashi, R., and Hidaka, H. (1994). Ca^{2+}/calmodulin-dependent protein kinase kinase cascade. *Biochem. Biophys. Res. Commun.* **203**, 694–701.

Sun, P. Q., Lou, L. M., and Maurer, R. A. (1996). Regulation of activating transcription factor-1 and the cAMP response element-binding protein by Ca^{2+}/calmodulin-dependent protein kinases types I, II, and IV. *J. Biol. Chem.* **271**, 3066–3073.

Tokumitsu, H., Brickey, D. A., Gold, J., Hidaka, H., Sikela, J., and Soderling, T. R. (1994). Activation mechanisms for Ca^{2+}/calmodulin-dependent protein kinase IV. Identification of a brain CaM kinase IV kinase. *J. Biol. Chem.* **269**, 28640–28647.

Tokumitsu, H., Enslen, H., and Soderling, T. R. (1995). Characterization of a Ca^{2+}/calmodulin-dependent protein kinase cascade. Molecular cloning and expression of calcium/calmodulin-dependent protein kinase kinase. *J. Biol. Chem.* **270**, 19320–19324.

Weekes, J., Ball, K. L., Caudwell, F. B., and Hardie, D. G. (1993). Specificity determinants for the AMP-activated protein kinase and its plant homologue analysed using synthetic peptides. *FEBS Lett.* **334**, 335–339.

Yano, K., Araki, Y., Hales, S. J., Tanaka, M., and Ikebe, M. (1993). Boundary of the autoinhibitory region of smooth muscle myosin light-chain kinase. *Biochemistry* **32**, 12054–12061.

Yokokura, H., Picciotto, M. R., Nairn, A. C., and Hidaka, H. (1995). The regulatory region of calcium calmodulin-dependent protein kinase I contains closely associated autoinhibitory and calmodulin-binding domains. *J. Biol. Chem.* **270**, 23851–23859.

Masatoshi Hagiwara*
Atsushi Shimomura*
Kazuhiko Yoshida†
Junko Imaki‡

*Department of Anatomy
Nagoya University School of Medicine
Nagoya 466, Japan

†Department of Ophthalmology
Hokkaido University School of Medicine
Sapporo 060, Japan

‡Department of Anatomy
Nippon Medical School
Tokyo 113, Japan

Gene Expression and CREB Phosphorylation Induced by cAMP and Ca^{2+} in Neuronal Cells

I. Introduction

Transcription of a number of eukaryotic genes is activated in response to an increase in the intracellular cAMP concentration. The stimulated genes have a common promoter element, the cAMP response element (CRE) (Montminy *et al.*, 1986), which is recognized by a specific binding protein, CREB. The binding of CREB to CRE does not itself induce transcription, because this requires phosphorylation of CREB at Ser133. In the case of the cAMP pathway, the activated catalytic subunit of cAMP-dependent protein kinase (PKA) translocates to the nucleus and phosphorylates Ser133 of CREB (Hagiwara *et al.*, 1993).

In the nervous system, signals transmitted across synapses are known to regulate gene expression in the postsynaptic cell. This process often involves membrane depolarization and subsequent elevation of intracellular Ca^{2+}. The transcriptional activation induced by such membrane depolarization

Advances in Pharmacology, Volume 36

and Ca^{2+} influx is mediated by a promoter element called the Ca^{2+}-responsive element (CaRE) (Sheng *et al.*, 1990). Recent studies of c-*fos* and proenkephalin gene expression have shown that CaRE and CRE are indistinguishable. In this chapter, we focus on the possible interactions between Ca^{2+} and cAMP signaling pathways in neuronal cells.

II. Ca^{2+}-Dependent Activation of CREB

As described previously, phosphorylation of Ser133 stimulates the ability of CREB to activate gene transcription, suggesting that CREB may mediate transcriptional induction by Ca^{2+} as well as by cAMP. The amino acid sequence of the CREB phosphorylation site is shown in Figure 1. Its Ser133 residue and the flanking amino acids constitute a phosphorylation site for both PKA and Ca^{2+}/calmodulin-dependent protein kinases (CaM kinase II and CaM kinase IV) *in vitro* (Fig. 2A). When cotransfected with a truncated CaM kinase IV (active form) expression plasmid into rat2 cells, the CRE-CAT reporter vector demonstrated as much activity as with the catalytic subunit expression vector of PKA (Fig. 2B). Essentially the same results have been reported by several groups (Sheng *et al.*, 1991; Enslen *et al.*, 1994).

We developed a polyclonal antiserum against CREB phosphorylated at Ser133 (Hagiwara *et al.*, 1993). This antiserum, No. 5322, can readily discriminate between phosphorylated and dephosphorylated forms, thereby enabling us to detect activation of CREB in response to extracellular signals *in vivo*. Using this antibody, we have assessed the physiology of CaM kinase-mediated CREB phosphorylation in neuronal cells.

III. Light-Induced c-*fos* Expression and Phosphorylation of CREB in the Neural Retina

The neural retina is the central nervous tissue best suited for studies of signal transmission because the vertebrate retina is a highly organized layer structure consisting of a limited number of cellular classes. The nuclei of neuronal cells in the ganglion cell layer (GCL) and the inner nuclear layer (INL) of the rat retina are immunopositive for anti-CREB antibody (No. 244) without any changes in light/dark conditions (Fig. 3). Light-

```
(127) Leu-Ser-Arg-Arg-Pro-Ser-Tyr-Arg-Lys (136)

PAK Consensus        Arg-Arg-Xaa-Ser/Thr

CaMKII Consensus   Arg-Xaa-Xaa-Ser/Thr-Yaa
```

FIGURE I   Phosphorylation site of CREB. Xaa, polar amino acid; Yaa, hydroxyl amino acid.

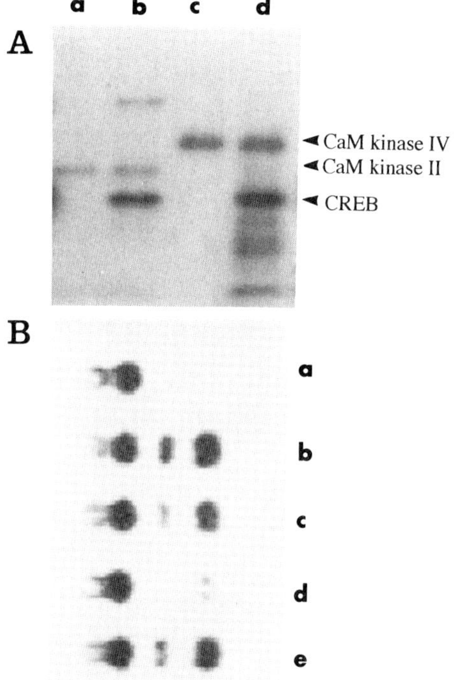

FIGURE 2   (A) Phosphorylation of CREB by CaM kinases *in vitro*. (a) CaM kinase II alone, (b) CREB + CaM kinase II, (c) CaM kinase IV alone, and (d) CREB + CaM kinase IV. (B) CaM kinase IV-induced gene expression in rat2 cells. (a) CAT basic, (b) RSV-CAT, (c) CRE-CAT + PKA catalytic subunit expression vector, (d) CRE-CAT alone, and (e) CRE-CAT + truncated CaM kinase IV expression vector.

induced stimuli are received by the photoreceptors lying outmost, processed through INL and GCL, and carried to the brain. Photosignals have been implicated in the regulation of levels of mRNAs coding for neurotransmitters and other proteins in retinal cells (Brann and Cohen, 1987; Korenbrot and Fernald, 1989).

To determine whether CREB is phosphorylated in retinal cells by changes in light/dark conditions, we have used an antiphospho-Ser133 CREB antibody (No. 5322) for immunocytochemistry. Phosphorylated CREB immunoreactivity (PCREB-IR) could not be detected in any area of the retina at the end of the dark period (08:00) (Fig. 4A). After 5 min exposure to sustained light, beginning at the end of the dark period, PCREB-IR was found in the nuclei in layers, the immunoreactive cells in the INL being in the inner border, as well as the outer half (Fig. 4B). After 5 min exposure of flashing light, PCREB-IR was also detected in the nuclei of the INL and the GCL. In the INL, immunoreactive cells were distributed in both the outer half and the inner border (Fig. 4C). When animals were examined

**Dark
Adapted**   **Flashing
Light**

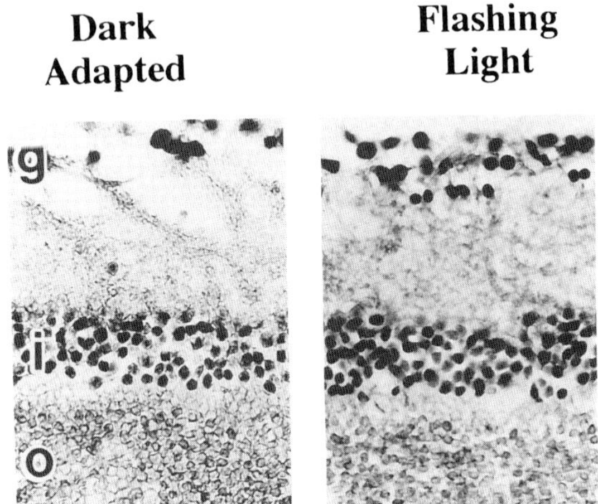

FIGURE 3   Immunohistochemical analysis of CREB in the rat retina. With No. 244 anti-CREB antibody, CREB immunoreactivity (CREB-IR) was observed in the nuclei of the INL (i) as well as the GCL (g) but not in the outer nuclear layer (o), both during the dark period (left) and after flashing light exposure (right).

after 30 min of either flashing or sustained light and 5 min, 30 min, and 2.5 hr after the onset of the dark phase, PCREB-IR was not present in any case (data not shown).

The c-*fos* gene was found to be expressed not in the INL and the GCL, but was weakly expressed in the outer nuclear layer (ONL) throughout the dark period under a 12:12 light:dark cycle (LD 12:12)(Fig. 4D). No hybridization of an antisense probe was observed in either the INL or the GCL. After 30 min of sustained light starting at the end of the dark period, c-*fos* expression was demonstrated by strong hybridization in the INL as well as in the GCL (Fig. 4E). Careful examination showed that the majority of silver grains were distributed in the outer half of the INL. Within the GCL, silver grains accumulated in only some of the cell bodies. As with sustained light exposure, 30 min of flashing light at the end of the dark period resulted in c-*fos* expression in the INL and in the GCL (Fig. 4F). In contrast to steady light exposure, however, flashing induced higher levels of silver grains at the inner border rather than at the outer half of the INL. In control experiments, a sense-oriented probe, generated by T3 polymerase, was applied to adjacent sections and failed to show any positive hybridization (data not shown). The consistency of the PCREB-IR and c-*fos* expression, except in the ONL, suggests that light-induced c-*fos* expression in the INL and the GCL is mediated by CREB.

IV. Ca^{2+}-Dependent CREB Activation in Rat Retina

An increase of cAMP in retinal cells after light exposure has not been reported. However, Kaneko (1970) showed that transient depolarization of

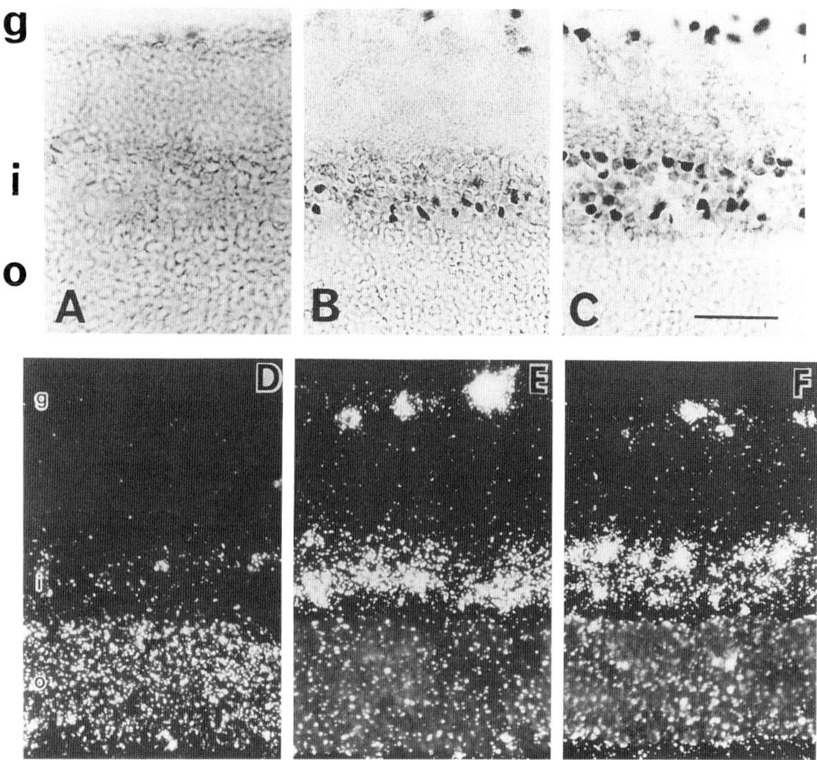

FIGURE 4   Effect of light exposure on CREB phosphorylation (A–C) and c-*fos* expression (D–F) in retinal cells. (A and D) At 08:00, the end of the dark phase of a 12 : 12-hr light : dark cycle: PCREB-IR is not detectable in any area of the retina. Autoradiographic grains are exclusively localized over the ONL (o). (B and E) After exposure to sustained light starting at 08:00, the end of the dark period: PCREB-IR is apparent in nuclei of cells in the INL (i) and the GCL (g). In the INL, the immunoreactive cells are more abundant in the outer half. c-*fos* mRNA expression in the ONL is decreased, but strong hybridization signals are evident in the INL as well as in the GCL. The majority of the c-*fos* mRNA appears to be distributed in the outer half of the INL. (C and F) After exposure of flashing light starting at 08:00, the end of the dark period: PCREB-IR is present in the INL and the GCL. In the INL, immunoreactive cells are located in both the outer half and at the inner border. c-*fos* mRNA is also expressed in the INL and the GCL with silver grains being distributed most densely at the inner border of the INL. Abbreviations: o, outer nuclear layer; i, inner nuclear layer; g, ganglion cell layer. Scale bar, 50 μm.

amacrine cells was induced by illumination and because these cells, located in the inner border of the INL, have voltage-sensitive L-type Ca^{2+} channels, we examined whether Ca^{2+} influx induced by Bay K 8644 [a novel L-type Ca^{2+} activator (Tsien *et al.*, 1987)] can trigger the phosphorylation of CREB in retinal cells. Fifteen minutes after vehicle treatment during the light period of an LD 12 : 12 cycle (examination time point 14:00, switch over at 08:00), the rat retina failed to show any PCREB-IR (Fig. 5A), but 15 min after Bay K 8644 application (1 mg/kg), it could be detected in the INL and the GCL

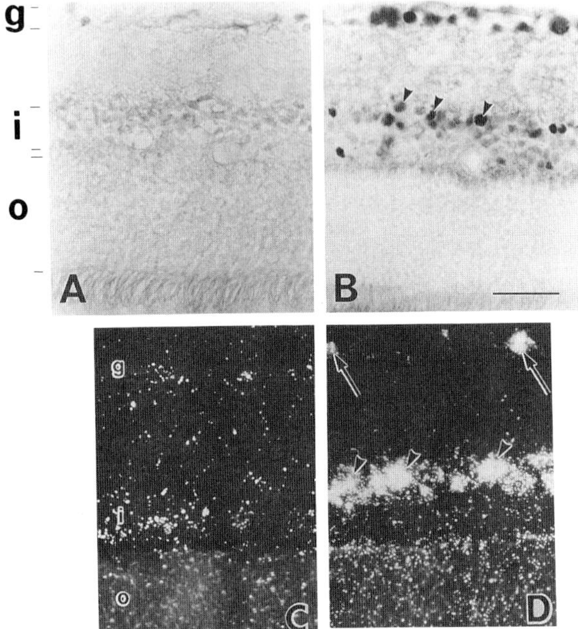

FIGURE 5   Effects of the Ca^{2+} channel activator Bay K 8644 on CREB phosphorylation (A and B) and expression of c-*fos* (C and D). (A and C) After intraperitoneal injection of vehicle during the light period (14:00), the rat retina fails to show PCREB-IR or any positive hybridization signals for c-*fos*. (B and D) After treatment with Bay K 8644 (1 mg/kg) at 14:00, PCREB-IR is apparent in the nuclei of amacrine cells at the inner border of the INL (arrowheads in B) and in GCL with c-*fos* gene expression becoming evident in photoreceptor cells, presumptive amacrine cells at the inner border of the INL (arrowheads in D) and ganglion cells (or displaced amacrine cells) of the GCL (arrows). Abbreviations: o, outer nuclear layer; i, inner nuclear layer; g, ganglion cell layer. Scale bar, 50 μm.

(Fig. 5B). In the case of INL, distribution was exclusively at the inner border, presumably localized in amacrine cells (Fig. 5B, arrowheads).

Next, we examined the induction of gene expression in the INL and the GCL after administration of Bay K 8644. Forty-five minutes after intraperitoneal injection of vehicle during the light period (14:00) of an LD 12:12 cycle, the rat retina failed to show any positive hybridization for c-*fos* mRNA (Fig. 5C). However, 45 min after treatment with Bay K 8644 (1 mg/kg) at the same time point, the c-*fos* gene was found to be expressed in photoreceptor cells, presumed to be amacrine cells at the inner border of the INL (Fig. 5D, arrowheads) and ganglion cells (or displaced amacrine cells) of the GCL (Fig. 5D, arrows), but not in the rod bipolar cells in the outer half of the INL. Enhanced somatostatin mRNA signals were also detected at the inner border of the INL and in the GCL after the injection of Bay K8644 (Yoshida *et al.*, 1996).

V. Differential Roles of CaM Kinases in the Neural Retina

As previously described, two types of Ca^{2+}/calmodium-dependent protein kinases, CaM kinase II and CaM kinase IV, are able to phosphorylate CREB at Ser133 (Enslen *et al.,* 1994). Immunocytochemical studies with antisera to CaM kinase II and CaM kinase IV revealed that both are expressed in the GCL and in the INL. CaM kinase II, however, is localized in the inner border of the INL (Fig. 6A), whereas CaM kinase IV is distributed in the outer half of the INL (Fig. 6B). Double staining with anti-CaM kinase II monoclonal antibody and polyclonal No. 5322 antibody in the same sections of rat retinas treated with Bay K 8644 demonstrated that most PCREB immunoreactive nuclei (Fig. 6D) in the GCL and amacrine cells in the INL are also immunopositive for anti-CaM kinase II (Fig. 6C), indicating colocalization of the Ca^{2+}-induced phospho-CREB and CaM kinase II.

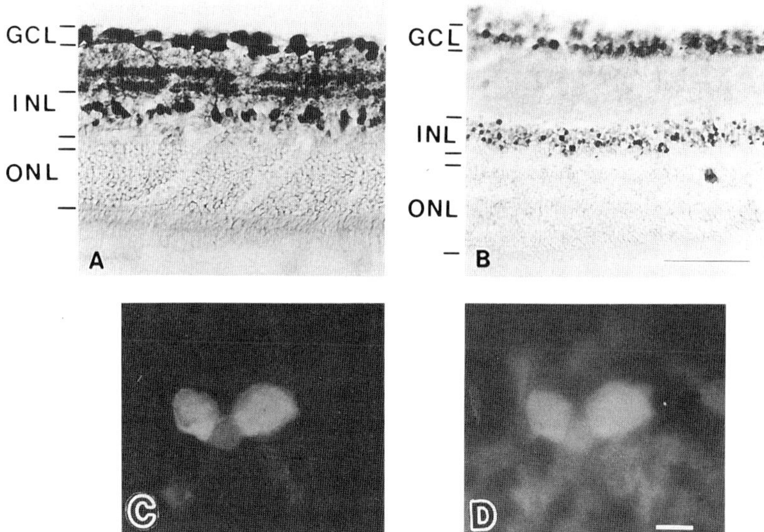

FIGURE 6   CaM kinase II (A) and CaM kinase IV (B) immunoreactivity in retinal cells and double staining of phospho-CREB (C) and CaM kinase II (D) after Bay K 8644 treatment. CaM kinase II is expressed in the GCL and in the inner border of the INL. CaM kinase IV is also distributed in the GCL and in the INL, although in contrast to the CaM kinase II case, immunoreactivity is predominantly in the outer half. The GCL, 15 min after treatment with Bay K 8644, demonstrates double staining with CaM kinase II monoclonal antibody, visualized with rhodamine-labeled anti-mouse IgG (C) and PCREB-IR visualized with FITC-labeled anti-rabbit IgG (D). Most PCREB immunoreactive nuclei in the GCL and amacrine cells in the INL are also immunopositive for anti-CaM kinase II monoclonal antibody. Scale bar, 50 μm.

The results thus suggest that sustained light activates only kinase IV in bipolar cells, whereas flashing light activates both CaM kinase IV and CaM kinase II in bipolar cells and amacrine cells. Taken together, our findings suggest that sustained light and flashing light differentially activate calcium/calmodulin signaling pathways and, in consequence, lead differentially to phosphorylation of CREB and expression of c-*fos*. Our finding that Ca^{2+} entry through L-type Ca^{2+} channels leads to phosphorylation of the transcriptional regulatory site, Ser133, of the CRE-binding protein CREB and expression of c-*fos* gene in the rat retina is the first evidence that a physiological Ca^{2+} signal transcriptionally regulates neural gene expression via CREB phosphorylation *in vivo*.

VI. Discussion and Conclusion

Hypothetical signal transduction pathways for induction of transcriptional activity through CREB, as discussed previously, are illustrated in Figure 7. In this figure, the pathway of Ca^{2+}-dependent CREB activation suggested by Impey *et al.* (1994) is indicated by a gray arrow. Currently, we have no clue as to whether Ca^{2+}/calmodulin-dependent (type I) adenylyl cyclase is involved in Ca^{2+} induction of CREB pathway. The nuclear localization of CaM kinase IV and an isoform (δb) of CaM kinase II provide support for the idea that Ca^{2+}/calmodulin-dependent protein kinases directly phosphorylate Ser133 of CREB and thus activate transcription.

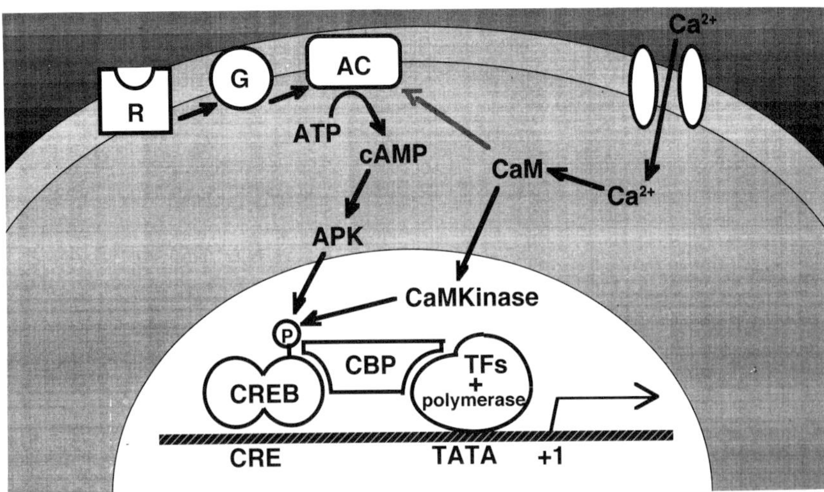

FIGURE 7   Schematic illustration of cAMP and Ca^{2+} signal transduction systems from the cell membrane to CREB in the nucleus. Abbreviations: R, receptor; G, G protein; AC, adenylate cyclase; CBP, CREB-binding protein; TFs, basic transcription factors.

Considering that dark/light adaptation in neural retina has some hallmarks of long-term memory, such as lasting several hours and being inhibited by cycloheximide, the signal transcription coupling described here may provide a basis for investigation of the memory system in the central nervous system. In ongoing experiments, we are now trying to clone genes whose expression is regulated by the dark/light cycle using the methods of subtraction cloning and differential display.

References

Brann, M. R., and Cohen, L. V. (1987). Diurnal expression of transducin mRNA and transaction of transducin in rods of rat retina. *Science* **235**, 585–587.

Enslen, H., Sun, P., Brickey, D., Soderling, S. H., Klamo, E., and Sodering, T. R. (1994). Characterization of Ca^{2+}/calmodulin-dependent protein kinase IV. Role in transcriptional regulation. *J. Biol. Chem.* **269**, 15520–15527.

Hagiwara, M., Brindle, P., Harootunian, A., Armstrong, R., Rivier, J., Vale, W., Tsien, R., and Montminy, M. (1993). Coupling of hormonal stimulation and transcription via the cAMP-responsive factor CREB is rate-limited by the nuclear entry of PK-A. *Mol. Cell. Biol.* **13**, 4852–4859.

Impey, S., Wayman, G., Wu, Z., and Storm, D. (1994). Type I adenylate cyclase functions as a coincidence detector for control of cyclic AMP response element-mediated transcription: synergistic regulation of transcription by Ca^{2+} and isoproterenol. *Mol. Cell. Biol.* **14**(12), 8272–8281.

Kaneko, A. (1970). Physical and morphological identification of horizontal, bipolar and amacrine cells in goldfish retina. *J. Physiol.* **207**, 623–633.

Korenbrot, J. I., and Fernald, R. D. (1989). Circadian rhythm and light regulate opsin mRNA in rod photoreceptors. *Nature* **337**, 454–457.

Montminy, M. R., Sevarino, K. A., Wagner, J. A., Mandel, G., and Goodman, R. H. (1986). Identification of a cyclic-AMP-responsive element within the rat somatostatin gene. *Proc. Natl. Acad. Sci. U.S.A.* **83**, 6682–6686.

Sheng, M., McFadden, G., and Greenberg, M. E. (1990). Membrane depolarization and calcium induce c-*fos* transcription via phosphorylation of transcription factor CREB. *Neuron* **4**, 571–582.

Sheng, M., Thompson, W. A., and Greenberg, M. E. (1991). CREB a Ca(2+)-regulated transcription factor phosphorylated by calmodulin-dependent kinases. *Science* **252**, 1427–1430.

Tsien, R. W., Hess, P., McClesky, E. W., and Rosenberg, R. L. (1987). Calcium channels: Mechanisms of selectivity, permeation and block. *Annu. Rev. Biophys. Chem.* **16**, 265–290.

Yoshida, K., Imaki, J., Matsuda, H., and Hagiwara, M. (1995). Light-induced CREB phosphorylation and gene expression in rat retinal cells. *J. Neurochem.*, **65**, 1499–1504.

Index

Contents of Previous Volumes

Volume 28

Volume 29A

Volume 29B

Volume 30

Pharmacologic Therapy of Obsessive Compulsive Disorders
Joseph DeVeaugh-Geiss

Mechanism of Action of Antibiotics in Chronic Pulmonary
Pseudomonas Infection
Niels Holby, Birgit Giwercman, Elsebeth Tvenstrup Jensen,
Svend Stenvang Pedersen, Christian Koch, and Arsalan Kharazmi

Quinolinic Acid in Neurological Disease: Opportunities for Novel
Drug Discovery
John F. Reinhard, Jr., Joel B. Erickson, and Ellen M. Flanagan

Pharmacologic Management of Shock-Induced Renal Dysfunction
Anupam Agarwal, Gunnar Westberg, and Leopoldo Raij

Autoantibodies against Cytochromes P450: Role in
Human Diseases
Philippe Beaune, Dominique Pessayre, Patrick Dansette, Daniel Mansuy,
and Michael Manns

Activation and inactivation of Gene Expression Using
RNA Sequences
Boro Dropulic, Stephen M. Smith, and Kuan-Teh Jeang

Therapy of Cancer Metastasis by Systemic Activation
of Macrophages
Isaiah J. Fidler

5-Hydroxytryptomine Receptor Subtypes: Molecular and
Functional Diversity
Frédéric Saudou and René Hen

Volume 31

Regulation of the Calcium Slow Channels of the Heart by Cylic
Nucleotides and Effects of Ischemia
Nicholas Sperelakis

Functional Adaptation to Myocardial Ischemia: Interaction with
Volatile Anesthetics in Chronically Instrumented Dogs
Patrick F. Wouters, Hugo Van Aken, Marc Van de Velde,
Marco A. E. Marcus, and Willem Flameng

Volume 33

Volume 35